Plant Nematology

Cyst and Root Knot Nematodes

Plant Nematology

Cyst and Root Knot Nematodes

Dr. K.K. Kaushal
Retired Principal Scientist
Division of Nematology,
IARI, New Delhi

2013
BIOTECH BOOKS®

ISBN 978-81-7622-270-9

Published by: **BIOTECH BOOKS®**
4762-63/23, Ansari Road, Darya Ganj,
New Delhi - 110 002
Phone: +91-011-23262132
E-mail: biotechbooks@yahoo.co.in

Printed at: **Chawla Offset Printers**
Delhi - 110 052

PRINTED IN INDIA

Preface

Recently, there has been a glut in the publication of so many books in plant nematology providing up to date information but the knowledge on various aspects of nematode study is so fast appearing in literature that it is becoming difficult to put that into black and white in one publication. The group is so large (an astronomical number of individuals, representing perhaps four out of every five multicellular animals on earth) and so ecologically, morphologically, and phylogenetically diverse that to attempt to discuss the diversity of the group in such a comprehensive way is practically futile. Efforts are being made to look for more fossil records to know exact time of their appearance in the universe. From the known fossil observations, their age is estimated to be 28000 years old. However, through application of various models of molecular evolution and molecular clock theory, estimates of the time of divergence of the nematodes from the rest of the animal groups appears to be about 1,177 ± 79 million years. Most authors consider the ultimate origin of nematodes to be from a marine ancestor that branched from the metazoan tree perhaps a billion years ago. Surprisingly, the relationship of Nematoda to other animal phyla remains controversial. However, molecular phylogenies place nematodes and insects together in a high-level taxon, named Ecdysozoa. Recent studies seem to support this grouping although some data remain contradictory. In light of this evolutionary closeness of nematodes with insects, similarities between insects and nematodes in their various interactions with plants become all the more interesting.

As we know that only a portion of the existing nematodes species has been described, still they pose an enormous economic and health threat not only to vast array of economically important agricultural crops but also to the world's human population. The members of the families, Heteroderidae and Meloidogynidae are in the forefront in causing huge losses to many life saving and economically important crops and other resources which provide shelter to people and feed to their domestic animals throughout the globe. In India also the root knot nematodes were the first to be known to the people for reducing the vigor of their plants and debilitating and deforming the roots. Incidentally, it was the turn of the cyst nematodes to show in reality what damage these tiny, hidden creatures can cause to the crop like wheat which is almost a staple food for most of Indians and other people of Nematodes of the world. They were hunted in many geographically different areas of the country for new species and few were described but I am very optimistic that there are atleast 10-20 more species of cyst nematodes and an equal number of root knot species is also waiting to be described. While working on this fantasting group of nematodes, efforts were made to establish 'An advance centre on Cyst Nematodes' that could not be materialized but certain ambitions don't die immediately and thus came the idea of writing about these nematodes. The Indian literature on these nematodes is so scattered that it was my interest to put the information at one platform and to make it available to the interested, be it teachers or students or any other worker.

As I was looking for such an opportunity, the Department of Science and Technology, Govt. of India (DST) gave me the go ahead along with requisite financial support to write about cyst and root knot nematodes of India. I am also indebted to my scientist colleagues in the Division of Nematology, IARI, who always encouraged me in this endeavor.

Dr. K.K. Kaushal

Contents

Chapter 1
History of Nematology

1.1 Introduction

B.G.Chitwood has very poetically described the development of study of the nematodes or roundworms, the beautiful little beasts, as cited,

'The grace of movement of some lowly soil inhabiting forms finds little equal among other living organisms, being comparable to the gliding of snakes, which Solomon noted as one of the four mysteries of life. The complex pattern of their body markings and of the head and other parts might well be used in designing ladies' dresses. None of their graces and beauty is suggested by a name that carries the stigma 'worm'.

It is a fact that if the beautiful patterns and other diverse markings on the body of the nematodes are adopted by some clothes designer of the day, it will revolutionize the fashion world. Each day they would have a new design for presentation.

According to the Russian philosopher and nematologist, A. Paramonov, nematodes are ancient group of organisms presenting a great degree of diversification, intensive speciation, diversity of life cycles and strategies, trophic specializations and occupation of various habitats. Nematodes are nonsegmented worms that generally lack external appendages. Most are vermiform, with tapering anterior and posterior ends, cylindrical in cross-section, and covered with a usually translucent, flexible, acellular cuticle secreted by an underlying cellular hypodermis. The cuticle may be smooth or ornamented with rings, longitudinal striations, spines or spikes, or it may have well-developed winglike structures called lateral alae that are very common on the externo-lateral surfaces. Some marine species have long setae, modified sensory papillae that are probably used in movement and in detecting their environment. In contrast to all other nematodes, some marine species have eye spots that enable them to detect light in their environment. The fine structure of the external body-cuticle is complex, which protects the animal from the external environment and allowing it to remain homeostatic inside. The cuticle can be extremely resistant, and depending on the nematode species and its life-history characters. On the other hand, nematodes may extremely be delicate, able to exist intact only within the osmotically balanced tissues of other animals (*e.g.*, members of the order Filaroidea), and if moved from the isosmotic solution to one with fewer salts, they may explode and die in a most amazing display.

Nematodes are called "pseudocoelomates" because their coelom (pseudocoelom) is not completely lined with mesodermally derived cells and they are triploblastic having three germ-layers-ectoderm, mesoderm and endoderm. The body cavity of nematodes is derived from the blastocoel of the embryo. Such a cavity is bounded externally, in part, by the somatic muscle cells which are of mesodermal origin, and by the ectodermal lining of the body wall, and internally by the gut cells which emerged from the primitive endodermal lining. Because of this mode of origin, the cavity is known as a pseudocoelom.The pseudocoelom may be absent in the pharyngeal regions of both secernentean and adenophorean nematodes. They do not possess circular body muscles, so the movement is accomplished by contraction and relaxation of longitudinal muscles against a hydrostatic skeleton. Because their internal pressure is high, causing the body to flex rather than flatten, and the animal moves by thrashing back and forth. However, to a lesser degree six other types of unusual and specialized locomotions in nematodes have been recorded (Adams and Tyler, 1980). They also donot have well defined circulatory and repiratory systems. The circulation is brought about by pseudocoelomic fluid and respiration is cutaneous. The body wall is flexible and very strong. All nematodes maintain their form because their body fluids are under a positive pressure relative to their environment. No cilia or flagellae are present. Nematodes have a complete digestive system with an anterior stoma just behind the mouth and usually a triradiate or triple muscle pumping type of esophagus that can be muscular or glandular in structure (or both). The intestine, tubular in form, is usually a single cell in thickness lined on the peritoneal side with a thin collagen-like material, internally lined with microvilli (Grassé, 1965; Maggenti, 1981, 1991). The tube extends from the esophagus straight to the anus or cloaca, sometimes showing out pouched cecae or diverticulae near the esophageal end. Males are usually smaller than females. Many nematode species are sexually dimorphic with separate sexes (diocious or amphigonus) and are oviparous; however, some are ovoviviparous and some are viviparous. In secernenteans there is one testis and two ovaries but in adenophoreans, there is two of each. Males usually have a cuticularized spicule or pair of spicules that are used to assist in the transfer of sperm to the females. In certain nematodes the length of spicule may reach 60 per cent of the body length as in *Odontospora cetiopenis* and in *Gongulonema ingluvicola* it nearly reaches 100 per cent of the body length. Many species also have a gubernaculum (an extension from the cloaca that guides the spicule during copulation and spicular eversion). In *Hoplolaimus* spp., a bilobed structure is present at the distal end of gubernaculums called titillae. Some species are hermaphrodytic; in these cases the nematode produces both sperm and ova from the ovotestis of the same individual at different times during ontogeny- the development of the organism (Maggenti, 1991; Malakhov, 1994). In the phylum Nematoda, it is thought that the presence of noncontractile axon-like myoneural processes or extensions that run from the contractile or body portion of muscle cells to the neural junctions of the nerve cords is unique in the world of organisms but this process already exists in many other phyla such as the Caphalochordata (Flood, 1966), Echinodermata (Cobb, 1967), Annelida (Rosenbluth, 1968), Coelenterata (Gulyayev, 1976) and Gastrotricha (Teuchert, 1977). The width of nematodes is usually less than 2 mm, even when extremely long. However, even here there are exceptions, with the giant kidney worm, *Dioctophyme renale,* attaining a size of 15 mm x 1,000 mm. The eggs of nematodes are also very similar in size and holds true for *Placentonema gigantissima* (8m) and *Paratylenchus* (0.25mm) with most ranging from 50 to 100 μm long by 20 to 50 μm wide, with an exception being the antarctic marine nematode *Deontostoma timmerchioi* (40 mm in length) that has eggs that range from 868 to 1,060 μm long by 24-35μm in width.

Nematodes occupy every conceivable niche where lives exist. Parasitic forms affect the health and well-being of most plants and animals and cost the world's crop and livestock industries tens of billions of dollars per annum in lost production. In addition, more than two billion people currently suffer poor health from nematode infections.There are deepwater marine forms that appear to consume mostly diatoms and others such as species of the genus *Draconema* that are mostly associated with marine algae. Nevertheless, not all nematodes are pests. Beneficial species are common and they contribute to the health of ecosystems through their role in nutrient cycling, by feeding on insects, weeds or plant pathogenic fungi or by preying on plant-parasitic nematodes. Some nematode species feed on fungi in the soil while others are trapped and are themselves consumed by different species of fungi. Still other nematodes such as *Mononchus* and *Clarkus* are predatory and hunt and eat other nematodes in the soil environment. Free-living bacterial feeding nematodes have been shown to be integral parts in the carbon and nitrogen cycles in healthy soils (Ferris *et al.*, 1998), and recent survey work has shown that undisturbed or natural soils in noncultivated habitats can have as many as 20 times more species than soil from similar areas that have been under cultivation (Al-Banna and Gardner, 1996).

This is a fact that there is no parallel in the biological science to these multicellular metazoans when we observe their virtual numbers, variation in size and the diversity of forms and structures.

1.2 World History of Nematodes

The research in animal science, as in most fields of science, is primarily based on the curious mind's continuous observations and in recording these observations. History of science is nothing but the summary of the personal experience of the scientists. The science of nematology also has its roots in recording such observations in Vedas (6000-4000BC), China Yellow Classic of Internal Medicines (2700B.C.), Egyptian Papyrus Ebers (1550B.C.), Middle East Bible 21:6-9, Fiery Serpents of Israelites (1250B.C.), Hippocrates (420 B.C.), Aristotle (350 B.C.) and Celsus (10 B.C.). There are frequent references in the Rig and Atharv Vedas (6000-4000 BC) to nematodes as 'Krimi' causing itching who readily entered in the body. Different types of symbolic descriptions available in Atharv Vedas, the words, 'Algandu' and 'Kururu' stand for roundworms and threadworms, respectively. Even, Charak (Charak Samhita, 3000BC) recognized twenty different 'krimis' which included not only nematodes but also arthropods and leeches. He described some of the large bodied intestinal helminths of human and animals. People in the Vedic era were very much aware of the ascarid in different parts of the human body and used to treat the affected persons with extracts of medicinal plants and herbs (*'I suppress by vedic chants all those krimis which enter into our bodies piercing through the skin and eat away flesh, dwelling in the intestine, head and sides of the body'*- Ath.2.31.4).

The plant parasitic nematodes and their symptoms have been described in thirteenth century Sanskrit anthology and refer to the well known root knot nematodes.The anthology in question is *Saringadhara paddhati* compiled by Sarugadhara, a courtier in the court of King Hammira of *Sakambhari desa* (Modern Bundelkhand) during 1283-1301. The original stanza 183,184 in section XIII of the *Upavanavinoda* dealing with *Tarucikitsa* (treatment of plants) are reproduced below:

दोषैर्यस्य विना प्रवालकुसुमम्लानिर्विरुढ़ं वपुः
मूले तस्य तरोर्भवन्ति कृमयो यत्नाच्चतानुद्धरेत।
गोमुत्राज्यविड़ङ्गसर्षपतिलैर्लिप्तः प्रणाष्टैस्ततः
सिक्तः क्षीरजलैरुदेति सहसा धूपैश्च धूपायितः ।। १८३
करञ्जारग्वधारिष्टसप्तपर्णत्वचाकृतः ।
उपचारः क्रिमिहरो मूत्रमुस्तविडङ्गवान् ।। १८४

183—One should do well to realize that worms (Krimyo) are at the roots of plants affected with tubercles or at plants for the paleness of buds and flowers of which no other cause can be assigned and one should do well to root out these worms with care. If now fresh urine of cow, clarified butter, Vidanga (*Embelia ribes*), mustard and sesamum are mixed together with milk and water they (the plants) grow.

184—All kinds of worms are destroyed if one applies to the roots of trees the bark of Karanja (*Galedupa arberea*), argvadha (*Cassia fistula*), arista (*Melia azadirachta*) and Saptaparna (*Alstonia scholaris*) pasted in the urine of cows together with Vidanga and musta (*Cyprus rotundus*).

There is evidence that the soybean cyst nematode was known to be present in China about 2500 years ago. Referring to the book 26 of *The Annals of Lu Buwei* compiled in China in 239 BC, rules were outlined for controlling three 'robbers', one of them was 'the land stealing the crops'. The prescription for this 'robber' was 'overworked soil need fallowing'. Since the symptoms in these fields described by Chinese farmers were 'fire burned seedlings' the problem might have been due to soybean cyst nematode.

Nematodes were noted early in human history because some serious human diseases were caused by relatively large vertebrate-parasitic nematodes. Humans have been parasitized by nematodes from the earliest times. Eggs of the pinworm, *Enterobius vermicularis* were found in human coprolites (organic fossils) from Hopug and Danger Caves, western Utah. The Caves were inhabitated by man from 10,000 B.C. to A.D. 1400 (Fry and Moore, 1969). Eggs of the whipworm *Trichuris trichuria* occur in coprolites dated to about 7,000 years old from dry areas of Peru (Weir and Bonavia, 1985), and eggs of the hookworm *Ancylostoma duodenale* have been reported from coprolites dated to around 7,230 years old. In eastern Brazil, *Ascaris lumbricoides* has been positively identified from human coprolites dated to about 28,000 years old (Ferreira *et al.*, 1983). The seeming dearth of other nematodes from human remains older than about 7,000 years old appears to be due to the fact that organic material comprising the coprolites themselves does not preserve well enough to last that long (Reinhard *et al.*, 1986). The earliest record, in which symptoms of intestinal worm, *Ascaris* have been narrated, appeared in China Yellow Emperor's classic of Internal Medicines (2700 B.C.). Incidences have also been noted in Papyrus Ebers records (1500 BC) of Egyptians physicians of intestinal worm, *Ascaris lumbricoides*; guinea worm, *Dracunculus medinensis* (*Dracunculus* is derived from Greek word, *'Dracontia micra'* means little snake) and of hookworm, *Ancylostoma* (Viglierchio, 1991). The mention of Fiery Serpents (1250 BC) in the Bible (21:6-9) may have a reference to the guinea worms (Chitwood and Chitwood, 1974) who also wrote about illness of many people caused by nematodes during the period of Alexander (180-146 B.C.). Hippocrates (420 B.C.) discussed about a nematode pinworm, *Enterobius* in *'Aphorisms'*. During 384-322 B.C., Aristotle (also known as Father of Zoology) in his *'Historia Animalium'* mentioned about many nematodes including *Ascaris* and pinworm and stated that "these intestinal worms do not in any case propagate their kind". A.C. Celsus (53 B.C.-7A.D.) was able to differentiate between round worms, nematodes and flat worms, cestodes. In 100 A.D., Columella observed an ascarid, *Neoascaris vitulorum* in calf. Till middle of sixteenth century, many workers recorded animal parasitic nematodes in different hosts -Vegetius (400AD)- ascarid in horse; Avicenna (1000AD) referred to parasitic worms like ascarids and dracunculids in his classic "Canon Medicinae", A. Magnus (1200-1280AD)-nematodes in birds; Caesalpinus (1579-1630)-*Dioctophyma renale* in kidney of a dog; and Vinegia (1547)-ascarid in falcon. In his,'De Animalibus Insectis' Aldrovandus (1623) reported dead grasshopper covered with long worms, possibly parasitic nematodes and considered them as responsible for the death of grasshopper and erected the name, *Vermes*. After the development of a microscope by Antonie van Leeuwenhoek (1650), workers were able to see and describe the tiny creatures, the nematodes in soil and plants. Perhaps, Borellus (1656) was the first person to see a nematode under a microscope and gave the description of the vinegar nematode, *Turbatrix aceti*. E. Tyson (1683) worked on the anatomy of nematodes and described many structures and F. Redi (1684) compiled the reports on vertebrate nematodes and their hosts. His first great publication was a study of parasitic worms, the *"Enterozoorum Sive Vermium Intestinalium Historia Naturalis"*. This is the first publication to describe the Nematoda. His second, the *"Synopsis cui accedunt mantissima duplex et indices locupletissima"* was the first work regarding the detailed study of the life cycle of important nematode parasites of humans, such as *Ascaris lumbricoides*. Copenhagen Academy of Sciences in 1780 offered a prize for the best essay on the intestinal worms indicating wide interest of biologists in nematodes.

The earliest mention of a plant parasitic nematode is a line "Sowed cockle, reaped no corn," by William Shakespeare (1594) in Love Labour's Lost', Act IV, Scene 3, most certainly referring to the blighted wheat caused by the plant parasitic nematode, *Anguina tritici*. About 150 years later, John Turberville Needham (1743), a clergyman discovered the same nematode recording the worms inside the cockles of wheat. He wrote to the President of Royal Society of London, '*upon opening the small black grains of wheat I saw soft fibrous tissues. When a portion of this was placed in a drop of water, it seemed to consisting of bundles of longitudinal fibers. After sometime when these fibers were separated they surprisingly took life and moved with a twisted motion which they continued to do so for 9-10 hours when they were thrown away. I was convinced that they were the species of aquatic animals and may be worms, eels or serpents with which they resemble*'. This note was published in 1744. Reaumur (1742) described a worm (later named as *Sphaerularia bombi*). Gould (1747) graphically described the emergence of worms, probably mermithids, from ants. During the middle of eighteenth century, Carl Linnaeus who was a physician, botanist and natural scientist pioneered a practical classification system in his *Systema Naturae* (1758) of organisms by his binomial system with a genus and descriptive species attributes. He named many animal and human parasitic nematode species including *Ascaris lumbricoides* and *Dracunculus medinensis* under the group Vermes (worms). He was the first to name the vinegar eelworm as *Chaos redivivus* (which leterally means raised from the dead by water after years of dessication). Linnaeus (1758) listed eight genera in the *Vermes Intestini*. J.A. Scopoli (1777) characterized this nematode and named it as *Anguina* (original spellings *Angvina*). Later on, Steinbuck (1799) named this nematode as *Vibrio tritici* and also described another species *V. agrostis* infesting seeds of bent grass. Goeze (1782) made the first serious study of nematodes under a microscope and described the vinegar eelworm. He began to distinguish between the various kinds of worms. He also described the emergence of mermithids from soil following a heavy rain. K. Rudolphi (1819) gave the name Nematoidea and produced a publication "*Entozoorum Synopsis*" with 350 species belonging to 11 genera. He is often considered as "Father of Helminthology".

Most of the workers, during this period, were involved in the study of animal parasitic nematodes rather than plant parasites. In 1835, R. Owen recorded the occurrence of *Trichinella spiralis* in the human muscles. C.von Siebold (1842) developed the concept of a life cycle involving different kind of hosts (animal parasites enter through the mouth). He also gave the name Gordiacea.

Joseph Leidy (1851) reported the first record of a free living nematode, *Anguilulla longa* in America which was later on redescribed as *Tribolus logus* by Cobb (1914). Miles Joseph Berkley (1855) reported the presence of galled roots in cucumber grown in greenhouse in England. The galled roots were produced by the root knot nematodes. This discovery was immediately followed by record of stem and bulb nematode, *Angulliula dipsaci* (*Ditylenchus dipsaci*) infesting teasel, *Dipsacus fullonum* (Kuhn, 1857). Monographs published during the middle and later half of the 19th century were milestones in the development of nematology. These were mainly concerned with taxonomy and difference in the forms of nematodes in various habitats. Dujardin (1845) was also a pioneer in the study of nematodes in insects in France. He described *Mermis nigrescens* in 1842 and *Mermis aquatilis* in 1845, Diesing (1851) and Schneider (1866) made headway in advancing the field of taxonomy by describing many new species and putting them in black and whiteb. Karl Diesing (1851) classified free living nematodes in two families- Cirrhosstomae and Anguillulidae on the presence or absence of cirri or setae in the head region. He published *Systema Helminthum*, with 175 insect nematode records and involving five entomophilic genera. He listed 118 species of *Gordius*, 17 of *Mermis*, including *M. nigrescens* and *M. albicans* and *Sphaerularia bombi* as members of the same suborder and tribe. Not surprisingly, he considered *Sphaerularia* as a genus inquirendum. He assigned twelve species to *Anguillula*, nine of which had been assigned to *Oxyuris* spp., and found in intestines of insects. Diesing recognized two genera, *Gordius* and *Mermis*, but kept them in the same suborder and tribe. In 1865, R. Leuckart and I.I.Mechnikov established the alternation of generation in the life history of animal parasitic nematodes and the obligate development of nematodes in the intermediate host. Henry Charles Bastian (1865) published a 'Monograph of the Anguillulidae' in which he described 100 new species of free living nematodes of which 23 species were new. He was able to differentiate animal parasitic nematodes from free-living and subdivided the latter into continental (soil and fresh water nematodes) and marine forms. Schneider (1866) provided one of the first classifications of nematodes. He accepted only *M. nigrescens* and named a new species *M. lacinulata*. He separated *Mermis* from *Gordius*, but left *Sphaerularia bombi* with *Gordius* in the same group. In the same monograph, he reveals his curiosity about *S. bombi* by including a chapter on its development. After this, Johann Adam Otto Butschli (1873) provided the first detailed morphological description of free living nematodes. He was not only a nematologist but a histologist whose embedding methods for thin tissue sectioning paved the way for all biological research and proved that nematodes could be useful study objects for embryology and genetics. He also set standard for nematode illustrations which stands even today. He is in reality "the founder of science of nematology" (Thorne, 1961).

Another taxonomic monograph on soil nematodes by Johannes Govertus de Man (1876) was published which was the beginning of nematology in The Netherlands and Belgium. He was one of the most interesting personalities ever engage in nematology (Thorne, 1961) and was known as 'the old master of nematode science'. He published 50 papers in nematology out of total 170 of his scientific publications showed his devotion to the subject. He published about 109 books on Nematoda and Crustacea, decapods and Stomatopode and edited many Dutch and foreign collections, including the Mergui Archipelago of decapods from the Indian Museum at Calcutta, extensive collections from the Indian Archipelago of the Museums in Amsterdam, Göttingen, Frankfurt, Lubeck, etc. His publications are of extremely high scientific quality. The descriptions are meticulous and extremely detailed, illustrated with drawings by its clarity and careful execution are often true masterpieces. Andrassy (1976) graded him as 'the first modern nematologist' and recommends his beautifully illustrated monograph (de Man, 1884) on nematodes in soil and fresh water as 'the bible of nematologists'. J.G. de Man's (1880) contributions in nematology in Europe are comparable to N.A. Cobb (1907-1918) in USA. He also contributed in the area of nematode anatomy. His famous de Man's formula for measurement of nematodes is still used (Coomans, 1992). In 1921, de Man gave us new genera like, *Ecphyadophora, Hemicycliophora, Psilenchus* and many new species. Thorne (1961) considered these publications as "beginning of the science of plant nematology".

P. Manson (1878) established mosquito as the intermediate host in transmission of the nematode, *Wuchereria bancrofti* causing 'elephantiasis' and it received much attention as it was the first case of biting insect as the intermediate host (Chitwood and Chitwood, 1974). Orley (1880) published a system of classification for 202 nematode species in 27 genera and proposed the family Tylenchidae. Vejdovsky (1886) separated Nematoda and Nematomorpha.

If somebody starts writing the history of nematology, he will never miss to write about the cyst nematodes of sugar beet, potato, soybeans, the root knot nematodes, discovery of DD, EDB, DBCP, the virus vectors, the burrowing nematodes, the stem and bulb nematodes, the pine wood nematode and above all the *Caenorhabditis elegans*- all have their own status to mark different phases of nematological developments. Plant parasitic nematodes all of a sudden came into the public limelight with the discovery of sugarbeet nematode. For many years during the second half of the nineteenth century, farmers of Germany were continuously suffering from severe decline in the production and productivity of sugar beet. Herman Schacht (1859) found that the causal organism of this disease commonly known as 'beet sickness', was a cyst nematode, named as *Heterodera schachtii* by A. Schmidt (1871). A detailed morphological study of this nematode was carried out by Strubell (1888). As the sugar beet industry was important for producing sugar, many scientists (led by Julius Kuhn) got interested in the problem and looked for it not only in Germany but also in other parts of Europe. This led to the discovery of the oat cyst nematode (Kuhn, 1874) and golden nematode of potato (Kuhn, 1881). At that time, these were called as potato or oat races of sugar beet nematode, respectively. He also established some of the management methods for nematodes especially for sugar beet nematode after studying the characters of the hosts, life history, habitats, distribution and etiology etc. His application of carbon disulphide in 1871 against sugar beet nematode was the first example of fumigation of soil for nematode control. He also used trap crops as a method of nematode control for the first time. The aim was to induce penetration of juveniles in the susceptible host and destroy it before the completion of nematode life cycle. Repeated use of this method was not only inapplicable but also cumbersome but crop rotation was most effective as also economical method of control. Four years later, when *H. schachtii* was described, an Italian Licopoly (1875) described small nematodes in galls of the roots of *Sempervivum tectorum* L. Later Cornu (1879) worked on the root knot nematode infesting a legume plant, sainfoin (*Onobrychis sativa*) and described it as *Anguilulla marioni*. He illustrated galls, some of them with secondary roots, eggs with developing juveniles, second stage juveniles and a root section with females and protruding egg masses. He described the females as cyst to express close relationship with *H. schachtii*; he thought it to be closer to *Anguillulina radicicola* Greeff, 1872 (=*Subanguina radicicola*).

Van Benden (1883) gave detailed description of meiosis in a parasitic nematode, *Parascaris equorum*. He also determined that egg and sperm contribute equal number of chromosomes during fertilization and Boveri (1893) did embryological studies in this nematode and reported that fate of cells is determined at the cleavage stage itself.

Muller (1884) found nematodes in galled roots of *Dodartia orientalis* and resembled them to *Heterodera* and renamed *Anguillulina radicicola* which had been described from the same host by Greeff (1872), as *Heterodera radicicola*. He described in detail the process of gall induction and development, followed by morphological observation of stylet, dorsal pharyngeal gland orifice, metacarpus, gonads and spicules. It was Muller who observed and illustrated for the first time a perineal pattern and briefly described the morphological differences between cyst and root-knot nematodes. Schoeyen (1885) illustrated and described a root galling nematode earlier recorded from Norway in 1879 as *Tylenchus hordei* without knowing it has already been described from Germany by Greeff (*Subanguina radicicola*). A German, Frank (1885) published many new hosts of *Heterodera radicicola* and gave the idea to use hosts for the reduction of root knot problem in the field.

The Dutchman, Treub (1885) described a root knot nematode a *Heterodera javanica* infesting sugarcane plants in Java (Indonesia) while studying the origin of Sereh-disease (stunted plants). His description was simple based on few unreliable female and eggs measurements. Beijerinck (1887) and van Breda de Haan (1899) considered *Heterodera javanica* identical to *Heterodera radicicola*. Haan also studied the life cycle of the nematode and speculated for the first time the parthenogenetic mode of reproduction and observed the presence of S-E duct and a pore and discussed the function of the egg sac.

In 1888, Neal, though he was trained in medicines, published an excellent discussion of the affects of root-knot nematodes (know as *Heterodera radicicola*, now *Meloidogyne* sp.) on peaches. This may be the first use of the term "root-knot". Neal (1889) published an interesting account of root-knot nematodes in U.S.A. He wrote, "In 1878, I found the root-knot prevalent over Florida and learnt from old residents that as far back as 1805 it had been known, and from time immemorial has been dreaded as foe to gardens and groves". He described the root-knot nematode as new species, *Anguillula arenaria* infesting groundnut, *Arachis hypogaea* but wrote 'it may belong to a new genus'. He described the tail spike, characters of swollen juveniles, and the loss of it after the last molt and illustrated a part of female rectal glands. He also studied the effects of abiotic factors and mentioned the role of temperature more than that of host or soil on nematode life cycle. He also published an excellent discussion of the affects of root-knot nematodes on peaches.

Earlier, C. Jobert (1878) from France had reported the occurrence of root knot nematode infestation on coffee in Brazil. Then, Goeldi (1887) worked in detail on the root-knot nematode of coffee in Brazil and described it as *Meloidogyne exigua* in the National Museum Archives, vol.8. He was the first person to use the generic name *Meloidogyne*. The nematode, the root knot of coffee, was the most widely spread in whole of Brazil. He also mentioned that the attack by root knot nematode was predisposing the coffee roots to root infecting fungi. At the same time Atkinson (1889, 1892) reported his work on disease complexes involving root knot nematodes and published a paper on 'Nematode Root-Galls'. He also proved that infection by root knot nematodes predisposes the host plant cotton to the infection by *Fusarium* wilt. During 1888-89, both George Francis Atkinson and J.C. Neal worked independently on the morphology, host range and losses caused by root knot nematodes to different crops (Atkinson, 1889; Campbell *et al.*, 1999). In the year 1889 itself, F. L. Scribner started his research work on the lesion nematode affecting potato in Tennessee. Leibscher (1892) described a new cyst nematode species, *Heterodera goettingiana* infecting pea in Germany as different from the sugar beet nematode, *H. schachtii*. Till that time all cyst nematodes were known as races of *H. schachtii*. In 1894, P. H. Rolfs wrote an account of root-knot on tobacco seedlings and illustrated the symptoms. Stone and Smith (1898) published details of development stages of root knot nematodes and illustrated coiling up of male in the juvenile cuticle. Dr. H. Harold Hume prepared a list of plants affected by root-knot in 1901.

The 1890s decade was the most productive from the point of view of advances made in taxonomy, biology, ecology, host ranges and also on management of the plant parasitic nematodes. During this period workers stared selecting and breeding resistant varieties against root knot nematodes. Neal (1889) suggested in a paper for using selected orange trees to control these nematodes. In 1897, Albrecht Zimmerman showed *Coffea arabica* to be more susceptible to root knot nematodes than *Coffea liberica* (Campbell *et al.*, 1999). H.J. Weber and W. A. Orton (1902) gave evidence of root knot and wilt resistance in cowpea and showed that the work was genuine and done for the first time as there was no resistant strain available for this kind of diseases. Orton also found that wilt free watermelon was free of root knot disease. Charles Bessey found certain type of grape cultivars particularly American lines, which were resistant to phylloxera were also resistant to nematode diseases (Campbell *et al.*, 1999). Bessey (1911) was able to publish a classical bulletin on root knot nematodes and clarified the confusing root knot taxonomy by grouping all of them in a single genus, *Heterodera radicicola*. He was for the coordination among workers to study nematode physiology and morphology quite seriously.

Lavergne (1901) reviewed the root-knot nematode records in South America and studied root-knot problems in Chilli and described a new species, *Anguillula vialae*. The first woman who published her work on root-knot nematodes was Kati Marcinowski (1909). Her extensive study on parasitic and semi parasitic nematodes included 24 pages on cyst and root-knot nematodes. According to her there were two species of cyst nematodes and only one species of root knot nematodes and synonymized *M. exigua*, *H. javanica*, *A. arenaria*, *A. marioni* and *A. violae* to *Heterodera radicicola* and reported 236 host plants. Two years later Bessey (1911) reported 480 host plants for *H. radicicola*. Between 1911 and 1919, three studies were conducted for the control of *H. radicicola* by Orton and Gilbert (1912), Bessey and Byars (1915) and Gilbert (1917). J. R. Watson (1915) began a rather long-term study of root-knot in Florida, especially control which lasted until his death in 1946. He found that the commercial product "cysnamidl" provided some measure of control as did calcium cyanamide, sodium cyanide, and mixtures of the latter with ammonium sulfate; he advocated rotations of susceptible crops with resistant plants such as velvet beans, *Crotalaria* spp., iron cowpea, etc., he first reported his work with organic mulches for the control of root-knot. In 1938, he also become active in breeding for resistance to root-knot; and in 1944, Dr. Watson first reported on the use of D-D soil fumigant in Florida (several workers reported on the use of D-D in 1944). From 1934 to 1936, Dr. G. R. Townsend, plant pathologist, did considerable research on temperature relationships of root-knot nematode.

Nathan Augustus Cobb often considered as the father of nematology, worked under renowned German zoologist Ernst Haeckel and joined USDA in 1907. He was a brilliant person who obtained his PhD degree from the University of Jena, Germany in 1889 only in 10 months after which he worked in Zoological Research Station, Naples, Italy. While in Naples he named a marine nematode as *Tricoma*. He then proceeded to Australia and joined New South Wales Department of Agriculture as a plant pathologist. Here he published his first paper on plant parasitic nematodes, '*Tylenchus* and root galls' unaware of the fact that root galls forming nematodes have earlier been named as *Heterodera* and *Meloidogyne*. Cobb (1893) produced a monograph, *Nematodes, mostly Australian and Fijian*, that contained many tylenchid nematodes including *Radopholus similis* which he described on the basis of males as *Tylenchus similis*, and the female of this nematode was described as a new species in this monograph. He did not remained confined to taxonomy and morphology but studied nematode ecology, their impact on farm animals as parasites and also how they affected the crop production. He prepared capsules of *Oxyuris* spp. and swallowed to show that these animals can survive in human beings. Being a keen observer, he noted that nematode (*Dorylaimus bulbiferus*) have parasites (probably *Pasteuria*). While in USDA working full time in nematology, he coined many new terms like amphid, deirid, phasmid and nema, a shorter name for nematodes. He published his first paper in 1913 on 'New nematode genera found inhabiting freshwater and non-brackish soils'. In 1914, he initiated a publication 'Contribution to a science of Nematology' for which he paid himself and in 1915 he brought another publication 'Nematodes and their Relationships' by which he tried to familiarized the science of nematology. His article 'Estimating Nema Population in Soil' in 1918, provided many of the conceptual methods and suggested methods of sampling nematode populations and for their extraction (Cobb's sieving and extraction technique) from the soil which are still in use. In 1919, he proposed the phylum, Nemata, and in 1932 gave proper diagnostic characters for it. In 1924, he erected a new genus *Caconema* for the root-knot nematodes but the name was generally not accepted. To facilitate successive examination of number of nematodes, his famous invention was that of a revolving table on which as many as eleven microscopes could be mounted. He had a unique interest and ability in developing complex instruments. He also developed a metal slide commonly known as Cobb's slide for mounting nematodes. He was known as 'skipper' among his students like G. Steiner, G. Thorne, J.R.Christei, B.G.Chitwood, M.B.Chitwood and A.L.Taylor. His concerted efforts to enhance the knowledge of nematodes make him 'father of plant nematology' among all the nematologists of the world. Initially, Thorne (in Salt Lake City, 1918) and Courtney (in Long Island, 1919) were working with Cobb in the fields outside Washington DC. Gotthold Steiner

(1920), Christei (1922) and Chitwood (1929) began to work with Cobb in USDA. He worked with all forms of nematodes. Steiner took Cobb's position (as Director of USDA, Division of Nematology) after 1932 upon his death. He was joined by Tyler (1932), Taylor (1936), Allen, Reynolds and McBeth (1937). Cobb's publication: "Plant parasitic nematodes the growers should know" (1949) is one of his good contributions which not only introduced nematodes to growers but also promoted applied nematology. Jenkins and Taylor (1967) in the introductary part of their book, 'Plant Nematology' wrote that, 'although *many workers have played important roles in development of plant nematology, none have had a greater impact, particularly in the United States, than N.A. Cobb. In 1913 Cobb published his first paper on nematology in the United States. From 1913 to 1932 he was the undisputed leader in nematology in this country. Through his efforts the widely renowned (USDA) nematology research program was initiated and developed. Many of his students and colleagues developed into the leaders in plant nematology in the 1930's, 1940's, and 1950's. In addition, during his productive career he contributed major discoveries in the areas of nematode taxonomy, morphology, and in methodology. Many of his techniques are still unsurpassed*'. Perhaps no one person has had as favorable an impact on the field of nematology as has Nathan Augustus Cobb.

In 1919, Kofoid and White reported a new nematode infection in man. In the stools of human beings they observed eggs of the nematode which had highly refractive hyaline bluish globules flattened asymmetrically at the two poles of the embryo as were seen in *Oxyuris* sp. So they named it *Oxyuris incognita* as the adult stage was unknown. Sandground (1923) reported that the eggs of *H. radicicola* could be demonstrated in the stools of man after the ingestion of parasitized roots and observed the same polar globules and egg morphology as Kofoid and White. This nematode was later known as *Meloidogyne incognita* (Chitwood, 1949). Earlier, Heinrich Micholetzky (1914) from Austria published a paper on nematodes and came with another publication in (1922), 'Die freilibende Erd-nematoden' giving an extensive key to species of diverse group of nematodes encountered in various habitats. Fuchs (1914, 1915) did lot of work on the taxonomy of the tylenchid parasites of bark beetles and published excellent papers on this group in 1929 (described many new species of *Tylenchus*) and 1937 (described many new species of *Aphelenchus*).

In Japan, Nagakura (1930) studied both cyst and root-knot nematodes in detail and listed 12 morphological differences between them. He observed three juveniles stages (first molt within the egg was seen for the first time by him) and three cuticular layers and six rectal glands in the female of root-knot nematodes.

Tom Goodey (1932) reviewed the nomenclature of root-knot nematodes and proposed that *Heterodera marioni* is the correct name for gall forming nematodes. He disagreed with Cobb to place them in a separate genus and synonymised *Caconema radicicola* with *H. marioni*, the name persisted till 1949. T. Goodey was the head of the department of Nematology which was established in 1947 at Rothamsted Experimental station in England. His son J.B. Goodey, D.W. Fenwick, M. T. Franklin and B.G. Peters provided full support in taxonomy and other fields and the department prospered. Later the department was joined by C.C. Doncaster (nematode behaviour), K. Evans (potato cyst nematode), J. J. Hesling (nematode populations), D. J. Hooper (Taxonomy), B. R. Kerry (Biological control), J. E. Peachey (Chemical nematicides), R. N. Perry

(Physiology), A. Shephered (cyst nematode hatching and ultrastructure), D. L. Trudgill (Heteroderidae), H. R. Wallace (Nematode ecology and locomotion), J. M. Webster (Nematode host-parasite relationships), T. D. Williams (cereal nematode management), A. G. Whitehead (nematode control) and R. D. Winslow (cyst nematode populations). During this time there were 45 nematologists working at the station on various aspects of nematology.

Tyler (1938) studied the asexual reproduction of root-knot nematode and role of temperature on nematode development and female egg laying capacity (up to 2000 eggs per female). Shebacoff (1940) reported differences in host plant interaction and found that in a root knot infested tomato field, cotton was not infested but in cotton infested field, tomato was heavily infested. Day and Tuft (1940) observed inconsistent reaction with the same peach varieties between two nurseries suggesting races of *H. marioni*. Christie and Cobb (1941) studied the life history of *H. marioni* and observed four molting.

Until 1968, *Meloidogyne* taxonomy was descriptive with a few exceptions. Taylor *et al.* (1955) studied the perineal pattern of eight species giving a key to these species. Goffart (1957) and Franklin (1957) published brief reviews of 11 described species of root knot nematodes. Gillard (1961) studied biology, distribution and control of root knot species and a critical review on species described by Chitwood (1949).

Nematological research in Latin America started much later in second half of the twentieth century when they received scientific and financial help from European and North American countries. G. Steiner of USA promoted nematological studies in Puerto Rico where he worked and trained young nematologists, A Ayala and J. Romain. L.G.E. Lordello, who was founder of nematological research in Brazil and Society of Brazilian Nematology, was also trained by Steiner. Investigations into nematode problems were initiated by A. Martin and continued by J. Franco and G. Gomez in Peru in early 1960s and by M. Costilla in Argentina under the guidance of J. Sasser and W. Mai of USA.

Among the first generation of true nematologists in USA, was Benjamin Goodwin Chitwood who wrote a classic book 'An Introduction to Nematology' along with Mabel B.Chitwood in 1937 (the book was revised in 1950 and 1974). This is a truly superb contribution to nematology not only in USA but also in the world as their detailed description of internal and external structures provided a strong foundation on which nematode systematics is based. He is commonly referred to as 'the greatest systematist of nematoda' or 'the most outstanding nematologist of all times' or even as the 'one of the earliest architects of the Science of Nematology'. He worked as a junior nematologist with N.A Cobb in USDA in 1928. In 1949, he reinstated the genus *Meloidogyne* proposed by Goeldi in 1887 (for *M. exigua* causing root knots in coffee) by putting five species, *M. incognita*, *M. javanica*, *M. arenaria*, *M. hapla* (n. sp. described by him) and *M. incognita acrita* under the genus in a paper,'Revision of the genus *Meloidogyne* Goeldi, 1887'. With this work, he established strong criteria for species identification and host races. Later on, in 1955 he was appointed as Chief Nematologist of the Florida State Plant Board where he established nematology section. He worked on various aspects including nematode parasites of vertebrates, invertebrates, plants, free living and marine forms. His revision of genus, *Meloidogyne*, establishment of two nematode classes (Secernentea and Adenophorea) and comprehensive books are some of the major contributions to nematology. Whitehead (1968) published a comprehensive account of 23 species of root-knot nematodes known at that time giving detailed morphological information on females, males and juveniles. He also proposed two keys including one for J2 for the first time. Franklin (1972) reviewed 32 species based mainly on the morphological characters of juveniles and perineal pattern. International *Meloidogyne* project started in July 1975 (1975-1984), under the supervision of J.N.Sasser in NCSU involving 100 scientists from 43 countries continued for eight years with a goal to assist developing nations to increase yields of economic food crops by promoting knowledge about root-knot nematodes and protecting the crops from their damaging effects. The project was funded by USAID. Three books were published during these eight years: Root-knot nematodes (*Meloidogyne* species) edited by Lamberti and Taylor (1979) and 'An advanced treatise on *Meloidogyne*' Vol. I and II, both edited by Sasser and Carter (1985). Triantaphyllou (1979 and 1985) studied 600 populations of 24 species of *Meloidogyne* cytogenetically and distinguished two groups in parthenogenetically reproducing populations. One diploid group with chromosomes no. of n=18 reproducing meiotic parthenogenesis and the other polyploid group with chromosomes no. ranging from 30-56 reproducing mitotic parthenogenesis. The latter group included the most important species like *M. incognita, M. javanica, M. hapla* and *M. arenaria.* Jepson (1987) published a book for the morphological identification of *Meloidogyne* species based on female, male and second stage juvenile characters. Esbenshade and Triantaphyllou (1985a, b and 1987) studied 300 out of 600 populations cytologically and biochemically listing four isozymes, esterase, malate dehydrogenase, superoxide dismutase and glutamate-oxaloacetate transaminase which proved very effective for routine identification because product of one female is sufficient for test and variability of these enzyme within a species is very low.

This was the ripe time to start writing self and experiences of other nematologisrs and their results in some form for consulation by budding workers. Thus came the first book on plant nematodes by Tom Goody 'Plant Parasitic Nematodes and the Diseases They Cause' in 1933, that was a land mark in the development of nematology. The book had nine chapters on various aspects of nematology including taxonomy, biology, host parasite relationship and management. Before his death he was able to complete another book 'Soil and Freshwater nematodes' in 1951. This book was enlarged and revised by his son, J. Basil Goodey in 1963. Tom Goodey was a man of many interests. He was a good actor and a renowned singer with the stage name as Roger Clayson not to confuge with his scientific personality. In 1934, Ivan Nicolaevich Filipjev, known as founder of Russian nematology published a book 'Nematodes that are of importance to Agriculture' in Russian which was later on translated in English and also 'Classification of the free-living nematodes and their relation to parasitic ones'. At age of 21, he published his first paper on the nervous system of nematodes and wrote about evolution of modern genetics in Russia. Then he shifted to U.S.A where he lectured on classification of nematodes and did nematode taxonomical research and grouped free living and plant parasitic nematodes in eleven orders. Filipjev based his classification of nematodes on the shape of amphids and considered other characters as supplementary. He also underlined the importance of embryology and physiology in the relationship of nematodes with related organisms. When writing about Filipjev in the book "Plant and Insect Parasitic Nematodes", Nickle and Welch stated "If Rudolphi is recognized as the Father of Helminthology, then Ivan Nikolaevich Filipjev (1889-1940) may certainly be named the father of Insect Nematology. Not only did he contribute indirectly through his work on nematode classification, but directly by bringing together the scattered information on insect nematodes and incorporated his own findings. This formed the second section (about 80 pages) of a monograph published in Russian in 1934 entitled "Nematodes that are harmful and useful in Agriculture" (Filipjev, 1934). Sections of this book were sent to Dr. Prof. J. H. Schuurmans-Stekhoven in Belgium who, as junior author, had the original Russian translated into French and then into English in the form of a book,' A Manual of Agricultural Helminthology' in 1941. This book had the motto, *'Concordia parvae res crescent'* which means 'Collaboration fosters small undertakings' and was dedicated to the memory of de Man, Cobb, Filipjev, Micoletsky and Ritzema Bos. This is a most comprehensive compendium of nematological information and is especially valuable as a reference source for plant parasitic, free-living, and insect parasitic nematodes described up to that time. Chitwood's book, 'An Introduction to Nematology' was published three years later than Filipjev's book (Chitwood and Chitwood, 1937) and this was revised in 1950 and then in 1974. In 1951, Hans Goffart summarized the plant parasitic nematodes in Europe and also Mary Franklin released her book 'Cyst forming species of *Heterodera*' in this year. Christie and Perry (1951) established the pathogenecity of an ectoparasitic nematode, *Trichodorus christie* which broadened the concept of parasitism. Christei also discovered the nematicidal potential of EDB (ethylene dibromide), which was marketed in the beginning of 1945. He also worked and produced classic study on the mermithid parasites of grasshoppers.

The first International Symposium held at Harpenden, U.K. (1951) was developed and organized by Tom Goodey and B.G.Peters. The second meeting was conducted at Copenhagen in association with Zoological Congress in 1953 and third larger meeting was held at Wegningen in Netherlands by P.A. van der Lan, M. Oostenbrink and J.W. Seinhorst where European Society of Nematologists was formed in 1955 and the trend for formation of professional societies started. The following year (1956) saw the publication of the journal, Nematologica (now Nematology). Ensuing this, the Society of Nematologists was formed in Europe in 1961 but it started the publication of the Journal of Nematology as late as 1969. Since that time many societies have come up (>30 professional societies) with Nematological Society of India (1969) publishing Indian Journal of Nematology (1971), Japanese Journal of Nematology, Pakistan Journal of Nematology and Chinese Journal of Nematology came into being in 1971, 1982 and 1996, respectively.

The Various Nematological Societies and Federations

- ☆ Afro-Asian Society of Nematologists (AASN)
- ☆ Australian Association of Nematologists (AAN)

- ✰ Brazilian Nematological Society (Sociedade Brasileira de Nematologia) (SBN)
- ✰ Chinese Society of Plant Nematologists (CSPN)
- ✰ Egyptian Society of Agricultural Nematology (ESAN)
- ✰ European Society of Nematologists
- ✰ International Federation of Nematology Societies (IFNS)
- ✰ Italian Society of Nematologists (Societa Italiana di Nematologia) (SIN)
- ✰ Japanese Nematological Society (JNS)
- ✰ Nematological Society of India (NSI)
- ✰ Nematological Society of Southern Africa (Nematologiese Vereniging van Suidelike Afrika) (NSSA)
- ✰ Organization of Nematologists of Tropical America (ONTA)
- ✰ Pakistan Society of Nematologists (PSN)
- ✰ Russian Society of Nematologists (RSN)
- ✰ Society of Nematologists (SON)

Nematology Journals and Other Publications (Publishing Society or Organization)

- ✰ Asian Journal of Nematology (AJN)
- ✰ International Journal of Nematology (ASSN)
- ✰ Nematologia Brasileira (SBN)
- ✰ Journal of Nematology (SON)
- ✰ Nematology Newsletter (SON)
- ✰ The Egyptian Journal of Agronematology (ESAN)
- ✰ Egyptian Society of Agricultural Nematology Newsletter (ESAN)
- ✰ Nematology News (ESN)
- ✰ Nematologia Mediterranea (Istituto di Nematologia Agraria of the C.N.R.)
- ✰ Japanese Journal of Nematology (JSN)
- ✰ Indian Journal of Nematology (NSI)
- ✰ African Plant Protection (NSSA)
- ✰ Nematropica (ONTA)
- ✰ Organization of Nematologists of Tropical America Newsletter (ONTA)
- ✰ Pakistan Journal of Nematology (PSN)
- ✰ Pakistan Society of Nematologists Newsletter (PSN)
- ✰ Russian Journal of Nematology (RSN)
- ✰ Nematology (Brill Academic Publishers)
- ✰ Nematology, Pakistan Journal of Nematology and Chinese Journal of Nematology coming into being in 1971, 1982 and 1996, respectively.

Plant nematology developed as the knowledge regarding all forms of nematodes as also in all aspects of the science increased. The workers started paying attention to interactive associations of nematodes with other biological agents like fungi, bacteria and viruses. The 'Friendly Fungi" by Duddington, (1957) was responsible for further enhancing the interest in studying the antagonists and parasites of nematodes as biological means of controlling them. This was the time when Hewitt, Raski and Goheen (1958) discovered that the plant parasitic nematode, *Xiphinema index* was responsible for transmission of grapevine fan leaf virus, thus, increasing the role of nematodes in disease etiology. Thereafter a series of books were published by leading nematologists in many countries, notable among them were: J.R Christie's (1959) - Plant Nematodes-Their Bionomics and Control; J.F.Southey's (1959)-Plant Nematology; Joseph N. Sasser and W.R.Jenkins's (1960)-Nematology-Fundamentals and Recent Advances with emphasis on Plant Parasitic and Soil Forms (This book was the result of series of lectures offered by leading scientist at the course organized at the North Carolina State University sponsored by Shell Chemical Company); William F. Mai and H.L. Lyon's (1960)-Plant parasitic Nematodes –a Pictorial Key to Genera (with five subsequent editions); Gerald Thorne's (1961)-Principal of Nematology; in Russia, Paramonov (1962, 64) wrote two volumes of his book. 'Principal of Plant Nematology'; Harry R. Wallace's (1963)-The Biology of Plant Parasitic Nematodes; J.B. Goodey's (1963)-Soil and Freshwater Nematodes; D.L.Lee's (1965)-The Physiology of nematodes; J.E.Peachy's (1969)-Nematodes of Tropical Crops; Neil A. Croll's !970)-The Behaviour of Nematodes; A.F. Bird's (1971)-The Structure of Nematodes; Zuckerman.B.M., W.F.Mai and R.A. Rhode's (1971)-Plant parasitic Nematodes (as eds) with third vol. in 1981; J.M. Webster's (1972)-Economic Nematology; and Harry R. Wallace's (1973)-Nematode Ecology and Plant Disease.

This was the period when nematologists also started developing formal and specialized teaching courses in nematology, setup nematology departments, formed societies which published journals and also paid attention to extension programmes for the awareness of farmers about nematodes. The first graduate level course in nematology was taught by Merlin W. Allen in 1948 in the University of California at Berkeley campus where he developed international level teaching programme. Allen then moved to Davis campus of the university in 1958 and along with Dewey J. Raski continued to develop the subject and its research activities. Chitwood also taught at Catholic University of Maryland from 1949-1952. During this time lot of trained personnel started taking interest in the subject doing not only teaching but research work also. Foremost in the line were J.N. Sasser, Hedwig Hirschmann at North Carolina State University, W.F.Mai at Cornell University, J. R. Christie, G. Thorne and V. G. Perry at University of Wisconsin, A. J. Overman,G. C. Smart and A. C. Tarjan at University of Florida.

A number of departments were formed solely for the science of nematology across various countries. This included the foundation of department at Rothamsted in U.K., the Landbougheschool, Gent, Belgium, the Institut voor Pflantezeictenkundig, Wegeningen and Division of Nematology at Indian Agricultural Research Institute, New Delhi, India in the early or middle sixties. In addition to the courses developed at USA and Europe, in 1961 F.G.W.Jones with the help of local scientists develop a course in plant nematodes in the Department of Botany at Aligarh Muslim University, Aligarh in India. With the detection of golden nematode of potato in Ooty in Nilgiris Hills in 1961 by him, interest in the plant nematodes increased further. Another international Course was organized by F.G.W.Jones with help of J. B. Goodey from Rothamsted and D. J. Raski from University of California, Davis at IARI, New Delhi in 1964. The first Congress of Nematology worldwide was sponsored by European Society of Nematologists, Society of Nematologists and the Organization of Nematologists of tropical America in Guelph, Canada, second congress by these societies again was organized in The Netherlands in 1990 and third in 1996 in Guadalupe. At this venue of third International Congress of Nematology, International Federation of Nematological Societies was also formed.

The important discoveries and documentation of crop threatening nematodes like potato and soybean cyst nematodes, the root knot nematodes, *Radopholus* sp., *Ditylenchus* sp. including their genetic diversity, pathogenic quality of ectoparasitic nematode and virus transmission by nematodes provided the much required impetus for development of plant nematology in 1940-1950s. The description of *Heterodera glycines* Ichinohe (1952) was also a part of this development as this nematode was discovered in USA in 1954 (Riggs and Wrather, 1992). With the discovery of D-D (1, 3-dichloropropene-1, 2-dichloropropane) in 1943 by Carter at Pineapple Research Institute in Hawaii, it was observed that D-D mixture controlled nematodes in pineapple more efficiently than chloropicrin. Additional discovery of EDB in 1945 by Christie which was equally effective against nematodes, a new phase of highly effective nematicides started which provided a reliable method for demonstrating the losses caused by the nematodes. Then in 1954, another very potent nematicide DBCP (Dibromodichloropropane) came into use (McBeth, 1954) that enhanced the utility of these highly effective chemicals mostly against the above nematodes. Soon these nematicides found usage throughout the world against many threatening plant parasitic nematodes. But due to ground water contamination and toxic effects on humans, now these chemicals are not in use.

The pioneering work of Oostenbrink (1966), Jones *et al.* (1967) and Seinhorst (1970) in modeling nematodes population dynamics and developing crop response models provided models for determination of crop losses caused by nematodes. Developments in precision agriculture including GPS (Global Positioning System) and GIS (Global Information System) have recently been used to improve the nematode associated crop losses and integrated pests' management. The monoxenic culturing of *Pratylenchus penetrans* (Mountain, 1958) and similar research work by others opened avenues for studies on nematode induced damage to plants and nature of nematode-host interactions. The characterization of cytogenetics and related work on plant parasitic nematodes by Triantophyllou (1985) and others contributed much to the studies on nematode biology and genetics. Effectively, the period during the 1960-1990's brought much of the awareness of understanding the ultra structure of nematodes, behavior, physiology, host parasitic relationships, biological control of insects with nematodes, molecular biosystematics, nutrient cycling, molecular level of understanding host parasitic interaction including resistance, cloning of resistant genes and study of whole nematode genomics. Few years back, the first CD-ROM came out that concerned with the root knot nematode taxonomic data base including the most important articles and book chapters on cytology, distribution, molecular biology, morphology and techniques (Karssen and Wiesenekkar, 1998).

By 1998, *Caenorhabditis elegans* has placed itself at everybody disposal and had become "a landmark in biology- determination of essentially complete DNA sequence of an animal genome". In 2002, Swedish newspaper had emphatically written headlines, 'Rundmask far Nobelpris" (Roundworms receives Nobel Prize) when three scientists (Sydney Brenner, Robert Horvitz and John E. Sulston) working with genomics of *Caenorhabditis elegans* got Nobel Prize. With its 959 cells and 20,000 genes *C.elegans* has helped world to sniff at the secrets behind many diseases like cancer, AIDS and Alzheimer's. As benefactor of mankind, it is no doubt that *C.elegans* is a unique star in the nematological sky.

1.3 The Indian History

Apart from numerous references about animal parasitic nematodes in Rig and Atharv Vedas and in Charak Samhita in the ancient times and the description of plant parasitic nematodes and their symptoms during 1283-1301, in section XIII of the *Upavanavinoda* dealing with *Tarucikitsa* (treatment of plants), the present day nematological work on plant parasitic nematodes started with beginning of the twentieh century in India. C. A Barber (1901) reported the occurrence of root knot nematode infesting tea plantation in Devala (Madras state) in South India. After that there were reports of several other nematode diseases from different parts of the country. The root knot nematodes on black pepper (1906) and 'Ufra' disease of rice caused by *Ditylenchus angustus* (1913) were reported by E.J. Butler. Hutchinson (1917) reported the occurrence of 'Tundu' disease of wheat caused by *Anguina tritici* in Punjab. During 1926-1934, Ayyar published several reports on the incidence of root knot nematodes in vegetables and other crops in south India. At the same time, Dastur (1936) reported the white-tip disease of rice caused by *Aphelenchoides besseyi* from the then Central Provinces. Thrumala Rao (1956) reported root knot nematode problem on citrus and occurrence of citrus nematode, *Tylenchulus semipenetrans* was reported by Siddiqi (1961). The organized research on plant nematodes in India could start only in the 1950s, first at the Aligarh Muslim University, followed by the Indian Agricultural Research Institute, New Delhi. The most contributing factor for the development of nematology in India was the discovery of 'molya' disease of wheat caused by *Heterodera avenae* in Sikar district of Rajasthan in 1958 by Vasudeva of Division of Mycology and Plant Pathology, I.A.R.I., New Delhi. Subsequently, it was reported from many areas in the same state. Then, the report of presence of potato cyst nematode from Nilgiris hills of Otacamund in Tamil Nadu by Jones (1961) was sufficient to strengthen the plant protection department by the government specially the work on nematodes as it resulted in sanctioning of ICAR scheme on potao cyst nematods at this place. Also in 1963, a coordinated scheme on nematodes infecting cotton was sanctioned in Coimbatore and Bangalore. This led to the organization of International Course in nematology in 1964 at I.A.R.I. with support from FAO and Rockfeller Fouindation under the guidance and supervision of Prof. D.J. Raski from

U.S.A., Drs F.G.W. Jones and J.B.Goodey from U.K. including many workers from India. Ultimately in December 1966, the Division of Nematology was established with full fledged research, teaching and extension programmes and was headed by Dr A.R. Seshadri with Dr S.K.Prasad as first Professor. In 1964-65, though nematology had just started taking its roots, potato cyst nemaode was detected in consignments received from West Germany, Ireland and Scotland.

This era (1967-68) also started with international collaboration in organizing several South Asian Nematology Courses of three months duration jointly organized by I.A.R.I. and Aligarh Muslim University, Aligarh in association with Dr Oostenbrink and his colleagues from International Agricultural Centre, Wegeningen, The Netherlands, which enabled many workers to get trained in nematology. In all, seven (1968 -1972 and 6th in 1975 and 7th in 1979) such training courses were organized at both the centres, the last one was organised at IARI, in 1979. In August 1969, the Division of Nematology organised the First All India Nematology Symposium which was inaugurated on August 21, 1969 by Dr B.P.Pal, then D.G., ICAR under the Chairmanship of Dr M.S. Swaminathan, then Director IARI, New Delhi leading to the formation of Nematological Society of India with its headquarters at Division of Nematology, I.A.R.I., New Delhi which subsequently started publishing Indian Journal of Nematology in 1971. Prof. D. J. Raski from USA, a well wisher of Indian nematology was also present on this day that greatly helped local nematologists in the formative years of the society. In 1969, MSc and Ph D teaching pogrammes in Nematology also started at IARI, New Delhi.

During the next few years, independent nematology departments or separate sections in entomology / plant pathology started functioning at various institutes in the country. In 1977, an 'All India Coordinated Research Project on Nematode Pests of Crops and their Control' started functioning with 14 centres in the country with its headquarters at Division of Nematology, IARI, New Delhi. At present, there are many agricultural research institutes, universities and many centres under AICRIP (Nematodes) where work on different aspects of nematodes including the study on transgenics and molecular characterization is being carried out.

Chapter 2
Classification and Nematode Identification

The father of modern nematology, Cobb (1919) published an article on the classification of the phylum "Nemates" although previously in 1917 he used the word "nemas," though, in the same article, the word "nematodes" was used. In 1931, in his presidential address before the American Society of Parasitologists he pointed out that the term "nematode" is a contraction or vulgarity for nematoid and a weaker term since it means 'thread like,' while nema means 'thread'. For thousands of years the word 'nema' meaning a nematoid or nematode has been a household word, the Greek word for 'yarn' or 'thread. In the same article he published various derivatives of the word Nema, *i.e.*, "nematicide," "nematosis," "denematize," and pointed out the value of their usage and said that'Nematologist' and 'Nematology' are established words. He preferred the word 'nema' than 'nematode'.

The word "nematode" is a corruption of the ordinal name Nematoidea of Rudolphi (1808). It was already in use in its German plural equivalent "Nematoden" at least by the time of Wiegmann (1835), who, however, continued "Nematoidea" for the taxon itself. Beginning with Burmeister (1837), the stem nematode was used by some workers for a formal zoological group in Burmeister's case for a Familie Nematodes. Subsequently, both R. Leuckart (1848) and von Siebold (1848) employed an Ordo Nematodes, which Diesing (1861) modified to Ordo Nematoda. Vogt (1851) first proposed the name Nematelmia for a Klasse to include the Ordnungen Nematoidea, Gordiacei, and Acanthocephala. This name was modified to Nemathelminthes by Gegenbaur (1859). As originally proposed by Rudolphi (1803, 1809), the order Nematoidea contained some species that were originally described as gordiids (horsehair worms) and that later workers placed in synonomy with gordiids. Von Siebold (1843) created an Ordnung Gordiacei, coordinated with the

Ordnung Nematoidea, but incorrectly associated the nemic genus *Mermis* with the horsehair worms of the genus *Gordius*. This continued to be true for much of the remaining half of the 19th century. Huxley (1864) promoted the horsehair worms provisionally to a Class Gordicea, coordinate with a Class Nematoidea in a category Scolecida. However, it appears probable, on the basis of later publications by him (Huxley, 1877), in which he reverted to a group Nematoidea of unspecified rank but containing as "Groups" Polymyaria, Meromyaria, and Holomyaria, in the last of which he included Gordius, that in 1864 he had no better concept of the Gordiacea than had von Siebold in 1843. Such was equally the case with Lankester (1877) who promoted the Nematoidea to the level of phylum, but included therein, as a "subdivision," the Gordiidae. The first clear distinction between the nemas and gordiids was realized by Vejdovsky (1886) at which time he named a group to contain the horsehair worms in the order Nematomorpha.

Cobb first proposed the Phylum Nemates in 1919 and gave a diagnosis in 1932 but Potts (1932) and Chitwood (1950) did not recognize Nemata Cobb, 1919, rather synonymized it with the Nematoda (as a phylum). The nematodes which are vermiform or worm-like, fall in the category of the heterotrophs mobile organisms in the kingdom Animalia under phylum Nematoda. There are about 30,000 named species but the true number may probably be 10 - 100 times this number. These animals are triploblastic (with three germ layers), bilaterally symmetrical (mirror image if the nematode is cut along the saggital plain), unsegmented (segmentation is superficial), pseudocoelomates (without true coelome). They are free living, parasites of humans and plants including our crops. They can live pretty much anywhere from deserts to deep oceans and from hot springs to antartica's icy lands. They have become very important in development studies, especially the species *Caenorhabditis elegans*, presumably due to its small size and constancy of cell number (eutely - 959 cells).

2.1 Nematode Fossils

Nematodes are soft bodied and mostly very small organisms. Any larger forms that existed were probably parasites of vertebrates; however, they left no fossil traces. The only fossil nematodes that are known are insect parasitic or plant parasitic forms that occur very rarely. Though eggs of the pinworm *Enterobius vermicularis* and the whipworm *Trichuris trichuria* occurred in coprolites (trace fossils) dated to about 7,000 years old from dry areas of Peru, and eggs of the hookworm *Ancylostoma duodenale* have been reported from coprolites dated to around 7,230 years old. In eastern Brazil, *Ascaris lumbricoides* has been positively identified from human coprolites dated to about 28,000 years old from caves in France, but this is the only oldest occurrence of a record. The seeming dearth of other nematodes from human remains older than about 7,000 years old appears to be due to the fact that organic material comprising the coprolites themselves does not preserve well enough to last that long. Because there are no fossil records of nematodes of Cambrian or Precambrian ages, estimates of the age of the nematodes have been only speculative, and without fossils, it is difficult to calibrate molecular clocks for the nematodes.

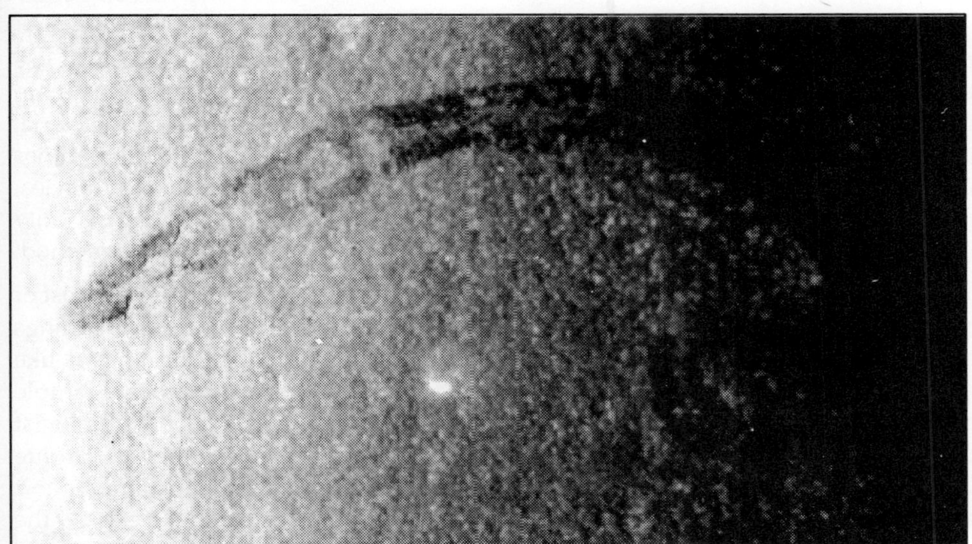

Figure 1: *Eophasma jurasicum*, A Fossilized Nematode (from Wikipedia)

However, through application of various models of molecular evolution and molecular clock theory, estimates of the time of divergence of the nematodes from the rest of the animal groups appears to be about 1,177 ± 79 million years (Wang *et al.*, 1999). Most authors consider the ultimate origin of nematodes to be from a marine ancestor (Maggenti, 1981; Malhakov, 1994).

2.2 Nematode Nomenclature and Classification of Living Organisms

Linnaeus (1758) is best known for his introduction of the method still used to formulate the scientific name of every species. His 1st edition of *Species Plantarum* and the 10th edition of *Systema Naturae* were chosen as starting points for the Botanical and Zoological Nomenclature (Bionomial nomenclature) respectively.

Whereas Linnaeus classified for ease of identification, it is now generally accepted that classification should reflect the Darwinian principle of common descent. Since the 1960s, a trend called cladistic taxonomy (or cladistics or cladism) has emerged, arranging taxa in an evolutionary tree. If a taxon includes all the descendants of some ancestral form, it is called *monophyletic.* Groups that has descendant groups removed from it (*e.g.* dinosaurs, with birds as offspring group) are termed *paraphyletic,* while groups representing more than one branch from the tree of life (science) are called *polyphyletic.*

A new formal code of nomenclature, the International Code of *Phylogenetic Nomenclature*, or *PhyloCode* for short is currently under development, intended to deal with names of clades. Linnaean ranks will be optional under the PhyloCode, which is intended to coexist with the current, rank-based codes. As advances in microscopy made classification of microorganisms possible, the number of kingdoms increased, five and six-kingdom systems being the most common.

Domains are a relatively new grouping. The three-domain system was first invented in 1990, but not generally accepted until later. The majority of biologists use the 5-Kingdom system, but a large minority uses the 3 domain system. One main characteristic of the three-domain method is the separation of Archaea and Bacteria, previously grouped into the single kingdom Bacteria (a kingdom also sometimes called Monera). Consequently, the three domains of life are conceptualized as Archaea, Bacteria, and Eukaryota (comprising the nuclei-bearing eukaryotes). A small minority of scientists adds Archaea as a sixth kingdom, but do not accept the domain method.

Though 6 kingdoms have been proposed by Cavalier Smith (2004) but most of the scientists agree for 3 kingdoms (Domains) of Woese *et al.* (1990). So living things are currently split into three domains-eukaryotes or coplex celled organisms like animals and plants and humans; bacteria and Archaea, the last two being simple celled microorganisms but recently Jonathan Eisen (2011), an evolutionary biologist of University of California, using a complicated gene sequencing technique (metagenomics) found that some of the genes did not fit into the three domains and could possibly had stumbled upon a whole new domain but he could not classify the new DNA.

Table 1: Classification of Living Organisms

Linnaeus, 1735 2 Kingdoms	*Haedkal, 1866 3 Kingdoms*	*Chatton, 1925 2 Empires*	*Copeland, 1938 4 Kingdoms*	*Whittaker, 1969 5 Kingdoms*	*Woese et al., 1977 6 Kingdoms*	*Woese et al., 1990 3 Domains*	*Cavalier Smith, 2004 6 Kingdoms*
	Protista	Prokaryota	Mychota	Monera	Eubacteria	Bacteria	Bacteria
				Protista	Archae-bacteria	Archae	Protozoa
		Eukaryota	Protoctista	Plantae	Protista	Eukarya	Chromista
			Plantae	Fungi	Plantae		Plantae
Vegetabilia	Plantae		Animalia	Animalia	Fungi		Fungi
Animalia	Animalia				Animalia		Animalia

Taxonomy (derived from Greek word 'taxis' means arrangement and 'nomos' means law- proposed by de Candolle (1813) is the systematic grouping of organisms according to their natural relationships. While based on thorough investigation of organism features and structures, opinions may differ among experts regarding the importance and significance of individual features as indicators of relationship. Additionally, the technologies like transmission and scanning electron microscopes, biochemical and molecular techniques, allow new interpretations of the relationships. Consequently, although guided by the International Rules of Zoological Nomenclature, the taxonomy of nematodes is seldom static. Changes in taxonomy are, of course, important indicators of the evolution of our understanding of the organisms but sometimes they can give rise to confusion when some characters are discussed. Competent traditional taxonomic work, *i.e.* based solely on morphological evidence, can be as revealing as any other modern behavioral, biochemical or molecular method as to the identity of a group (Mayr and Ashlock, 1991; Wang *et al.*,

2008). However, the reasoning through which a decision about identity is made inherently involves a certain level of subjectivity, for it is the experience of the taxonomist that matters most in making these decisions.

Above the level of the order, confusion reigns relative to the classification and systematic arrangement of the nematodes. Maggenti (1991) is usually followed in this regard, and his analyses followed corroborated historical analyses in recognizing the two main classes: the Secernentea and the Adenophorea. Recent work showed that these groups are substantiated both in morphological and in molecular analyses, although competing phylogenetic hypotheses and associated classifications have also been proposed (Dorris *et al.*, 1999; Blaxter *et al.*, 1998; Brooks and McLennan, 1993; Adamson, 1989).

2.3 Molecular Level Diagnostics for Nematodes

The use of molecular techniques for nematode identification is in its infancy though some species of cyst nematodes (*H.australis* and *H.pratensis*) have been described based on differences in DNA analysis. DNA analysis through RFLP and RAPD is very useful for overcoming problems associated with morphological diagnosis in identifying sibling species, races, or pathotypes. These techniques are highly valuable in statutory control or quarantine where presence of even a single nematode is significant. Technology for utilization of DNA probes is evolving rapidly at present and further advances are sure to occur soon. Some of these will involve development of methods that are technically simpler, less expensive, and more capable of automation. In addition to identification, molecular tools will provide valuable insights in resolving some of the complex aspects in plant nematology such as development of resistant cultivars, definition of nematode-host interactions, nematode population dynamics, development of useful biological control agents, and design of new nematicides. Molecular characters can be used in identify, diagnose and delimit species in the same way as the morphological characters. Potentially sensitive, objective and powerful, molecular characters can be dangerous, if used carelessly. Species occupy an important position in all aspects of biology, therefore the species concept matters. It determines, for example, the outcome of biodiversity assessments and distribution patterns. However, several problems remain to be solved before a consensus will be reached about the choice of a concept. As of current trends, nematode taxonomy in the future will inevitably become increasingly molecular. The practicality of molecular methods will work to the advantage of nematology, as the use of these methods will encourage the inclusion of this diverse nematode group in biodiversity and environmental studies. Such an inclusion would be an important step towards a better understanding of the biology and ecology of nematodes. Therefore, molecular methods will help us realize our goals of understanding the biological diversity and ecological role of nematodes. Nonetheless, despite some difficulty associated with the use of morphology in the taxonomy of some groups of nematodes, the extensive amount of information accrued on various aspects of nematode biology is currently linked with morphology. Consequently, this makes the integration of morphology in taxonomic studies still relevant and essential. It is rightly said that DNA sequences alone are not sufficient to characterize a species, but their unique reproducibility helps to guard against duplicate descriptions (Tautz *et al.*, 2002). Despite its obvious advantages and the fact that it is possible to predict a certain level of gene expression from sequences, our current inability to predict ecological functionality of entire species from DNA sequences alone invalidates such a speedy jump to abandon morphology altogether. In short, molecular methods are best supplemental tools, as are other methods (Mallet and Willmott, 2003). With all its limitations, morphology is as good as any of the other taxonomic tools in delimiting species (Mayr and Ashlock 1991; De Ley, 2006; Agatha and Strüder-Kypke, 2007). Future research in nematode systematics should be balanced one comprising well-focused taxonomy based on a combination of classical and modern methods in a way that can raise the interest of young scientists.

The modern day technologies like molecular phylogenetics, bioinformatics and digital communication have significantly altered the dynamics of nematode systematic. There are many advantags in employing molecular methods which are applicable in high throughput systems, *viz.*,

1. Genetic information in the form of DNA sequences of a few genes can be acquired for many taxa in a few days. Therefore, these methods offer development of high throughput systems which will make nematodes attractive to environmental studies.
2. Homology of genes is simpler to predict and test than homology of morphological characters. In addition, both types of data can be used to infer phylogenetic relatedness and thus create a systematic framework in an evolutionary context. However, because of their sheer number and simple methods for predicting homology (Schwarz, 2005), molecular characters rapidly outnumber morphological ones. Inferring phylogeny from DNA sequences, similar to morphology, has its own recognized methodological limitations that may affect the conclusions. Alignment of sequences using computer algorithms may introduce biases, especially when they are subsequently modified by eye. Deciding positional homology of DNA sequences is a more consistently applicable process than similar decisions for morphological characters.
3. DNA sequences are already digital and can be communicated across laboratories without interpretation.
4. DNA sequences are inherently genetic and therefore overcome many problems of apparent similarity from which morphology suffers, especially with microscopic organisms.

Amplification and sequencing of diagnostic regions of nematode DNA have become the major source of new information for advancing our understanding of evolutionary and genetic relationships (Hajibabaei *et al.*, 2007; Meldal *et al.*, 2007). DNA sequences have unveiled cryptic species among individuals considered conspecific based on morphology alone (Eyualem and Blaxter, 2003; Hoberg *et al.*, 1999; Chilton *et al.*, 1995; Nadler, 2002; Fonseca *et al.*, 2008; Derycke *et al.*, 2008; Bhadury *et al.*, 2008). On the other side of the spectrum, despite its demonstrable and tangible advantages, there is anxiety among biologists, especially taxonomists, that

2.3.1 Advantages, Limitations, and Applications of Molecular Methods in Nematode Identifications (Abebe *et al.*, 2011)

Method	Use	Limitations	Application Example
Allozymes	- First biochemical method in nematode taxonomy - Provides comparable Data between geographical isolates	- Lack of confirmed genetic basis - Cannot be used for single nematode studies - Life Stage specific - Lacks data on specific characters - Requires non-denatured proteins	Nematode genera identified using this method includes: *Meloidogyne, Heterodera, Pratylenchus, Ditylenchus* and *Aphelenchus*. (Esbenshade and Triantaphyllou 1990; Navas *et al.*, 2001).
Restriction fragment length polymorphism (RFLP)	Suited to differentiate between closely related taxa based on presence/ absence of restriction fragment bands	- Lacks homology of characters - Requires large amounts of PCR products to use for different restriction enzymes.	- Detection of species or populations within species (Curran *et al.*, 1985, 1986; Powers *et al.*, 1986; Xue *et al.*, 1992; Fargette *et al.*, 1996).
Amplified fragment length polymorphism (AFLP)	- Suited to assess variation among individuals of the same species	Lengthy procedure	- Detection of species in *Heterodera avenae* group (Subbotin *et al.*, 1999), molecular characterization of *Pratylenchus* species (Waeyenberge *et al.*, 2000) and the study of intraspecific variation in *Radopholus similis* (Elbadri *et al.*, 2002).
Random amplified polymorphic DNA (RAPD)	- Unlike AFLP this method doesn't require a restriction step - Simple and rapid	Not suitable to all organisms and lacks reproducibility - Sensitive to variations in primer and DNA concentration	- Detection of *Globodera rostochiensis, G. pallida, Meloidogyne incognita, M. javanica,* and *M. arenaria*. (Fullaondo *et al.*, 1999; Zijlstra *et al.*, 2000)
Real-time PCR	- Qualitative and quantitative detection of species - Rapid and precise	Requires species-specific primers - Multiple species detection requires lengthy optimization	- Detection of potato cyst nematode (*G.pallida*) and beet cyst nematode (*H. schachtii*), quantification of *Meloidogyne* spp. from tomatoes, *Pratylenchus thornei* and *Pratylenchus neglectus*, detection of *Bursaphelenchus xylophilus* (Madani *et al.*,

Method	Use	Limitations	Application Example
			2005, Stirling *et al.*, 2004, Hollaway *et al.*, 2004).
Multiplex PCR	- Detects more than one species at a time - Cost effective - Rapid	Involves optimization of reagent concentrations, variation in primer annealing temperatures, primer interactions, and amplification of non-specific products - Requires species-specific primers	- Identification of *Globodera pallida* and *G. rostochiensis* (Fullaondo *et al.*, 1999), *Pratylenchus penetrans* and *P. scribneri* (Setterquist *et al.*, 1996), *Meloidogyne chitwoodi* and *M. fallax* (Petersen *et al.*, 1997), and *P. coffeae* and *P. loosi* (Uehara *et al.*, 1998).
DNA sequencing	- Preferred from the above mentioned methods as factors that affect patterns in these methods (*e.g.* variation in primer and DNA concentration, DNA template quality, gel electrophoresis, and the type of DNA polymerase) can be controlled and the sequencing step can be optimized. - Fast and accurate	- Difficult in finding an ideal gene for taxonomic identification as well as phylogenetic inference, that works in all nematode groups	Widely used in recent years; in various groups including family *Hoplolaimidae*, (Subbotin *et al.*, 2007), order *Tylenchida* (Subbotin *et al.*, 2006), suborder *Criconematina* (Subbotin *et al.*, 2005), suborder *Cephalobina* (Nadler *et al.*, 2006) , *Pratylenchus* (Al-Banna *et al.*, 1997), *Acrobeloides* (De Ley *et al.*, 1999), *Steinernema* (Stock *et al.*, 2001), *Meloidogyne* (Castillo *et al.*, 2003), *Longidorus* (Rubtsova *et al.*, 2001; Mekete *et al.*, 2009).
Denaturing gradient gel electrophoresis (DGGE)	- Easy and straight forward approach for nematode community analysis - Used to differentiate multiple species in a sample	- Choice of marker is still an open issue - DNA extraction for DGGE problematic - Not useful for quantification of species richness in marine communities (Cook *et al.*, 2005)	- Community analysis in different ecosystems (Foucher *et al.*, 2004; Griffiths *et al.*, 2006; Shi-Bin *et al.*, 2008; Waite *et al.*, 2003.

Source: After Abebe *et al.*, 2011

molecular barcoding will replace taxonomy as we know it and will reduce the biological complexity of an entire organism to a small fraction of the organism – a gene (Lipscomb *et al.*, 2003; Will and Rubinoff, 2004).

Recently, pyrosequencing has emerged as a new sequencing technology (Creer *et al.*, 2010). This technique is a widely applicable alternative technology for the detailed characterization of nucleic acids. Fundamentally different from DNA sequencing, pyrosequencing occurs by a DNA polymerase driven generation of inorganic pyrophosphate, with the formation of ATP, and the ATPdependent conversion of luciferin to oxyluciferin. The generation of oxyluciferin causes the emission of light pulses, and the amplitude of each signal is directly related to the presence of one or more nucleotides. It is a nonelectrophoretic real-time DNA sequencing method in which enzymatic reactions yield detectable light, proportional to the number of the incorporated nucleotides (Diggle and Clarke, 2004; Edwards *et al.*, 2006). One of the most important applications of this technology is the ability to identify large numbers of species from complex communities and recently it has been widely used to generate the genome sequences of complex environmental samples (Edwards *et al.*, 2006; Joseph *et al.*, 2009). This methodology has shown encouraging results for the analysis of nematode diversity from metagenomic nematode samples (Creer *et al.*, 2010).

It is imperative that the specimens used for sequence generation are correctly identified, preferably by a trained taxonomist using morphological and morphometrical characters, before the sequence data is submitted to public databases. Luc *et al.* (2010) highlighted the limitations of using DNA barcodes from sequences deposited in the GenBank database that had been incorrectly identified. To avoid incorrect labeling, the DNA barcode must be taken from one of the paratypes or paralectotypes when the nominal species-group taxon has been established on many specimens. If such material does not exist, or cannot be used, specimens should be sought in the type locality or a practicable place as near to the original type locality as possible; for parasitic species, it should be from the same host species and the identity confirmed by a specialist in the group in question. Alternatively, DNA can be extracted from archived specimens. For example, successful PCR amplification and sequencing of the 18S rRNA fragment was achieved for marine nematodes that had been fixed and stored for up to 20 years using a modified hot lysis protocol (Bhadury *et al.*, 2007). To improve the resolution and reliability of nematode systematics, studies should ideally combine morphological with molecular data (De Ley and Blaxter, 2002). Thus, molecular approaches should not be considered in isolation but a balanced molecular and morphological taxonomic approach is required.

Meanwhile, morphology based taxonomy as such suffered a decline. Poor descriptions, too much routine work and low citation rates hampered it. As a result, the discipline became less attractive. With only a small fraction of the biodiversity known, this situation will lead to serious problems in the future in all those fields of nematology depending on a correct identification of species. Phylogenetic analyses of nematodes have been mainly based on morphology and supplemented with developmental characters. Phylogenetic estimates derived from independent data may provide new insights in character homologies through reciprocal illumination. Classifications of nematodes were often biased according to the expertise of the author and were only recently based on the principles of phylogenetic systematic (Coomans, 2002).

The nematode systematics has a long history of constant revision, and there may be as many classifications as there are nematode taxonomists. One of the first phylum wide classifications was proposed by Chitwood and Chitwood (1933). They divided the phylum into two classes, the Phasmidia and Aphasmidia, later renamed to Secernentea (*'secretors'*) and Adenophorea (*'gland bearers'*), respectively. This was based mainly on the fact that the Secernentea share several characters, including the presence of phasmids, small sensory organs on the tail. Although it was already recognized at the time that the Adenophorea did not form a natural group, the division of the Nematoda into these two groups persisted for a long time (Maggenti 1963; De Coninck 1965; Siddiqi 1983; Maggenti 1983). The first to propose a tripartite system was Andrássy (1976) who divided the Adenophorea into the Penetrantia (pocket like amphids) and the Torquentia (spiral amphids). However, for various valid reasons, few people accepted this system. The first person to apply cladistic principles to nematode systematics was Lorenzen (1981). He also recognized that the Adenophorea were not a monophyletic group, but could not provide an alternative. Also at lower taxonomic levels (order, family and genus level), systematics was far from stable (De Ley and Blaxter 2002). The main reason for this is that though nematodes are ecologically and physiologically very diverse, their conserved morphology and small size resulted in a paucity of observable, phylogenetically informative characters. Furthermore many characters display a convergent evolution. In recent years DNA sequence data have brought a revival to the field of systematics (Blaxter *et al.*, 1998; Mullin *et al.*, 2005; Subbotin *et al.*, 2006; Meldal *et al.*, 2007). The first major classification to incorporate both morphological and molecular phylogenetic information is that of De Ley and Blaxter (2002).

The Phylum Nematoda

The Phylum *Nematoda* consists of two classes: *Secernentea* and *Adenophorea.*

Characters of Secernentea

1. Pore-like or slit-like amphid apertures - labial.
2. Deirids present in some near nerve ring.
3. Phasmids present, generally posterior.
4. Excretory system tubular.
5. Cuticle 2 to 4 layers, striated; lateral field present.
6. Esophagus varies but has 3 esophageal glands.
7. Male generally with 1 testis.
8. Caudal alae common.
9. Sensory papillae cephalic only, although may be caudal papillae in males.
10. Almost exclusively terrestrial, rarely freshwater or marine.

Characters of Adenophorea

1. Amphids post-labial, variable shape, pore-like to elaborate.
2. Deirids not seen.
3. Phasmids generally absent.

4. Hypodermal glands present.
5. Simple non-tubular excretory system, when present.
6. Male generally has 2 testes.
7. Caudal alae rare.
8. Sensory papillae in cephalic region and along body.
9. Marine, freshwater, terrestrial.

The classification of organisms above the genus and species levels is not governed by the International Rules of Zoological Nomenclature. In essence, higher classification schemes represent expert opinion based on available evidence. Several classification systems have been proposed for nematodes. New opinions on the relationships among taxa are emerging from molecular data.

1. Classification Based on Morphology

Phylum Nematoda

Class–Adenophorea		*Class–Secernentea*		
Enoplia	Chromodoria	Rhabditia	Spiruria	Diplogasteria
Enoplida	Araeolaimida	Rhabditida	Spirurida	Diplogasterida
Isolaimida	Chromadorida	Strongylida	Ascarida	Tylenchida
Mononchida	Desmodorida		Camallanida	
Dorylaimida	Monhysterida		Drilonematida	
Stichosomida				
Triplonchida				

2. Classification Based on SSU-RNA (De Ley and Blaxter, 2004)

Class–Enoplea		*Class–Chromodorea*
Sub-Class–Enoplia	*Sub-Class–Dorylaimia*	*Sub-Class–Chromodoria*
Enoplida	Dorylaimida	Rhabditida
Triplonchida	Mermithida	Plectida
	Mononchida	Araeolaimida
	Dioctophymatida	Monhysterida
	Trichinellida	Desmodorida
	Isolaimida	Desmoscolecida
	Muspiceida	Chromodorida
	Marimermithida	

Based on small subunit ribosomal DNA (SSU rDNA). The small subunit is the 16s component of the rDNA.

2.4 Keys to the Classes and Subclasses of Nematoda

1. Amphids always post labial; hypodermal glands present; hypodermal cells have one nuleus; 3 cuadal glands; male with two testes and with supplementary median ventral glands in a single row; generally have 5 esophageal glands .... **Class–Adenophorea**
2. Amphid apertures on lips, often difficult to see; no hypodermal glands; hypodermal cells are multinucleate; no caudal glands; male with a single testis and with paired pre-anal supplements, subventral in 2 rows; 3 esophageal glands .... **Class–Secernentea**
3. Body without annules; amphid aperture not elaborate, may be large, oval, or stirrup shaped; 5 esophageal glands .. **Sub-class–Enoplia.**
4. Body usually has annules; amphids elaborate, spiral; three esophageal glands. .... **Sub-class–Chromadoria**
5. Three-part esophagus (corpus, isthmus and postcorpus), valve in postcorpus; corpus and postcorpus in 12 cells .... **Sub-class–Rhabditia**
6. Wine bottle esophagus; large nematodes (many >2.5 cm); all animal parasites .... **Sub-class–Spiruria**
7. Three-part esophagus with valve in metacorpus (median bulb) and postcorpus glandular .... **Sub-class–Diplogasteria**

2.4.1 Characters of Chromadorea (after De Ley, 2006)

These animals have evolved from at least four lineages that have attained greater habitat space than all Enoplia and most Dorylaimia. Chromadorians are common in marine sediments, but they have also flourished in terrestrial habitats *e.g.*, those subject to frequent episodes of rapid de- and rehydration, such as mosses and lichens or extremely xeric and/or cryogenic soils. Throughout Chromadoria, cuticular structure has undergone a wide range of evolutionary modifications, sometimes resulting in strikingly decorative ornamentations. A key adaptation within the order Rhabditida was the development of a chemically impermeable cuticle, which clearly contributed to their success as parasites and colonizers. Another subject of striking modification in Chromadoria is the pharynx, which is structurally much more diverse in this subclass than in Enoplia or Dorylaimia. This diversity revolves mostly around the evolution of one or more rounded muscular bulbs, which has apparently allowed for more compact body designs. Free living Chromadoria on average are smaller than Enoplia and Dorylaimia, correlated with a greater dominance of rapidly reproducing bacterial feeders. Several chromadorian lineages have independently evolved curved teeth in can-opener-like arrangements. These are used to pry open the diatomaceous algae, or to slice the cuticle of other nematodes. Some of these species are among the smallest known predatory nematodes. The most distant relatives of *C. elegans* that can presently be efficiently cultured with *C. elegans*-like methods are certain bacterivorous species of the orders Monhysterida and Plectida (De Ley and Mundo-

Ocampo, 2004). Within the order Rhabditida, a major factor has been the development of a modified juvenile stage specifically adapted to long-term survival. A true dauer stage is rarely reported outside of the suborder Rhabditina (which includes *C. elegans*). True dauers are often capable of seeking out and hitching rides phoretically on larger animals, which has in turn allowed multiple invasions of the internal organs of other animals. At least three Rhabditida lineages have independently evolved major zooparasitic animals. A fourth lineage has not only given rise to zooparasitic species, but to the most diverse group of plant parasites and fungal feeders among nematodes. These are the tylenchs, equipped with a protrusible stomatosylet that is convergent with, but clearly different from, the odontostyle of dorylaims and the onchiostyle of trichodorids. SSU sequences have confirmed the previously unpopular hypothesis that their closest relatives are the morphologically very dissimilar cephalobs (Siddiqi, 1980). Both groups are therefore now united in the suborder Tylenchina - another example of a more drastic change induced by the new phylogenies.

Parthenogenesis appears to be much more common in Tylenchina than hermaphroditism, and Goldstein *et al.* (1998) speculated that this could in fact be linked to the mechanism of sex determination. Within the suborder Rhabditina, diversity segregates into four major groups: strongylids, diplogasterids, bunonematids and rhabditids (*sensu stricto*). The latter group includes *C. elegans* as well as many other species that differ morphologically in details of the male and female reproductive systems. Strongylids were traditionally placed in their own order, on the basis of their importance as animal parasites and their morphological complexity as adults. However, they actually arose from within rhabditids, as is clear from SSU sequences (Sudhaus and Fitch, 2001), juvenile morphology and male genital characters. Compared to rhabditids, diplogasterids are characterized by a shift of pharyngeal pumping function to the median bulb, with concomitant muscle and valve reduction in the basal bulb. This arrangement is superficially very similar to the pharynx of Tylenchina, but ultrastructural and molecular data strongly indicate that the resemblance is purely convergent (Blaxter *et al.*, 1998; Baldwin *et al.*, 2001). The group includes the "satellite model" *Pristionchus pacificus*, in a colorful array of bacterivores, fungivores, animal parasites and "can-opener" type predators (Fürst von Lieven and Sudhaus, 2000). The closest relatives of diplogasterids appear to be certain species that exhibit the morphology typical of rhabditids (Sudhaus and Fitch, 2001), but also the truly perplexing bunonematids (Fürst von Lieven, 2002). The latter include some of the most unusual anatomies among nematodes, with complex cuticular ornamentations arranged in dorsoventral symmetry, *i.e.* at right angles to the bilateral symmetry of the internal organs. They appear to be specifically adapted for life along surfaces within decomposing material.

2.4.2 Characters of Enoplea (after De Ley, 2006)

The phylum Nematoda occurs in an incredibly wide spectrum of ecological habitats and natural histories, ranging from *e.g.*, deep sea sediments to arid deserts, or from interstitial bacterivores to obligate parasites with multiple intermediate hosts. Several of its constituent clades cover a large subset of this ecological spectrum, but interestingly none of them appears by itself capable of covering the full ecological range of the phylum, especially so within the class Enoplea. This suggests that the evolution of ecological adaptations within each nematode taxon was constrained by limitations on the rates of change in genes and ecophysiology, or by competitive exclusion from habitats previously colonized by other taxa, or both.

The non-marine occurrence and the present diversity of Dorylaimia both suggest that these could have been the first nematodes to conquer freshwater and terrestrial habitats. Such an early origin could explain the great diversity within this subclass, which includes not only the mostly freeliving Doryaimida and Mononchida, but also remarkable animal parasites such as Mermithida and Trichinellida. Mermithids are highly unusual among metazoan parasites in that they actually leave the host before reaching adulthood - a property otherwise only found in the phylum Nematomorpha. The present existence of such exceptional biologies hints at much greater past diversity. The most successful surviving clade within Dorylaimia, however, is the order Dorylaimida. This includes many species of large predators/omnivores, as well as the plant-parasitic family Longidoridae, of which some species transmit plant viruses. The evolutionary radiation of dorylaims appears to have resulted in large part from functional diversification of the odontostyle, a protrusible, hollow and often needle-like tooth used for puncturing and emptying food items. Although predation or plant feeding are well documented for larger dorylaims, the food sources of smaller species (with much smaller odontostyles) remain unknown.

Enoplia are especially diverse in marine habitats, but multiple lineages are also found in freshwater sediments and/or moist soils. Enoplia are especially interesting phylogenetically because of the occurrence of features that are presumably ancestral within nematodes, such as a highly indeterminate mode of development (Justine, 2002) and retention of the nuclear envelope in mature spermatozoa (Lee, 2002). No Enoplia are known to have adapted to terrestrial environments subject to extreme temperatures, nor are there any surviving lineages that parasitize animals (with the possible exception of a few enigmatic taxa of uncertain position). The one enoplian order that has clearly undergone extensive evolution in soils is the order Triplonchida, which includes plant parasites such as *Trichodorus*. These are convergent with dorylaims in a number of respects, *e.g.*, they also have a protrusible tooth for feeding (called an onchiostyle) and several species are known to act as virus vectors. Molecular data have shown that the triplonchid clade includes free living nematodes such as *Tobrilus* and *Prismatolaimus*, even though these are morphologically quite divergent from trichodorids.

For the origin of these nematodes, the molecular data published to date confirm the presence of three early nematode lineages, corresponding to the previously recognized subclasses Chromadoria, Dorylaimia and Enoplia (Lorenzen, 1981; Inglis, 1983). The exact order of appearance of these three lineages is not yet resolved. It seems likely that Enoplia appeared first, and it is even possible that Dorylaimia and/or Chromadoria could have originated within Enoplia. On the other hand, SSU data also allow for the possibility that Dorylaimia diverged first, which is an intriguing possibility because all known Dorylaimia are absent from marine habitats. A "Dorylaimia first" topology would therefore imply that the ancestor of all nematodes was perhaps a freshwater organism, and not a marine animal as more commonly assumed (De Ley and Blaxter, 2004). Whereas the plant parasitism evolved independently in three traditional nematode orders, Dorylaimida Pearse 1942, Triplonchida Cobb 1920 (both Adenophorea), and Tylenchida Thorne 1949 (Secernentea). In the Tylenchida, ribosomal RNA analysis suggested that plant parasites have evolved within clades that include many free-living nematodes (Baldwin *et al*, 2004). Insight from molecular phylogenies based on the nuclear small subunit ribosomal RNA indicated that the adenophorean Dorylaimida and Triplonchida lack a common ancestor. Tylenchida shares a most recent common ancestor with Cephalobina (primarily bacteriovores Cephaloboidea). Triplonchida appears to be the sister clade to the Enoplida Filipjev, 1929 (primarily marine omnivores and bacteriovores), and Dorylaimida are embedded within a clade that includes insect parasites (Mermithida), vertebrate parasites (Trichinellida), and predators (Mononchida, and other Dorylaimida). It cannot be assumed that plant parasitism arose only once in each order (Baldwin *et al*, 2004).

2.4.3 Systematic Position of *Heterodera* and *Meloidogyne*

Class:	Chromodorea		
Subclass:	Chromodoria		
Order:	Rhabditida		
Suborder:	Tylenchina		
Infraorder:	Tylenchomorpha		
Super family:	Tylenchoidea		
Family:	Hoplolaimidae	and	Meloidogynidae
Subfamily:	Heteroderinae		Meloidogyninae
Genus:	*Heterodera*		*Meloidogyne*

Chapter 3
Diversity and Distribution of Nematodes

Biological diversity is the variety of life forms at all levels of biological systems (*i.e.*, molecular, organismic, population, species and ecosystem)". It may be defined as "the variability among living organisms from all sources, including terrestrial, marine, and other aquatic ecosystems, and the ecological complexes of which they are part: this includes diversity within species, between species and of ecosystems". This definition is used in the United Nations Convention on Biological Diversity, 1992. Nematodes are biologically so diverse in form and structure that they live in all possible niches wherever life can sustain and probably they are just behind the bacteria in this respect. Nematodes are unusual among all invertebrates having highly diverse forms and are able to combine endless variation in morphology with deceptively simple underlying structures. Basic pattern of body organization is very much similar in all nematodes, but these tiny creatures are structurally very diverse and have in many respects, amazing ways of life. Foremost among these is their numerical superiority, which surpasses all imaginations.

Nathan Angustus Cobb amassed a huge amount of knowledge and came to have a deep appreciation for the immense number of species of nematodes that existed. With scientific knowledge based on keen observational skills, he understood the nature of both the great numerical density and species diversity of nematodes in all habitats of the globe that he examined. Thus, he wrote the following: '*if all the matter in the universe except the nematodes were swept away, our world would still be dimly recognizable, and if, as disembodied spirits, we could then investigate it, we should find its mountains, hills, vales, rivers, lakes, and oceans represented by a film of nematodes. The location of towns would be decipherable, since for every massing of human beings there would be a corresponding massing of certain nematodes. Trees would still stand in ghostly rows representing our streets and highways. The location of the various plants and animals would still be decipherable, and, had we sufficient knowledge, in many cases even their species could be determined by an examination of their erstwhile nematode parasites. We must, therefore, conceive of nematodes and their eggs as almost omnipresent, as being carried by the wind and by flying birds and running animals; as floating from place to place in nearly all the waters of the earth; and as shipped from point to point throughout the civilized world in vehicles of traffic'.*

Nematodes are aquatic animals for all of their activities so must exist in an aqueous environment. This environment includes soils, muds, sands, plants, and animals. They can be found living in soils with moisture contents as low as 5 to 10 per cent, but in the majority of these cases, the nematodes are associated with plants. It is evident that the environment in which nematodes live constrains their ultimate size. Soil-dwelling species live in the water film of the interstices of soil particles. Nematodes must live and carry out all functions of life in these spaces, thus free-living or plant-parasitic soil-dwelling forms are usually extremely small. Morphological diversification of nematodes can operate only within the constraints of their life history parameters and evolutionary pathways for nematodes living in an interstitial-soil or sediment/sand environment are limited. Thus, these forms have limited abilities to diversify into the nonaquatic regions of the earth and their size.

3.1 Numerical Diversity and Trophic Groups

Cobb (1914) wrote, "*Nematodes are extremely widespread, and to be found in most unexpected places; they are also inconceivably abundant. A thimbleful of mud from the bottom of river or ocean may contain hundreds of specimens. The nematodes from a 10 - acre field, if arranged single file, would form a procession long enough to reach around the world*". Not only do they occur in huge numbers as parasites of all known animal groups (15 per cent), but also they are found in the soils, as parasites of plants (10 per cent), and in large numbers in the most extreme environments, from the Antarctic dry valleys to the bottom of the ocean as freeliving (25 per cent) and marine nematodes (50 per cent). They are extremely variable in their morphological characteristics, with each group showing morphological adaptations to the environment that they inhabit. Nematodes are one of the major synanthropic (associated with man) metazoans, with some species such as pinworms having coevolved with humans and their relatives since the beginning of the lineage of the primates.

They are most numerous metazoa and in sheer numerical density of individuals in any given environment, they (two to 20 million individual per square meter of soil, usually smaller than one millimeter) exceed even the mites and beetles combined. More than 90,000 nematodes were recorded from a single decomposing apple, and one report showed that 1 cc of marine mud contained 45 nematodes representing 19 species. Nematodes in marine estuaries occur at high numerical densities with reports of 4,420,000/m^2 in surface mud and 527,000,000/acre in the top 3 inches of sand on the Massachusetts coast. Counts and extrapolations for relatively moist soils 10 to 70 per cent moisture content) worldwide show that in the uppermost levels, nematodes occur in mind-boggling abundance: 7 to 9 billion/acre in undisturbed soil in North China; from 800,000,000 to > 1 billion/ acre (representing just 35 species) in Utah and Idaho, and around 3 billion/acre in low-lying alluvial soils of Europe and other areas of North America. In one soil sample, weighing not more than 1 kg from a locality in Malnad tracts of Karnataka, India, 60 genera of nematodes were recorded. This alone speaks volumes about tremendous diversity of group as also their wide distribution. A single grain of wheat parasitized by well known wheat-gall nematode, *Anguina tritici* may contain up to 100,000 of its juveniles and can remain in viable condition for more than 35 years under anhydrobiosis.

In various soils, nematode populations differ not only in trophic groups but also in different species of the same trophic group. Johnson *et al.* (1972) identified 178 species representing eight orders in 18 forest sites. Baird and Bernard (1984) reported 100 species in two wheat-soybean fields in Tennessee. Coomans (1989) listed 228 terrstrial species from Belgium.Within the plant parasitic nematodes there may be 4 to 40 species within a given habitat (Norton, 1987). Wheeler (1990) considered the phylum Nematoda to be a potential megataxon, with many thousands of species yet to be described. Even in the most intensively study regions new species are frequently found. Sibling species normally morphologically indistinguishable but not interbreeding may be far more common than we realize (Huettel *et al.*, 1984).

Soil dwelling nematodes can be functionally separated into five broad trophic groups (Wasilewska, 1971): i) microbivores ii) fungivores iii) plant parasites iv) predators and v) omnivores. These species are not mutually exclusive, some species always placed in one category may in fact have developmental stages or generations that fit another category. Juvenile stages of several species of order Mononchida may feed on bacteria (Yeates, 1987) and the phaenopsitylanchid genus *Beddingia* has alternating fungivorous and insect- parasitic generations (Bedding, 1967, Remillet and Laumond, 1991). At least one bacterial feeder, *Rhabditus orbitalishas* has three types of third stage juveniles of which one is obligatory parasitic in the conjunctival sacs of rodents (Schulte, 1989).A bacteria feeding cephalobid nematode, *Acrobeloides buetschlii* (de Man) Steiner and Buhrer, depresses N_2 fixation in the pea by feeding on bacteroids within nodules, enlarging and partly destroying the nodules (Westcott and Barker, 1976).

3.2 Animal Parasitic Nematodes

Parasitic nematodes infect 3 billion people worldwide, leading to considerable morbidity; they are a problem for livestock and domestic animals. Some 5,000 species of nematodes are estimated to be parasites of vertebrate animals and humans. These species are often characterized in a larger group of worm parasites as helminths. The nematode parasites of humans cause a variety of disease conditions and symptoms, ranging from lack of energy and vigor to blindness and malformations. Pinworms, hookworms, and roundworms are extremely common intestinal helminth infections of humans; worldwide, roundworms are probably the most common, but in the U.S., pinworms predominate. Pinworm transmittal generally occurs through ingestion of fecal-contaminated material, and infection occurs commonly in children. Some of these nematodes are highly host-specific, surviving and reproducing successfully only in host individuals comprising a single species or perhaps a closely related group of species. Other nematodes show less specificity, being much more likely to jump from one suitable vertebrate host to another during opportune times during their life history (Brant and Gardner, 2000). Within a host, many different types of habitats may be occupied by nematodes. For example, in a human, nematodes can occur as juveniles in muscle tissues (usually smooth muscle such as the diaphragm or the tongue (*Trichinella*) and in the mucosa of the intestine (*Strongyloides*); as migrating forms in blood and lungs (*Ascaris*); as microfilaria in blood or lymph (filarioids of humans); and as adults in the small and large intestines (*Ancylostoma, Necator, Ascaris*), in mesentaries and subcutaneous tissues (*Onchocerca, Loa, Wuchereria*), and in the large intestine and cecum (*Enterobius*).

Numbers of Common Nematode Infections in Humans Worldwide (Crompton, 1999):

- ✰ *Ancylostoma duodenale* and *Necator americanus* 1,298,000,000- Worldwide
- ✰ *Ascaris lumbricoides* 1,472,000,000 -Worldwide
- ✰ *Brugia maylayi* and *B. timori* 13,000,000 -South Pacific, SE Asia, India
- ✰ *Dracunculus medinensis* 80,000- Sub-Saharan Africa and Yemen
- ✰ *Loa loa* 13,000,000 -West and Central Sub-Saharan Africa and Yemen
- ✰ *Onchocerca volvulus* 17,660,000- Central and South America and Sub-Saharan Africa
- ✰ *Strongyloides stercoralis* 70,000,000 -Temperate regions
- ✰ *Trichuris trichiura* 1,049,000,000- Worldwide
- ✰ *Enterobius vermicularis* 400,000,000 -Temperate regions

The parasitic nematodes in humans attack the muscles, intestines and other bodily tissues. The modes of entry include bite by an infected vector, unintentional ingestion of eggs through consumables and lastly, penetrative entry via skin. *Ascaris lumbricoides* also known as round worm or giant intestinal worm is an intestinal parasite, pinkish colored, a large nematode species (35 cm in length). Infection occurs after ingesting contaminated food or through soil. A whip-like species, whipworm (*Trichuris trichiura*) is transmitted from contaminated soil, and remains asymptomatic in mild intestinal infections. Of the different nematode species, pinworm (*Enterobius vermicularis*) causes the highest number of infections in the United States. Small in size (about 13 mm in length), pinworm infection occurs after ingestion of eggs. There are two hook worm species of parasitic nematodes (*Ancylostoma duodenale* and *Necator americanus*). As per medical data, they are the second leading intestinal parasites found in humans. At the time of infection, the larvae penetrate human skin. Infection of trichina (*Trichinella spiralis*) is called trichinosis. This parasite nematode is present in pork meat and infection of the same is resulted after consumption of undercooked pork meat. The larvae from the stomach later migrate to the muscular tissues.

Pinworms, nematodes of the order Rhabditida, superfamily Oxyuroidea show high level of host specificity, and it is well-known that almost all species of rodents and primates have one or more species specific pinworms (for instance, *Homo sapiens* harbors both *Enterobius vermicularis* and *E. gregorii*). Both recent and historical studies have shown that pinworm nematodes exhibit high levels of coevolution, *i.e.*, concomitant host-parasite speciation (and host specificity) with their primate and rodent hosts (Hugot, 1999). At the present time, slightly more than 716 species of Oxyuroidea have been described, with the vertebrates hosting about 496 species and invertebrates about 217 species. The greatest diversity in the Oxyuroidea to be found in the future is expected to come from examination of the Arthropoda, especially beetles, cockroaches (4,000 species described and more than 20,000-30,000 expected to be found), and millipedes (17,000 species described and more than 60,000 species expected to be found). This gives huge numbers of pinworms occurring just in the cockroaches and the millipedes if each species harbors its own species of pinworm. At least two species of pinworms (genus *Thylastoma*) occur in laboratory colonies of *Periplaneta americana*.

3.3 Nematodes Diverse Habitats

In contrast to the small sizes of plant-parasitic, free living marine, freshwater, or soil nematodes (some adults can be small like *Criconemoides* with an adult length of around 250 µm but a free living nematode, *Micronema micronema* is as small as 80 µm whereas *Greefiella minutum* is 100 µm), species that occur as parasites of animals are free from many of the physical constraints on their size, so the bodies of species in mammals are, therefore, relatively large. In fact, the largest nematode thus far recorded is *Placentonema gigantissima* from the blue whale with a length of more than 8 m (Maggenti, 1981). The ratio between the smallest plant parasitic adult nematode to the largest animal parasitic nematode can be calculated as 0.250 mm (*Criconemoides*)/ 8,000 mm (*Placentonema*) = 0.000031, compared to the ratio between a shrew and a blue whale (50 mm/24,4000 mm = 0.12), indicating that the size differences in nematodes are three orders of magnitude greater than the mammals. In the marine-benthic environment, Lambshead (1993) has reported that there may be as many as 100 million species of marine nematodes. As exorbitant as that estimate seems, deep-sea nematodes have been shown to be extremely rich in species diversity. Nematodes occur in tissues and organs of all species of vertebrates that have been studied, and some, such as *Physaloptera* spp. live, feed, and reproduce in the strongest stomach acids of mammals, birds, and reptiles.

Desiccated specimens from both the Arctic and Antarctic have been rehydrated to form viable colonies, and living nematodes have been found in the limited melt-water in some of the dry valleys of Antarctica, an area that is probably one of the most extreme biotopes on the Earth. Nematodes are thought to make up 90 per cent of all life on the sea floor and are found in the deepest ocean trenches, where the pressure is 100 times greater than at the surface (Poage *et al.*, 2008). Three species of nematodes, the microbial feeders *Scottnema lindsayae* (feeding on bacteria and yeast) and *Plectus antarcticus* (a bacterial feeder), and the omnivore *Eudorylaimus antarcticus*, that presumably feeds on individuals of the other two species (Freckman and Virginia, 1991) are found in the McMurdo Dry Valleys of Antarctica, one of the harshest environments on earth, where temperatures reach –60 °C in the winter and wind speeds exceed 320 km/h (200 mph), stripping away almost all moisture. Species of another nematode, *Acantholaimus* spp are fairly distributed in Antartica and about 18 species have been found coexisting at one station. This nematode is considered a typical deep sea genus within the free living marine nematodes (Mesel *et al.*, 2006). Nematodes can survive in little drops of moisture within rocks, consuming bacteria. The nematode, *Halicaphalobus mephisto* (specific name in honor of Faust's demon Mephistopheles-a light-hating demon of the underworld) lives nearly a mile (1.3 km) underground in fluid filled rock fractures near South African goldmines (Borgonie *et al.*, 2011). The water in which the nematode lives is at least several thousand years old. The nematode is 0.02-inch-long (0.5 millimeters), tolerates high temperature, reproduces asexually and feeds on bacteria. This discovery expand the known metazoan biosphere and demonstrate that deep ecosystems are more complex than previously accepted. The discovery of multicellular life in the deep subsurface of the Earth also has important implications for the search for subsurface life on other planets in our Solar System.

Two *Cryonema* species permanently live at freezing point inside lacunae in arctic ice; one of them preys on other lacunary nematodes (Tchesunov and Riemann, 1995). Another biological extreme, Death Valley in California, has recorded some of the highest temperatures in North America, and the soils of the valley are teeming with nematodes, many of which have been discovered to possess similar adaptive traits to those found in the cold deserts of Antarctica. Individuals of some species of Nematoda are capable of resisting extended periods of dessication, for example, 2nd stage juveniles of *Anguina tritici* have been dried for more than 35 years in a state of anhydrobiosis in which all metabolic activities are shut down (Maggenti, 1981). These nematodes have been shown to have specialized proteins that fold into stable/ preserved structures as the organism dries. In this state of cryptoor anhydrobiosis, individual nematodes can remain viable through incredible extremes of temperature, desiccation, hypoxia, and even synthetic nematicides designed to kill nematodes. When water again becomes available, the animal comes back to life when the molecular structures rehydrate and the proteins and enzymes spring back into normal operation. Rizvi (2010) has reported the presence of two species, *Cervidellus vexilliger* (de Man, 1880) Thorne, 1937 and *Chiloplacus demani* (Thorne, 1925) Thorne, 1937 first time from cold and dry Ladakh region of J.&K. India, and also recorded another species *Acrobeloides nanus* (de Man, 1880) Anderson, 1968 which is a first record from this area.

Organisms have evolved various types of survival strategies against adverse conditions. Some species move away from sites with adverse conditions to conditions more favorable for growth and reproduction. Dehydration is one of the most serious stresses in both aquatic and terrestrial organisms. Water is the major component of living organisms. The average amount of body water in invertebrates is around 70 per cent (from 17 per cent to 90 per cent) and water makes up 95–99 per cent of the total number of molecules (Hadley, 1994). Most organisms have only limited ability to survive water loss (Wharton, 2002). Humans do not survive 14 per cent loss of their body water and most organisms die when they lose 50 per cent of the body water at the individual, organ or cellular level. The ability of different species to tolerate adverse conditions varies enormously and one of the key factors implicit in desiccation survival is the ability to control the rate of water loss. The majority of soil-dwelling nematodes show little intrinsic ability to control water loss, being dependent on the environmental conditions of high relative humidity within soil pores to slow down or prevent water loss. In general, nematode anhydrobiotes can be grouped into those that rely on environmental factors to control water loss and those that have intrinsic abilities to control water loss. Perry and Moens (2011) termed these groups external dehydration strategists and innate dehydration strategists, respectively. Both groups require controlled drying in order to survive, the rst group to prolong the time to lethal low water content and the second group to enable biochemical changes to take place to facilitate long term survival. Control of the rate of drying is the rst phase; successful entry into long-term anhydrobiosis depends on subsequent biochemical and molecular adaptations.

Longivity Records in some Anhydrobiotic Nematodes

- ✰ *Tylenchus polyhypnus* 39 years Steiner and Albin, 1946
- ✰ *Anguina tritici* 32 years Fielding, 1951; Norton, 1978; Womersley, 1980
- ✰ *Globodera rostochiensis* 25 years Norton, 1978
- ✰ *Ditylenchus dipsaci* 23 years Perry, 1977; Norton, 1978; Wharton, 1996
- ✰ *Panogrolaimus sp.* 8 years Aroian *et al.*, 1993
- ✰ *Heterodera glycines* 6 years Norton, 1978
- ✰ *Heterodera avenae* 5.5 years Norton, 1978
- ✰ *Anguina agrostis* 4 years Norton, 1978; Preston and Bird, 1987
- ✰ *Ditylenchus triformis* 2.5 years Norton, 1978
- ✰ *Aphelenchus avenae* 2.2 years Crowe and Madin, 1974; Higa and Womersley, 1993

3.4 Nematodes Surviving Extremes

Nematodes evade the biotic and abiotic obstacles by employing a combination of behavioral and physiological survival strategies. Nematodes of the subfamily Stilbonematinae are covered with a dense "fur" of species-specific ectosymbiotic sulfur-oxidizing bacteria, allowing them to thrive at the redox boundary layer in sulphur-rich marine sediments (Nussbaumer *et al.*, 2004). *Oncholaimus* mates by traumatic insemination: males inject sperm through the female cuticle; females then develop specialized internal structures for sperm transfer to the reproductive system - or for evacuation of excess sperm into the intestine (Coomans *et al.*, 1988). The entomopathogenetic nematode, *Steinernema tami* produces dimorphic sperm, with 50–100 μm wide megaspermatozoa functioning as self-propelled spermatophores carrying the 2μm wide microspermatozoa on their surface (Yushin *et al.*, 2003). *Mehdinema alii* uses male crickets as vectors for transmission between female cricket hosts, female nematodes give birth to fully formed dauers while males have a motile copulatory claw extruding through a separate postcloacal opening (Luong *et al.*, 2000). The millipede gut inhabitant *Zalophora* is an intra-intestinal predator and cannibal of other gut nematodes (Hunt and Moore, 1999). Microbivorous fungal and bacterial feeding nematodes are also extremely abundant, with species specializing in their feeding habits on bacteria, fungi, diatoms, and other microscopic organisms.

Some, such as juveniles of species of the genera *Heterorhabditis* and *Steinernema*, carry bacteria of the genus *Photorhabdus* in their digestive system. The 3rd stage juveniles wait in the soil until an unwary insec passes nearby. The nematode then homes in on and penetrates the hapless insect, making its way into the hemocoel where it releases the bacteria, which proliferate, killing the host. The nematode then feeds on the bacterial colony and reproduces in the insect, eventually again producing 3rd stage juveniles that leave the carcass of the insect and disperse into the soil, waiting there for another insect to invade. Bacterial-feeding forms occur in the soil in extremely high numbers. In soils that have not been disturbed and that have a good layer of organic matter, large numbers of all kinds of nematodes occur. One of the most well-known groups is the Rhabditida, or the rhabditid nematodes. These forms feed on bacteria and yeasts growing in the soil and are typically found in high numbers in moist soils with high organic content. The most famous of these forms are members of the genus *Caenorhabditis*, of which the complete genome of *C. elegans* has been sequenced. One form of nematode is entirely dependent upon fig wasps, which are the sole source of fig fertilization. They prey upon the wasps, riding them from the ripe fig of the wasp's birth to the fig flower of its death, where they kill the wasp, and their offspring await the birth of the next generation of wasps as the fig ripens.

A newly discovered parasitic tetradonematid nematode, *Myrmeconema neotropicum*, apparently induces fruit mimicry in the tropical ant, *Cephalotes atratus*. Infected ants develop bright red gasters, tend to be more sluggish, and walk with their gasters in a conspicuous elevated position. These changes likely cause frugivorous birds to confuse the infected ants for berries and eat them. Parasite eggs passed in the bird's feces are subsequently collected by foraging *Cephalotes atratus* and are fed to their larvae, thus completing the life cycle of *Myrmeconema neotropicum* (Yanoviak *et al.*, 2008)

3.5 Spread and Distribution of Plant Parasitic Nematodes

One might think of soil as a safe environment, but to a microscopic nematode it is a hostile world filled with danger. A nematode must contend with voracious predators, changes in soil temperature and moisture, and the death of its host plant. For a nematode population to survive, it must be able to circumvent these obstacles. Nematode survival is not affected only by biotic factors, but also by abiotic ones such as temperature and water availability. The onset of winter or the drying of the soil can be disastrous for a nematode. Interestingly, many nematodes are well adapted to abiotic stress and are capable of cryptobiosis (hidden life) i.e. the ability to enter a state of suspended metabolic activity during unfavorable environmental conditions (drying, heat, cold). While not all nematode are capable of cryptobiosis, the ones that are can often survive for years in a cryptobiotic state awaiting favorable conditions that will trigger their revival. The ability of nematodes to undergo cryptobiosis is one reason some nematode species are very difficult to eradicate from a field.Among the plant parasitic nematodes, the most striking feature of their distribution and damage within a field is the irregularity of infested areas. Damaged crops will appear as irregular patches or streaks that may vary in size, shape, and number. These variations usually reflect the compounding of nematode stress on a plant by such other factors as physical soil differences and irrigation and drainage patterns. Previous patterns of cropping, initial introduction, and soil movement through cultivations are also important factors. Cysts containing viable eggs are introduced by contaminated harvesting or cultivation equipment, tractor or truck wheels, human feet, or animal hooves; by irrigation water; or by windblown in from an adjacent infested area. Following introduction, at a single site or at separate sites within the field, the nematode becomes established during host crop development and, over a period of years, increases population levels. As the population increases, the sites gradually enlarge in area and, when the nematode population reaches a damaging level, the sites show as unthrifty or damaged plant areas within the field. In later years, under short rotations, these areas enlarge further and coalesce until large areas of the field become uniformly infested. The spread of nematodes within the field from these original sites of introduction is mediated largely through the movement of cysts floating on water used in flood or furrow irrigations. Similarly, cattle grazing on crop tops at harvest can distribute infestation by their hooves. In this example, processes related to initial introduction and subsequent spread of infestation are the major determinants of fieldwide distribution.

Nematodes usually are brought to new areas by movement of soil, water, and plant material. Agricultural practices such as clearing or leveling land, cultivation, and irrigation are important in moving nematode-infested soil. The ability of nematodes to form environmentally resistant stages makes their dissemination even easier, since dried nematodes can be blown with the wind or plant debris over large geographical regions. Even migrating birds are suspected to be able to carry nematodes along their flight paths, assisting the nematodes getting established in new areas. Nematodes are established frequently in a new area with plant material. Plant propagation material (seeds, bulbs, tubers, cuttings, and transplants) can conceal numerous kinds of nematode pests. Naturally, cuttings and transplants rooted in soil greatly increase the chance of pests including nematodes introduction. Wind and animals also may move nematode pests.

In one such example of introduction of nematode distribution in imported consignments, Dr. F. G. W. Jones, Head, Nematology Department, Rothamsted Experimental Station, Harpenden, U.K., on a personal trip to Ootacamund in 1961, first detected the golden nematode of potato in India from a field situated at an elevation of 2125m MSL. This nematode has certainly been introduced in India from Europe in imported material. Realizing the importance of this problem in potato, the Indian Council of Agricultural Research (ICAR) and the Government of Tamil Nadu launched the *'Golden Nematode Scheme'* at Ootacamund in 1963 (Seshadri and Sivakumar, 1962). Large-scale inspection of potatoes in the marketing mandies at Mettupalayam (trading centre and rail-head at the foothills of Nilgiris) and field-to-field surveys were undertaken (Seshadri, 1970). These studies indicated presence of cysts in potato consignments meant for transportation to Bombay, Calcutta, Cuttack and Poona (Seshadri, 1978). Detailed surveys conducted in other major potato growing areas of Assam, Himachal Pradesh, Karnataka, Punjab, Tamil Nadu and Uttar Pradesh indicated this nematode to be restricted to the Nilgiri (Gill, 1974) and Kodaikanal hills (Thangaraju, 1983) of Tamil Nadu. Presence of cysts of PCN in several potato growing villages was also recorded in Karnataka (Singh and Krishna Prasad, 1986). Later, G. *pallida* was observed to be associated with potato at the Devikulam locality in Idukki district of Kerala (Ramana and Mohandas, 1988). Species composition, distribution and intensity indicated 57 per cent PCN population in the Nilgiris to be that of *G. pallida* and 43 per cent was that of *G.rostochiensis.*

So the nematodes are diverse metazoans with an estimated total number of a million species (Lambshead, 2004). They are arguably the most numerous metazoans in soil and aquatic sediments. From an environmental point of view, nematodes are part of nearly all ecosystems in their roles as bacterivores, herbivores, parasites of animal and plants, and consumers of dissolved as well as particulate organic matter. They have critical roles in the flow of energy and cycling of nutrients. From an anthropocentric point of view, they parasitize humans and plants, domestic and wild animals and they can serve as indicators of environmental change.

Chapter 4
Economic Importance of Nematodes

The importance of these tiny creatures had been the talk of the people suffering from the ailments inflicted by various animal parasitic nematodes since the time these nematodes came to be known to them. The knowledge about plant parasitic nematodes is recent in comparison to animal parasitic nematodes so there economic importance was also realized late with the discovery of sugarbeet nematode, *Heterodera shachtii* in the second half of nineteenth century when huge crop losses were suffered by the sugarbeet growers in Germany followed by the damage caused by poato cyst nematode in whole of Europe and America during/around this time. Plant parasitic nematodes are unique in their ubiquitous nature and persistence in soil. The confusing nature of their attack allows their presence to often pass unnoticed while crops slowly decline in vigor and yield. Rarely is any crop free from attack of these tiny and microscopic pathogens. They are present every where, in fields, glasshouses, orchards, home gardens, and yet their presence is generally not felt till the concerned individual is baffled by the continuous decline of the crop in spite of best agrono-mic practices. The losses, as a result of nematode attack do not consist of yield reductions only but other aspects like lesser ability of infected roots to utilize fully the available nutrients from soil or the necessity to grow uneconomical rotational crops, in an effort to control the noxious nematodes, are some of the effects which follow as a result of nematode infestation. Considering their impact on crops, McCarter (2009) estimated a global total loss of $118 billion for 2001, of which nearly half was related to only two crops, rice and maize. This being so, it is remarkable they are among the least studied, with only close to 30,000 (< 3 per cent of estimated) species described to date (Hugot *et al.*, 2001; Hallan, 2007). Overall average annual loss of the world's major crops due to damage by plant parasitic nematodes was estimated to be 12.3 per cent. For the 20 major crops which serves as the man's primary food source, annual yield loss of 10.7 per cent was estimated, while for another group of 20 crops mainly of commercial importance, a 14 per cent annual yield loss was assessed. Developing countries had a crop loss of 14.6 per cent compared to 8.8 per cent in developed countries (Sasser, 1989).

Another way to consider the impact of plant-parasitic nematodes is through the management strategies employed in their control. In 1982, 50,000 tonnes of nematicide active ingredients were applied to crops in the United States, at a cost exceeding $US1 billion (Landels, 1989) and between 1986 and 1990 in The Netherlands, nematicide application was more than three times the combined total of chemicals needed to combat insects, fungi and weeds on experimental and sustainable farms (Lewis *et al.*, 1997). However, in recent decades issues such as ground water contamination, toxicity to mammals and birds, and residues in food have caused much tighter restrictions on the use of agricultural chemicals, including suspension of nematicides in many countries (Thomason, 1987).

Nematodes are second only to insects in the number of species in the animal kingdom. One cubic foot of soil may contain millions of individual nematodes belonging to several different taxonomic groups. Amongst the soil inhabitants, they constitute one of the most important groups of organisms which quite often play a vital role in the deterioration of the health of the standing crops. With intensification in agriculture and also with continued and uninter-rupted new introductions in countries, it is obvious that there is likelihood of hitherto unknown problems which may become serious in the near future. The golden nematode problem in Long Islands, U.S.A. and in the southern part of India is one of the most outstanding examples of such type.

Considerable information has been generated on the losses caused by the nematodes to staple and other crops in many countries where well established infra structure and man power is available in the subject. At the same time this is also a fact that there is still no reliable information on the exact losses caused by the nematodes to various crops. Most of the reports are approximations based on indirect evidences, interpreted on the basis of pot trials and chemical trials under field conditions. The validity of the results on yield increases, as a result of chemical applications, is open to question since the total effect on biological flora of the soil is not fully understood and therefore, it might not be justifiable to interpret results of field trials with nematode damage alone. The quantitative aspect of the nematode damage to the crops is not only affected by the size of the nematode population but also on other variables such as the species of the host plant, temperature and other environmental conditions, the flora and fauna of the soil, which singly or jointly may contribute to the overall expression of disease syndrome. A range from death of seedlings/plants, in some cases, to an absence of recognizable damage may not be uncommon in nematode infestations. One of the most important factors influenc-ing yield of crops is the size of the nematode population in soil and also availability of a susceptible host. Measurable damage occurs only when the population density exceeds a certain limit.

The root-knot and cyst nematodes are the most common and destructive nematode pathogens. They produce some of the most dramatic symptoms and can substantially reduce crop yields. The root-knot nematodes belonging to the genus *Meloidogyne*, which have a wide host range, are widely prevalent all over the world. They produce conspicuous galls on the roots and the infestation can be recognized easily in fields. The wide distribution of these nematodes through tropics, sub-tropics and temperate regions and their occur-rences in cultivated fields and greenhouses as well, make them as one of the most common and economically important agricultural pests.

4.1 Losses Due to Root Knot Nematodes

More than 100 species of root-knot nematodes have been described from throughout the world. In India, out of thirteen species known to be present only four species, *M. indica, M. lucknowica, M. piperi* and *M. triticoryzae* are described where *M. lucknowica* has been synonymised with *M. javanica* (Fortuner *et al.*, 1987) and the fate of *M. triticoryzae* is also hanging between a new species and *M. graminicola*. Economically, four species *viz., M. incognita, M. javanica* (both on vegetables), *M. arenaria* (on groundnut) and *M. triticoryzae* (on rice and wheat) are important as they all cause significant losses to crops they infest (Gaur and Sharma, 1998; Swarup and Sosa-Moss, 1990). *M. incognita* can cause poor seedlings establishment and reduce yield of upland rice upto 60 percent if the population of nematode is near 8000 juveniles/dm^3 of soil at the time of sowing (Babatola, 1984) and Dasgupta (1962) estimated losses to the tune of 60 percent in vegetable crops in Delhi. A large number of field trials using different management schedules have shown 3-4 times increase in yield of brinjal, okra, tomato and many other vegetable crops at various location in India (AICRP Nematodes, 1990). In tobacco, Hussaini (1983) observed that in the untreated field the root knot infection reduced the yield of green and cured leaves by 36 and 34 per cent in comparison to treated plants whereas Rich and Barker (1984) reported delayed flowering in rootknot infected plants of tobacco. Under Indian conditions, an initial level of even two larvae of root knot nematode per gram of soil results in overall yield reduction of 42.5 per cent (Krishna Prasad, 1993). Root knot nematode infection on tubers manifests as pimple-like blisters which greatly reduces the marketable quality of potatoes. Mere presence of two infected tubers per bag of 80 kg seed potato is sufficient to reject for export from Himachal Pradesh, a state which follows potato seed certification (Krishna Prasad, 1986).

They are mostly severe on vegetables but can cause appreciable damage to cereals also though economic importance of root-knot nematodes in cereals is limited to few species only. *M. naasi* seriously affect wheat yield in Chile (Kirkpatrick *et al.*, 1976) and Europe (Person-Dedryver, 1986). On barley, it has been known to cause 75 percent damage in California (Allen *et al.*, 1970), France (Caubel *et al.*, 1972), Belgium (Gooris and D'Herde, 1977) and Great Britain (York, 1980). The losses (90 per cent) caused to wheat due to *M. artiellia* have been reported from Italy (Di Vito and Greco, 1988), Greece and Israel (Kyrou, 1967; Mor and Cohn, 1989). It is especially a serious problem with vegetable growers as also to cotton and tobacco growers. Groundnut (*Arachis hypogaea*) is an important food crop in India, Africa, China and USA where it is mainly attacked by at least by five different kinds of nematodes including two species of root knot nematodes, *M. arenaria* and *M. hapla*. *M.arenaria* is an aggressive pathogen of this crop causing extensive yield loss at relatively low population levels (1.0 larvae /100cc of soil) (McSorley *et al.*, 1992). Prasad (1997) reported 33.1 per cent yield loss of groundnut due to *M. arenaria* and *R. reniformis* found together in the fields of Uttar Pradesh.There are many estimates of losses due to *M arenaria* causing 39 per cent loss in grain yield of chickpea in Gujarat (Patel, 1997) and Mhase *et al.* (1997) estimated 42 per cent yield loss due to root-knot nematode in Maharashtra. Information gathered from different sources on avoidable yield loss due to root-knot nematode, mainly *M.incognita* in chickpea indicated 17-60 per cent losses in Bihar, 17-56 per cent in Gujarat, 8 per cent in Haryana, 35-43 per cent in Maharashtra, 22 per cent in Punjab, 17-60 per cent in Rajasthan and 22-23 per cent in Uttar Pradesh.

Rice root-knot nematode, *Meloidogyne graminicola* has gained considerable attention during the recent years due to its damage potential to cause significant loss in rice-wheat cropping system in South East Asia. A yield loss of about 21 per cent in rainfed and well drained soil, and 16-32 per cent in upland rice has been recorded due to nematode infection in India (Prasad *et al.*, 1986; Jairajpuri and Baqri, 1991; Bridge, 1990; Prot and Matias, 1995; Padgham *et al.*, 2004; Pokharel *et al.*, 2007).

4.2 Losses Due to Cyst Nematodes

Among the cyst nematodes, the Schachtii group of Mulvey's grouping of cyst nematodes, contains most devastating species, the glaring example is that of *H. schachtii* which was responsible for almost ruining the economy of Germany and other European countries in late nineteenth century and in USA in the first half of the twentieth century. Similarly, *Globodera rostochiensis* and *G. pallida* were extremely devastating in the reduction of potato yield in Europe during this period. In India the potato cyst nematodes may result in 65 per cent loss of the crop (Krishna Prasad, 1993). Overall damage due to these species has been estimated at about 9 per cent of crop production in United Kingdom (Evans and Stone, 1977). The soybean cyst nematode, *Heterodera glycines*, is of great economic importance in U.S.A. and has neces-sitated the development of intensive nematode management programmes there as well in other countries where it is a serious problem *H. glycines* is a major pest of soybeans in Asia also. In Japan, yield loss was estimated to be 10-75 per cent (Inagaki, 1977; Ichinohe, 1988), whereas in the USA, Riggs (1977) calculated a loss of 85 million USD from five states only.

Cereal cyst nematode, *Heterodera avenae* complex is acknowledged as a global economic problem on wheat production systems, both in many regions of the developed world (Europe, USA and Australia) and also in developing countries (North Africa, West Asia, China and India). In India, this nematode has been known to damage the wheat crop in the states of Rajasthan, Haryana, Punjab and Delhi. The damage caused used to be so severe that it was difficult to get back even the sown seed as yield. It has been associated with economic damage exclusively in light soils. However, it can cause losses irrespective of soil type when the intensity of cereal cropping exceeds a certain limit (Kort, 1972). Yield losses due to this nematode are 15 to 20 percent on wheat in Pakistan (Maqbool, 1988); 40 to 92 percent on wheat and 17 to 77 percent on barley in Saudi Arabia (Ibrahim *et al.*, 1999); and 20 percent on barley and 23 to 50 percent on wheat in Australia (Meagher, 1972). Staggering annual yield losses of $4.5 million in Europe, $72 million in Australia and $9 million in India have been calculated as being caused by *H. avenae* (Wallace, 1965; Brown, 1981; Van Berkum and Seshadri, 1970). In India, *H. avenae* has been reported to destroy 70-80 percent of wheat crop in very severely infested fields in northern India and the losses are being neutralized with the introduction of tolerant wheat varieties or by growing resistant barley cultivars. Swarup *et al.* (1976) have recorded 4 and 16 times increase in yield of wheat and barley respectively after nematicidal treatments. The losses in Australia also are now greatly reduced due to control of the disease with resistant and tolerant cultivars. The *Heterodera avenae* group of several closely related species has been documented to cause economic yield loss particularly on rainfed wheat production systems. The species like *H. filipjevi*, *H. iri* and *H. latipons* which have been implicated in economic losses. The species like *H. mani*, *H. arenaria* and *H. hordecalis* seem to persist with their hosts apparently causing insignificant damage. Little is known about the economic importance of the species *H. latipons*, even though it was first described in 1969 (Sikora, 1988). Field studies in Cyprus indicated a 50 percent yield loss on barley (Philis, 1988).

The pigeon pea cyst nematode, *H.cajani* has been reported to cause yield losses of over 30 per cent in plant growth (Sharma *et al.*, 1993). The symptoms of nematode injury include stunting, reduced leaf lamina size and yellowing on cotyledonary leaves has been observed (Gaur and Singh 1977). Yield loss estimates in nematicidal trials have indicated 20-43 per cent loss in yield of blackgram, french bean, green gram and pigeonpea due to pigeonpea cyst nematode (Sharma *et al.*, 1992). *Heterodera cajani* has been reported to infect pigeonpea to such an extent that it may cause the loss of grain yield upto 30 per cent under field conditions (Saxena and Reddy, 1987). The maize cyst nematode (*Heterodera zeae*) is another nematode problem which has gained an economic importance not only in India but as well in USA. The sorghum cyst nematode, *H. sorghi* is reported to present in many states in India but damage due to it is quite uncertain.

The rice cyst nematode *H. oryzicola* has been reported to infest plants which exhibit root browning and chlorosis causing 7-29 percent yield reductions (Usha Kumari and Kurien, 1981). Jayprakash and Rao (1984) have found that the nematode infestation is compounded by the infection of a fungus, *Sclerotium rolfsii* and if the fungus comes subsequent to the nematode infestation then the fungal effect is enhanced resulting in seedling blight.

The turf grass cyst nematode, *H. graminis* is a great menace in golf areas in many countries including India causing visible symptoms of patchy and dry areas particularly in 'greens'.

4.3 Losses Causd by Other Nematodes

The estimated yield-loss in potato by plant parasitic nematodes around the world is 12.2 per cent (Sasser, 1989). There are many species of plant parasitic nematopdes which are responsible for causing economic losses to the crops they infest. For example, in plantation crops such as coconut, arecanut, black pepper, banana and beal vine in India, the most important nematode problem is the root rot disease caused by burrowing nematode, *Radopholus similis* causing huge losses in these crops (Koshy *et al.*, 1980).

Involvement of various ecto-parasitic nematodes, that feed ectoparasitically on plant roots, besides the other endo or semi-endoparasites, notably *Pratylenchus species* (lesion nematodes), *Tylenchulus semipenetrans* (citrus nematode) and *Rotylenchulus reniformis* (reniform nematode) in the plant disease, adds to the dimensions of losses caused by these noxious nematodes to economi-cally important crop plants (Rebois *et al.*, 1978, Nath *et al.*, 1979 and Varaprasad, 1982). It is estimated that losses to tune of 12-24 per cent due to *Pratylenchus thornei* and 11 per cent due to *Rotylenchulus reniformis* occur in Madhya Pradesh. The lesion nematodes, in general, and particularly *Pratylenchus coffeae* on coffee in southern parts of India (Palanichamy, 1973), Indonesia (Toruan-Mathius *et al.*, 1995) and Brazil (Kubo *et al.*, 2002); *P. loosi* on tea in Sri Lanka (Whitehead, 1968) have attracted great deal of attention from the growers as well as the scientists as these are the most destructive species. Many of these may not cause conspicuous damage immediately but with continued root feeding, the root growth may stop and the affected plants may express unthriftiness and stunted growth. Sharma *et al.* (1981) has recorded many nematode species of the genera like *Ditylenchus* (*D. destructor* and *D. myceliophagous*) and *Aphelenchoides* (*A. composticola*, *A. sacchari* and *A. asterocaudatus*), *Aphelenchus* sp.and *Paraphelenchus* sp.and many species of saprophagus nematodes which are associated with huge damage that is caused in mushroom cultivation due to them. However, not all the soil nematodes are harmful to the plants. Some of dorylaims and mononchs prey on nematodes, rotifers and soil micro-organisms. These and some predacious fungi, which attack nematodes, may well be maintaining a natural balance in the soil fauna in favor of the plant growth.

Even though a very large number of plant parasitic species have been recorded and documented, only 15 have been identified, so far, as world problems. Besides these major nematode problems, there are several other important nematode diseases which cause heavy losses in certain parts of the world and under specific conditions. For example, the ear-cockle disease of wheat (*Anguina tritici*) could lead to as high as 80 per cent or more crop loss in certain pockets of India (parts of eastern U.P. and Bihar) as contaminated seed is regularly used for sowing by the poor farmers who do not have access to the certified seed. The 'ufra' and white tip diseases of rice, caused by *Ditylenchus angustus* and *Aphelenchoides besseyi*, respectively, often are responsible for serious crop losses in south-east Asian countries including India.

The red ring disease of coconut caused by *Rhadinaphelenchus cocophilus* in Surinam and British Guiana and on oil palm in Venezuela is rated as one of the most devastating disease throughout the world.

4.4 Nematodes as Disease Aggravators

Some of the most important root diseases are caused by nematodes in combination with other soil organisms involving disease complexes. The nematode-fungi and nematode-bacteria interactions are quite numerous and varied. The weak fungal and bacterial parasites can cause significant amount of damage once they enter the root tissue through the avenues created by the feeding nematodes. They are also involved in lowering the resistance of the plants to the diseases like vascular wilt caused by fungi and bacteria. In Japan, *H. glycines* interacts with the fungus, *Phialophora gregata* on *Vigna angularis* and when the nematodes are controlled, the fungus causes no damage (Ichinohe, 1988).

4.5 Virus Transmission by Nematodes

The economic importance of nematodes also lies in the transmission of many viral diseases. The Longodorids, Xiphinematids and Trichodorids are involved in the transmission of diseases caused by various ring spot, black ring, grapevine fan leaf; pea early browning and tobacco rattle viruses. It has been estimated that only 1 per cent of all the described species of plant parasitic nematodes are virus vectors. Of about 350 species of longidorids, only 17 species (seven longidorids, one Paralongidorid and nine Xiphinematids) are natural virus vectors of 13 of the 38 known nepoviruses and 13 trichodorids (nine Paratrichodorids and four Trichodorids) of 70 species described are involved in viral transmission of three tobraviruses (Pea early browning, tobacco rattle and pepper ring spot).

4.6 Entomopathogenic Nematodes

Entomopathogenic nematodes of the family Steinernematidae (65 species of *Steinernema* and one species of *Neosteinernema*) and Heterorhabditidae (16 species *Heterorhabditis*) are involved in controlling insect pests belonging to many families particularly in lepidopteron and coleopteron orders. Most of these insect-pests are found in confined areas where the EPNs can be applied in high concentrations. The EPNs are particularly well suited for the integrated management approach since they are resistant to many insecticides. In addition certain compounds are applied specifically on the host and the EPNs are applied in the soil, so the two control methods can compliment each other. *Xenorhabdus* spp and *Photorhabdus* spp of entomopathogenic bacteria are symbiotically associated with the above entomopathogenic nematodes. The genus *Xenorhabdus* appears more diverse than the genus *Photorhabdus*, and the occurrence of symbiotic bacterial genotypes is related to the ecological distribution of host nematodes.

4.7 Predatory Nematodes

Predatory nematodes belong to four major taxonomic groups: Mononchida, Dorylaimida, Aphelenchida, and Diplogasterida. They feed on plant-parasitic nematodes when they come in physical contact. Each group has a different type of feeding apparatus, feeding mechanism, and food preferences. The feeding apparatus of mononchids has a large buccal cavity, which is equipped with a large dorsal tooth and sometimes with smaller ventral teeth and denticles. The mononchids feed on nematodes, protozoans, rotifers, tardigrades, and small oligochaetes. They may swallow the whole prey if the size is smaller or rupture the body of the prey to suck the body contents and can be reared on prey species (*e.g.*, *Prionchulus punctatus* and *Mononchus aquaticus*). The feeding apparatus of dorylaimids is a hollow odontostyle (nygolaimids have needle-like feeding apparatus). Generally dorylaims are omnivorous. They feed on nematodes, oligochaetes, protozoa, rotifers, and eggs of nematodes and other invertebrates. The species of *Aporcelaimus, Aporcelaimellus, Dorylaimus, Eudorylaimus, Mesodorylaimus, Discolaimus, Labronema,* and *Nygolaimus* feed on plant-parasitic nematodes. They feed either by piercing the body of the prey or inject hydrolytic enzymes and suck out the predigested contents. The Aphelenchids are capable of feeding on larger preys by injecting toxins, which cause rapid paralysis of prey. Species of the genus *Seinura* are predatory and have a typical aphelenchid stylet. They have high reproductive potential, and can be easily and rapidly cultured on fungivorous nematodes. The stoma of diplogasterids is armed with teeth. They feed on other nematodes and bacteria. Diplogasterids are generally abundant in decomposing organic manure and can be easily reared on simple nutrient media containing bacteria. The predation of nematodes depends on size, motility, thickness, and degree of annulation on the cuticle of the nematode. The predatory nematodes are mostly omnivorous and predation on nematodes is coincidental (Stirling, 1991). Much of the information on the predatory nematodes is based on agar plate experiments. The use of these nematodes as biocontrol agents of plant-parasitic nematodes may be impractical, but they may be playing an important role in reducing parasitic nematode populations in nature.

4.8 Role of Nematodes in Organic Matter Decomposition

Saprophytic nematodes comprise 52 per cent of the total nematode genera (Wharton, 1986). They are less studied than the parasitic species. The quality and quantity of organic matter affect distribution and abundance of these nematodes. The organic matter provides an energy source for growth of bacteria and the bacteriophagus nematodes feed on the bacteria. The chemotropic bacteria convert the excretory product of nematodes (*i.e.* ammonia) to nitrate and nitrite compounds. The predominance of bacteriophagus nematodes in agricultural systems indicates faster rate of mineralization, decomposition, and nutrient turn over (Freckman, 1988). The bacteriophagus nematodes have an unarmed open stoma. These nematodes (*e.g. Rhabditis* sp) can consume up to 1.018 mg bacteria (Tietjen and Lee, 1977). The bacteriophagus nematodes have a short life span and generation time varies from few days to few weeks depending on the species; for example, *Mesodiplogaster lheritieri* completes its life cycle in 4 days and *Acrobeloides* sp in 11 days (Anderson *et al.*, 1981). They have very high fecundity rate. An *Acrobeloides nanus* female produces about 2 to 3 x 10 nematodes within a month (Wasilewska *et al.*, 1975). They are more abundant in undisturbed soils than in heavily-managed agricultural systems (Yeates, 1979). The bacteriophagus nematodes affect organic matter decomposition by (1) feeding on microbes and regulating the rate at which organic compounds are degraded into inorganic ions (2) dispersing the soil microorganisms which adhere to the cuticle of the nematodes and are dispersed in soil. The nematodes defecate 30–60 per cent of ingested bacteria in viable conditions and help in dispersal of rhizobia, algae, mycoplasma, and fungal spores (3) feeding on saprophytic and plant-pathogenic bacteria and influencing the composition of the microbial community (4) serve as a prey and a source of nutrients for microflora and fauna such as soil nematophagus fungi and (5) affecting the distribution and function of soil and plant symbionts.

4.9 Nematodes as Ecological Indicators

Nematodes have certain qualities that make them useful as ecological indicators (Freckman, 1998; Neher and Campbell, 1994). Various kinds of disturbances to soils such as addition of synthetic fertilizers (Wasilewska, 1989), cultivation (Hendrix *et al.*, 1986), liming (Hyvonen and Persson,1990) and accumulation of heavy metals (Bongers *et al.*, 1991) affect the species richness, trophic structure and successive nematode community formations. Because they reflect changes in the soil structure and functuion related to these disturbances, indices of nematode community structure show promise for monitoring the ecological condition of soil (Bongers, 1990; de Goede, 1993; Ettema and Bongers, 1993 and Freckman and Ettema, 1993). There are many methods of measuring nematode community structure but the maturity and trophic diversity indices are capable of differentiating sites better and more effectively than measured based on populations or ratios of individual trophic groups. Maturity index is a measure of successional status and trophic index measures food web structure.

4.10 Role of Nematodes in Nutrient Cycling

4.10.1 Carbon Cycle

The decomposition of plant material can be studied by measuring weight loss, directly or by monitoring the $^{14}CO_2$ produced from ^{14}C labeled material (Lynch, 1983). The total nematode community respires comparatively lesser of the total carbon input in natural ecosystems (0.6–0.9 per cent) than in agroecosystems (1.3–2.0 per cent) (Sohlenius *et al.*, 1988). The bacteriophagus nematodes indirectly stimulate the rate of organic matter decomposition by grazing (Freckman *et al.*, 1987). In a microcosm study on *Pseudomonas* sp. and *Mesodiplogaster* sp., the microcosms with nematodes and bacteria had greater CO_2 output than those with bacteria alone (Anderson *et al.*, 1981). Similarly, free-living marine nematodes (*e.g.*, *Diplolaimelloides* spp.) may increase the rates of carbon mineralization by 300 per cent (Findlay and Tenore, 1982).

4.10.2 Nitrogen Cycle

Nematodes have been demonstrated to affect plant productivity by increasing nutrients availability through regulation of mineralization process. Shoot biomass and nitrogen content of plant shoots grown in the presence of protozoans and free living nematodes, are greater than those grown without mesofauna (Verhoef and Brusaard, 1990; Yeates and Wardle, 1996). Nematodes and protozoa regulate mineralization processes. Evidence suggests that between 30 and 50 percent of the nitrogen present in crop plants was made available by the activity of bacteria-consuming nematodes (Ingham,1996). Research in Denmark has indicated that nematodes convert about as much energy as earthworms in certain forest soils (Dropkin, 1980). The basis of this relationshipis is that grazing on microbes by mesofauna releases and mineralizes nutrients immobilized in microbes, subsequently converting nitrogen from organic to inorganic forms that plants can utilize (Seastedt *et al.*, 1988 and Sohlenius *et al.*, 1988).

It is estimated that 60 per cent of nitrogen ingested by the total nematode community comes from bacteria, 15 per cent from fungi, and 25 per cent from plant roots. The bacteriophagus nematodes and amoeba together account for over 83 per cent of nitrogen mineralization (Hunt *et al.*, 1984; Griffiths, 1990). The nematodes have been related to increased plant production owing to increased nitrogen availability (Ingham *et al.*, 1985). Nematodes contribute directly to nitrogen mineralization by excretion of nitrogenous wastes, mostly as ammonium ions (Anderson *et al.*, 1983 and Hunt *et al.*, 1987). In addition to serving as a stimulatory force in net mineralization of nutrients nematodes also promote nutrient immobilization because their bodies constitute resvoirs of nutrients. When nematodes die nutrients immobilized in their tissues ae mineralized and subsequently become available to plants.

4.11 Nematodes as Models

Cell biologists believe that the basic life processes proceed along similar paths in animals. This idea has given impetus to use of nematodes as models for studying aging in animals. In some important aspects, age-related changes in nematodes have been shown to closely parallel that in mammals (Zuckerman, 1987). Effects of nutritional supplements can be directly related to effects on post-embryonic and adult development or aging and it is not influenced by changes due to cell turn over. The free-living nematodes *Caenorhabditis* sp, *Turbatrix* sp, and *Panagrellus* sp have been exclusively used for these studies. *Caenorhabditis elegans* is widely used in developmental biology, neurobiology, gerontology and genetic control of various life processes, cell lineage leading to organ formation, nutrition and physiology. One of the most attractive features of *C. elegans* is that it can be handled like a microbe. Large numbers of worms can be maintained inexpensively on bacteria growing on standard agar plates, but viable cultures can be stored as frozen stocks and then revived when required. Unlike other animals, *C. elegans* does not have male and female sexes. Instead, there are males and hermaphrodites. A single hermaphrodite can self-fertilise and produce over 1000 eggs per day. If crosses need to be carried out, hermaphrodites can be mated with males. The life cycle of *C. elegans* is two weeks long, similar to that of the fruit fly, so it is very amenable to genetic analysis.

The rise of the nematode worm as a model organism began in 1963, when Sydney Brenner chose it for his studies of the role of genes in development and the nervous system.

The *C. elegans* genome is 97 million base pairs in length and contains about 20 060 genes. Many of these genes appear to have functional counterparts in humans, and whole pathways are often conserved. This makes *C. elegans* a useful model for human diseases. For example, the insulin signalling pathway is fully conserved between humans and nematodes so mutant worms impaired for insulin signalling are useful models of type II diabetes. Due to their microbe-like properties, these nematode mutants can be screened with thousands of potential drugs to identify compounds that return the insulin-insensitive disease physiology to normal. *C. elegans* mutants provide models of many other diseases including neurological disorders, congenital heart disease and kidney disease. The molecular basis of cell death has been studied extensively in *C. elegans* and in the future this worm may even hold the key to counteracting the effects of ageing in humans.

Pristionchus pacificus is another nematode that has been established as a model organism in evolutionary developmental biology (Sommer *et al.*, 1996; Hong and Sommer 2006). It shares many advantageous features with *C. elegans*, in that it can be grown easily under laboratory conditions by feeding on *Escherichia coli* OP 50, it has a short generation time (4 d at 20°C), and it is a self-fertilizing hermaphrodite, which makes it amenable to forward and reverse genetics. The genome of *P. pacificus* was recently sequenced in a whole-genome shotgun approach with 10-fold coverage. The calculated genome size is 169 megabases (Mb) with a total number of 29,000 predicted protein-coding genes and a minimal gene content of 23,500 genes, as inferred from RT-PCR analyses (Dieterich *et al.*, 2008). Many of these genes share no sequence similarity with already known genes in other nematodes and different phyla ("pioneer" genes). In comparison, the genome of *C. elegans* is completely assembled, consisting of a 100-Mb genome with 20,060 encoding genes (The *C. elegans* Sequencing Consortium 1998; Dieterich and Sommer 2009). *P. pacificus* and *C. elegans* belong to the same phylogenetic clade which provides an ideal evolutionary distance for comparison of their proteome structures (Borchert *et al.*, 2010). The overall mutation rate in *P. pacificus* is 7.6×10^{-8} per site per generation and is less than one order of magnitude different from estimates in *C. elegans* and *Drosophila* (Ruxandra *et al.*, 2011)

Grass-inhabiting free-living nematodes, *Panagrolaimus* spp. were found to biomineralize lead heavy metal. Cultures of two isolates of *Panagrolaimus superbus*, were exposed to 500 ppm lead for an exposure time of 20 min. Control worms were exposed to deionized water. TEM sections taken from the oesophageal region revealed the presence of metal particulates in lead exposed nematodes only. X-ray elemental analysis confirmed the presence of lead in the particulates. This is the first account of heavy metal biomineralization in nematodes (Williams and Seraphins, 1998).

The human parasitic nematode, *Trichinella spiralis* has been used as host resistance model for detection of immunotoxocity in humans (Luebke, 2007).

The discovery of virus-infected nematodes (in wild strains of *C.elegans* and *C. briggsae* by Félix *et al.*, 2011) in nature proves the worm's value as a model system for studying viral infection. In addition to enabling exploration of defenses against viruses at all stages of their life cycle, the worm will help researchers understand how anti-viral immunity evolves as well as how hosts and pathogens coevolve with each other. And with a wealth of genetic tools already in place, this tiny, transparent model organism will surely help uncover more mechanisms of innate immunity. All eyes in the world of molecular immunology will be watching to see what discoveries this system yields next (Meadows, 2011).

Recently a dye, basic yellow1, also known as Thioflavin T (ThT), extended lifespan in healthy nematode worms by more than 50 percent and slowed the disease process in worms bred to mimic aspects of Alzheimer's (Alavez *et al.*, 2011). The research could open new ways to intervene in aging and age-related disease. The study highlights a process called protein homeostasis — the ability of an organism to maintain the proper structure and balance of its proteins, which are the building blocks of life. Genetic studies have long indicated that protein homeostasis is a major contributor to longevity in complex animals. Many degenerative diseases have been linked to a breakdown in the process. The study also highlights the value of curcumin, the active ingredient in the popular Indian spice turmeric, which had a significant positive impact on both healthy worms and those bred to express a gene associated with Alzheimer's.The authors stated that 'People have been making claims about the health benefits of curcumin for many years. Maybe slowing aging is part of its mechanism of action' (Alavez *et al.*, 2011).

Chapter 5
Methodology and Techniques

5.1 Sampling Techniques for Nematodes Assessment

Nematodes are microscopic and the most abundant multicellular organisms organisms that live in soil and water. To ascertain the involvement of a particular nematode with a crop requires careful sampling. The sampling has become increasingly important in modern agriculture as the concepts of Integrated Pest Management (IPM) and Integrated Crop Production are developed and utilized. Scientists concerned with nematode populations have improved methods for their assay; however, data from the best extraction methods are of limited value if the sample is not representative of the area. Diagnosis of nematode damage and management requires assessment of nematode infestation levels in the soil samples collected from the field or a plant where problem exists. Effective diagnostic sampling may involve rating plant roots (*e.g.* galls caused by *Meloidogyne* spp.), bioassays, or visually assessing above-ground growth for effects of foliar pathogens in addition to collecting soil and root samples for nematode counts.

The major purpose of sampling for nematodes includes population estimation for research, advisory or some predictive estimates, disease diagnosis and general surveys for nematode identification. The degree of required accuracy depends on the sampling purpose. In case of qualitative studies, such as diagnosis of diseased plants or taxonomic studies, the required accuracy is relatively low and a simple sampling plan is sufficient. In case of an ecological study to establish the species composition of a nematode population, sampling needs to be done more accurately, as rare species need to be included in the study. The required level of accuracy in sampling, further, increases if its purpose is to advise on the need for control measures or crop choice, and for the determination of biomass. A very high level of accuracy is needed if sampling is done to issue a phytosanitary certificate as part of legal requirement and inspection regulations, which mandate that plants for multiplication or export purpose are free of nematode infection (so-called 'zero tolerance').

5.1.1 Sampling Scheme

Sampling for nematodes is an increasingly important component of plant disease diagnosis. Without confirmation through sampling, poor plant growth because of nematodes may be misinterpreted as nutrient deficiencies or other maladies. The art and practice of collecting representative root and/or plant tissue as well as soil samples is a critical component of diagnosing the malady. For example, sampling from adjacent, healthy-appearing plants may be just as important as collecting samples from the most severely affected ones. For identification of nematode involved in damage function of a particular crop and management on an as-needed basis requires assessment of nematode infestation levels in the representative soil sample collected from the targeted field or a small plot. Roots should be included in the sample for the recovery of endoparasitic nematodes like cyst, root knot and lesion nematodes which complete a part or whole of their life cycle in the root tissue. However, the problem of accurately identifying the nematode involved and determining the level of its infestation will largely depend upon the thoroughness of the sampling method, handling and storage of the sample as well as the biology of the nematode species involved.

The plant parasitic nematodes are not uniformly distributed in any infested field even years after their establishment. Their distribution has best been described by a negative binomial model where sample variation (variability between samples) is greater than the sample mean (sample average).This clumping distribution is further complicated in many nematode species like root knot and cyst nematodes which deposit large numbers of eggs in the egg-sacs or in the cyst and their hatching or emergence takes place at the spot where they are laid (aggregated hatching and emergence). Soil population of plant parasitic nematodes varies greatly during the growing season of the crop; generally it is low when the crop is just sown and high at the harvest time of the crop. Soil population assessed at the time of planting is more useful for relating it with performance of the crop and potential yield losses. However, it is most convenient to sample at harvest time or just prior to it when population is at its peak. Knowledge of the life cycle of the nematodes to be sampled especially if the nematodes are plant parasites, is required to choose the 'optimum' time of sampling. Sampling should be done when the soil is close to its field capacity and it should be avoided when the soil is too dry or wet.

5.1.2 Number of Samples and Size

To manage the uneven horizontal distribution of nematodes, it is pertinent to take many composite soil samples from the field or the area to be surveyed. Obviously, higher the number of soil samples taken per unit area, higher accuracy will be obtained in assessing the nematode infestation levels. If a portion of the field exhibits unthrifty growth of the plants or there is lot of soil type variability in the field, soil samples should be drawn from such areas separately. Sample size depends on the area and crop being sampled, the crop's value and any supplementary information to be gathered during sampling. Sample size is measured in terms of the number of cores of soil constituting the sample. Very large fields should be broken down to smaller plots on the basis of a grid, and representative areas with uniform soil and crop histories should be selected randomly to provide a basis for sampling. The precision of the resulting nematode estimates can be improved by increasing the number of soil cores in the sample. This is also less expensive than increasing the actual number of samples. If the area is less than 500 sqm, 8-10 samples; 500-2000sqm, 25-35 samples and if the area is between half hectare to 2.5 ha. then 50-60 soil samples are required to estimate the nematode population. Aa a matter of fact, minimum of four composite soil samples should drawn from an acre fields and each composite sample is suggested to consists of at least 15 sub-samples or more, which is always better. The pattern of sampling to be done may include random, stratified random, systematic or two stage sampling. The specific type of spatial pattern and relative density of the target nematode species should be considered in selection of the sampling procedure. One should follow a predetermined soil sampling design for collecting sub-samples and preferably it should in the form of either 'V', 'W', 'X' or 'Z' in the given field. The sampling done in the form of 'W' will give best precision estimate of nematode population in the field or even while sampling nematodes around tree trunk. This can further be extended to sampling in done in row crops though in this case hexagonal pattern is the best. To determine, whether nematodes are involved in a decline or poor plant growth problem in an established crop, the symptoms should be assessed and their distribution pattern should be noted down (random or uniform and severity of symptoms). Samples should be drawn from different symptoms pattern from the root zone of affected plants and also from healthy plants collected both root and soil samples. In case of aerial plant infection samples should be taken from the affected palnt parts. Sampling depth varies with the host crop, turf: 0-10 cm; field crops, pastures, vegetables: 0-20 cm; trees, vines, perennial ornamentals: 10-30 or even 60 cm and discard the surface soil to minimise the influence of dried top soil, weeds and cover crops.

Using a spade, 'khurpi'- a narrow blade shovel or a sampling probe, a small volume of soil (200-300 gms) should be collected from the ploughed layer at each sub-sample location and after mixing few sub-samples a composite soil sample of 400-500 gms of soil should be placed in the polythene bag of proper size (2kg size). The top 3-5 cms layer of the soil should be discarded before collecting the soil and labeled properly giving location, milestone mark, crop, farmer's name and date of collection etc. Use a pencil or a permanent waterproof marker for writing on the label or bag, to avoid decoloration and bleaching.

These nematodes are highly sensitive to high and a low temperature as well to drying, so proper handling during storage is very important. It is extremely important to preserve the soil or root samples in as close a physiological active state as possible until they are processed for nematodes or bioassay with a susceptible plant. It is equally necessary to prevent exposure of the collected samples to direct sunlight or to heat by placing them in a shaded cool place or better in the refrigerator (10°C) if the samples are to be stored for few days before processing. But it is always good to process them as soon as possible (within a day or two) to get the best results. For processing the samples to extract a variety of nematodes, Cobb's sieving and decanting method is given preference over all other methods being simple, easy to handle and without any complications.

Estimate of nematode communities in the sample should be exercised at the appropriate time *i.e.* immediately after extraction, though the precision of nematode population is dependent on relative density, degree of aggregation, type of sample and reliability of extraction methods. Most nematode assays are based on numbers or density per unit volume or weight of soil. In addition, one may also find how many specimen/ species are in a sample relative to the total mean number of nematodes for all species. The relative density is calculated as:

$$\text{Relative Density} = \frac{\text{No. of individuals of a species in a sample}}{\text{Total no. of individuals in a sample}} \times 100$$

The sum of all relative densities for a community equals 100 per cent .

For calculating relative biomass we have:

$$G = \frac{a^2 b}{(1.6)(1{,}000{,}000)}$$

where,

G = biomass in µg; a = the greatest body width (µm); b = body length (µm) 1.6 = constant for correcting volume of nematode and 1,000,000 is a factor for converting μm^3 to µg.

Frequency is a measure of rate of occurrence (Nortan, 1978). Two types of frequencies may be calculated:

$$\text{Absolute Frequency} = \frac{\text{No. of samples containing a species}}{\text{No. of samples collected}} \times 100$$

and,

$$\text{Relative Frequency} = \frac{\text{Absolute frequency of a species}}{\text{Sum of frequencies of all species}} \times 100$$

Absolute frequency is an independent calculation for rate of occurrence for each species. In contrast relative frequency gives the total frequency of all species on the basis of 100 per cent .Therelationship of population density and frequency can be expressed as prominence value.

$$\text{Prominence Value} = \text{Density} \sqrt{\text{Frequency}}$$

These determinations are based on the absolute density and frequency per species. The importance of a nematode species may be estimated by calculation of the importance value:

Importance Value = relative frequency + relative density + relative biomass.

The sum of all relative biomass values for a community also equals 100 as with relative frequency and density. An 'index of similarity' may be determined for comparison of nematode infestations of two areas:

$$\text{Index of Similarity} = \frac{2c}{a+b}$$

where,

c = no. of species that the two samples or areas have in common; a and b = no. of species in a sample of areas a and b respectively.

Simulation models and a number of computer programs are available for evaluation of various sampling procedures and for fitting nematode population data to standard distributions as well as mapping nematode distributions over space and time. Three dimensional distribution maps can be developed from nematode population data by computer.

5.2 Processing of Nematodes from Soil and Plant Tissues

A rapid, reliable and highly efficient method for the extraction of viable nematodes from soil is of particular importance not only for field diagnostics but also laboratory research. Most of the extraction methods are indirect, making use of a number of properties using the size, density and movement of the nematodes to separate nematodes from the surrounding medium.

5.2.1 Weight and Rate of Settling

In water, nematodes are separated from particles that settle faster and can subsequently be poured off (decanted). This principle is the basis of a number of applications, such as the use of an undercurrent of water that keeps nematodes afloat while other particles settle (elutriation) and use of a liquid with higher specific gravity than nematodes, which keeps them afloat while other particles (with a higher specific gravity than the liquid) sink to the bottom. This is applied in centrifuge floatation techniques. Dried cysts contain air bubbles, making them to float on the water surface, which separates them from sinking particles.

5.2.2 Size and Shape

Because of their size and elongated shape, nematodes can be separated from other (soil) particles by using a set of sieves with different mesh size.

5.2.3 Mobility

Living nematodes can be separated from other particles due to their mobile nature. When a sample is placed on a sieve with a moist filter paper, placed in a shallow water-filled tray, nematodes will crawl from the sample into the water where they can be collected as a clear suspension.

5.3 Extraction Methods

An ideal extraction method would make it possible to extract all stages of all nematode species at 100 per cent efficiency, irrespective of temperature and soil type, and at low costs (labour, equipment, water) (McSorley, 1987). Unfortunately, none of the existing methods comply with this ideal. There are many methods for extraction of nematodes from soil. We are seldom concerned with the extraction of eggs but mostly look after juveniles and adults including females of root knot nematodes and cysts in the technique. Among the various extraction techniques of sugar floatation, elutriation, centrifugation and decanting, the most commonly used is the method of decanting and sieving the soil material in different steps (Cobb's decanting and sieving method, 1918). In decanting, the nematodes are less dense than debris and other mineral particles which sink much faster than the nematodes. This phenomenon is a result of Stokes' Law which states that the settling velocity is directly proportional to the difference in density but proportional to the square of the diameter of the particles. Thus large heavy particles sink much faster than the less dense and smaller diameter nematodes. Decanting relies upon pouring the water onto the sieves (840, 250, 45 and 28μ-sieve openings). The Cobb decanting and sieving method is one of the most commonly used methods for extraction of nematodes from soil. This method depends on the differences in size and specific gravity between nematodes and soil components. In this method, the soil sample (*e.g.*, 100 cc or 250 cc) is soaked in a bowl containing one liter water for 10 minutes. The soil and water are mixed by hand and stirred thoroughly. The soil suspension is immediately poured on a 20-mesh sieve (840 μm pore size) into another bowl. While still holding the 20-mesh sieve over the bowl, the material on the sieve is washed with a jet of water. The soil suspension or slurry, which passes through the sieve, is collected in another bowl. The residue on the sieve is discarded and the sieve is washed thoroughly. The slurry in the bowl is stirred again by swirling the bowl and slowly poured on a 60-mesh sieve (250 μm pore size). The residue on the sieve is washed and collected in a 250 ml beaker (marked 60 mesh) and the soil suspension in water which passed through the sieve is collected in a clean bowl for further processing through 325-mesh (45 μm pore size) or preferably 400-mesh (38 μm pore size) sieve. The suspension is collected in a beaker marked 325 or 400 mesh. Since the method is not dependant on nematode movement, sluggish nematodes are collected as effectively as active ones.

Deng *et al.* (2008) has given another method of extraction of nematodes from mainly roots with the use of iodixanol based on an iso-osmotic density gradient medium but it is of less use.

5.3.1 Extraction of Cysts from Dry Soil

Extraction of cysts from dry soil can be done using Fenwick Can. The Fenwick can is a widely used apparatus for extracting nematode cysts (Fenwick, 1940) and suitable for soil samples of up to 500 cc. The apparatus is generally 30 cm high and made of brass. Firstly, the can is filled with water. The soil sample (500 cc) is slowly poured on top of the wet sieve (pore size 840 μm) placed on the funnel of the apparatus and is washed with a strong jet of water from above. The soil along with dry cysts as well as some organic material overflows rapidly into the collar and passes down to the collecting sieves (pore size 840 μm, 250 μm, and 45 μm). The upward current of water carries the cysts that are trapped during sedimentation. The residue collected on 250 μm sieve is examined for nematode cysts under a stereoscopic microscope.

5.3.2 Extraction of Nematodes from Plant Tissues

For extraction of nematodes from plant tissues Seinhorst mist technique is the best. The basic principle of spraying a mist of water onto infected plant tissues on Baermann funnel like supports gives the advantage of ease of automation and yield of large numbers of active and clean specimens. The root incubation technique is suitable for immature stages and males of sedentary parasites. The roots (1–5 g) are washed free from soil and debris and cut into small pieces of 5–10 mm. These pieces are placed either in closed petri dishes or polyethylene bags and incubated at 25–30°C. If the roots are immersed in water, it is advisable to pour off the water and rinse the roots after 24 h, retaining the original water and washings for examination. Most of the nematodes are recovered within 4–7 days but much depends upon the nature of the roots and kind of nematodes present in the root tissues. The maceration filtration method is quicker and more efficient than the incubation technique and is used for the recovery of migratory endoparasites (*e.g. Pratylenchus* spp). About 5 g roots are washed and chopped into small pieces of about 1 cm. The pieces are then macerated in an electric blender at full speed for 5 seconds. The suspension is poured on the tissue paper supported on a wire mesh sieve, and immersed in a dish containing water. The nematodes move down and are collected in the dish after 24–48 h. The sedentary nematodes such as *Meloidogyne* spp, *Heterodera* spp, and *Globodera* spp can either be examined in situ by staining or can be removed from roots by maceration and centrifugation.

5.4 Collection of Eggs

The eggs of endoparasitic nematode species can be obtained from root tissue by the centrifugation methods. Egg masses of *Meloidogyne* spp can also be collected by shaking the roots in dilute sodium hypochloride (NaOCl) solution. Hussey and Barker (1973) observed that eggs collected by dissolving the gelatinous matrix or egg masses with NaOCl gave better inoculums for infesting roots than those recovered by maceration and centrifugation, although long exposure to NaOCl could be detrimental (Vrain, 1977). The roots are properly washed and cleaned to remove the adhering soil particles. These roots are shaken vigorously in 200 cc of 1.05 per cent NaOCl in a 500 cc bowl for 4 minutes. The NaOCl solution is passed through a 200-mesh sieve (pore size 75 mm) and over a 500-mesh sieve (pore size 35 µm). The 500-mesh sieve is rinsed with water to remove the excess NaOCl. The roots are also washed again with tap water to remove additional eggs, which are collected in a beaker. There are various methods for separating out different stages of nematodes. But generally, the cyst and large nematodes (*e.g.*, longidorids) are trapped on 60-mesh sieve (pore size 250 µm) while most of the other nematodes are collected on 325-mesh or 400-mesh sieves.

In soil samples where the population of nematodes particularly of root knot or cyst nematode juveniles is barely detected, bioassays could be required. In this technique the incidence of root knot nematodes may be indicated by the relative extent to which the roots are galled. In case of cyst nematodes, the presence of white females on the roots will be the indicator of their existence.

5.5 Killing and Fixation

After obtaining the clear nematode suspension, the nematodes can be killed by bringing the temperature of nematode suspension to 60-70°C on a hot plate. Solution of 2-4 per cent formaldehyde (5-10 per cent formalin) is generally used for 24 hrs to fix the nematodes. Fixation of samples in the field is an option if samples cannot be processed directly. In general, formalin (4-10 per cent) is added until the sample is just submerged and the mixture is then stirred or shaken (Elmiligy and de Grisse 1970; Freckman *et al.*, 1977). In wet samples, especially samples of sediments, the concentration of formalin sharply decreases due to dilution by the water present in the sample. Fixation of the sample is incomplete if the final concentration drops below 4 per cent. In fixed nematodes, much of the internal body contents especially the gonad structures may be obscured by the granular appearance of the intestine which can be cleared by processing them in lactophenol (lactic acid 20ml; phenol 20ml; glycerine 40 ml and distilled water 20ml). It is good clearing agent and widely used to process the nematodes to glycerine.

5.6 Preparation of Mounts

5.6.1 Water Mounts

Living nematodes can be easily identified and studied in a water mount. In this type of mount, a number of structures like the spear, the oesophagal lumen, and the excretory pore can be seen more easily than in dead and fixed nematodes. Water mounts are usually made on common glass slides after the nematodes are killed.

5.6.2 Permanent Mounts

The permanent mounts can be prepared either with slow or rapid glycerin methods (Seinhorst, 1962). As slow method takes lot of time, rapid method is usually followed. In this method, the fixed nematodes are transferred to Seinhorst Soln.1 (S1) which contains 20 ml of ethanol, 1 ml glycerine and 79 ml of distilled water in a small cavity block which is then kept on a platform in a vessel containing ethanol for a minimum of 12 hrs at 40C. The nematodes are in ethanol and little of glycerin. The cavity blocks is removed and with care the supernatant is pipetted out and Seinhorst Soln.2 (S2) (5 parts of glycerin and 95 parts of distilled water) is added in the cavity block containing nematodes. The cavity block, partially closed, is placed in an oven at 40C for 3 hrs or till the ethanol evaporates. The nematodes are now in pure glycerin and can be mounted in a drop of anhydrous glycerin on to a glass slide using round cover slip (19mm). As the processed nematodes are too soft, support of glass-wool pieces (equal to diameter and size of nematodes) should be provided before placing the cover slip on the slide. The slide is then sealed with a ring of 'Zut' (glyceel) to prevent it from drying. Clear nail polish can also be used in the absence of any other sealing material. The other method is that small three pieces (of the size of a drop of glycerine) of paraffin wax of 60-65C melting point are placed around the drop of nematodes in glycerin and the cover slip is placed on the pieces. The slide is then heated just to melt the wax which spreads and fill the space between the slide and cover slip holding the nematodes in the centre. There is no need of sealing the cover slip as the wax acts as sealing medium.The nematode specimens can be studied for morphology and morphometric requirements by using camera lucida mounted on a microscope. Generally two types of microscopes are used: the "dissecting microscope" and the "compound microscope". The "dissecting microscope" is a low-powered stereoscopic microscope for small magnification rates (10-50x), offering a wide-angled vision. This microscope is mainly used for analysing nematode suspensions and for dissection work. The "compound microscope" is a stereoscopic microscope offering magnifications ranging from 40 to 100x. Working at a magnification of 100x requires an oil immersion lens. This type of microscope is used for identification of nematodes and other types of detailed observations. Usually the material has first been mounted under a dissecting microscope. The image analyser can be used for easy recording different dimentions of the nematode body.The automated image analysis of digitized images of nematodes is described by Been *et al.* (1996).

5.7 Staining the Roots

The study of root knot and cyst nematodes usually involves penetration and development of juveniles in the plant tissues which may either be required for host parasite related (histopathological) investigations or for general information on their life history. For studying penetration and development, many staining procedures have been developed. The acid-fuchsin-lactophenol method (Mcbeth *et al.*, 1941) has been the most widely used procedure before Byrd *et al.* (1983) gave the Sodium-Hypochlorite-Acid-Fuchsin method which is much improved technique as it stains only nematodes and not the root tissues. The advantage of this method is also the elimination of toxic substances like phenol from the procedure. Since root tissue is cleared in sodium hypochlorite (NaOCl), it does not get heavily stained as in other procedures, only nematodes are stained which can be seen under stereoscopic microscope and counted easily. The time required for staining is also shortened and frequent destaining is avoided. The procedure involves washing the infected roots and cutting them to small pieces before they are put in a beaker having 50 ml water and bleached by adding 20 ml of 5 per cent sodium hypochlorite solution (for hard roots this quantity can be increased). Then soak the roots for 1-4 minutes (depending upon their hardness) and agitate occasionally. Wash the roots in running water for few seconds before putting the roots in a beaker containing tap water. Change the water 2-3 times after every one hour to remove the residual bleach which may affect the staining. Prepare acid fuchsin stock solution containing 3.5 gm acid fuchsin, 250 ml of acetic acid and 750 ml of distilled water. Put three ml of stock solution of acid fuchsin in the beaker containing bleached roots in 30 ml of water (the stock soln. stain and water may be in the ratio of 1:10 or more depending upon the thickness of the roots). Boil the roots for 30 seconds on a hot plate and cool them to room temperature before rinsing them in tap water. Place the roots in glycerin in a petri-dish for sometime. Replace the stained glycerin with pure glycerin in the petri dish after a day for direct observation of nematodes in the roots under stereoscopic microscope.

5.8 Preparation of Perineal Pattern (for identification of root knot nematodes)

The perineal pattern, the pattern of striations around anus and vulva, is basic to the identification of root knot species although morphological and morphometric variations of juveniles and males are also as important. The stylet length and the position of excretory pore from the anterior end is sometime of taxonomic significance while describing root knot species. The differences in these characters are now being supplemented with various biochemical and molecular level characterizations. For preparation of the pattern, the swollen matured female of any root knot nematode is taken on a drop of water preferably on a plastic slide, is given a small cut in the middle of the body to prevent it from bursting if the female is directly cut in the center which may destroy (tear-off) the pattern. The small cut will give way for the intestinal material to come out while pressing from one side. Then the female is cut into two halves. The posterior half is slowly cleared of the intestinal and other internal material. It is trimmed further till only a portion of the skin is left around the perineal pattern. The vulval area is cleared of the muscles attached to it with utmost care without causing any injury to the perineal area. This cut portion of perineal pattern is then stained with few granules of acid fuchsin stain for few minutes. On a clean slide place a small drop of lacto phenol or glycerine and place the stained cut portion of perineal pattern in the center of the drop while taking care that it is placed upside down and is in a proper position (there should not be any shrinkage, tilted or bend portion). The perineal pattern when placed in the lactophenol gets folded due to change in the medium (from water to lactophenol or glycerine), therefore, care should be taken to immediately unfold it and bring it to proper position. Place a cover slip on the slide and possibly seal it with glyceel or nail polish before labeling it properly for permanent record.

5.9 Preparation of Cone-Rop Structure (for identification of cyst nematodes)

As in case of the root knot nematodes, the identification is mostly based on the characters of the perineal pattern found in the vulval regions of the matured females, the identification of cyst nematodes is also relied upon the cone top structures (a variety of structures on the top of the cone) present in and around the vulval region of the cysts. For preparation of the cone top structure, a cyst is taken in a drop of water preferably on a plastic slide and with a sharp blade an incision is made in the center of the cyst to facilitate the exclusion of eggs and juveniles from the body of the cyst to make it empty. The body is cut into two halves and the cone top half is trimmed further until sufficient skin is left around the cone. With the help of a fine and soft 'pick' the outer and internal parts are cleaned properly without disturbing the main structure. The cone top is then placed in absolute alcohol taken in a small cavity block which is covered with lid and kept for 3-4 minutes for dehydration before placing it in clove oil for ultimate cleaning. Any muscles remained attached with cone top gets obscured in clove oil. The structure can be kept in this oil for a long time till mounted on a slide. A standard drop of Canada balsam or euprol is taken on a glass slide and the cone top structure is placed on it and lowered to the bottom of the drop on the surface of glass slide with the help of fine sharp 'pick' and covered with a cover slip. In the event of cone top being large, very small pieces of crushed cover slip can be placed on the three sides of the drop before placing the cone top structure to avoid it tilting towards one side. There is no need of sealing as Canada balsam itself acts as a sealing material. Label the slide properly for records.The slide with cone top is now ready for studying under compound microscope.

5.10 Nematode Measurements

The nematodes are characterized by the differences in the body measurements of various structures and ratios derived from the measurements of body parts. Measurements have to be accurate and be taken very carefully and patiently and for that proper procedure is to be followed for processing and mounting of nematodes. The workers have to remain aware of the developments that take place in the observations, identification and interpretations of the characters. In case of root knot and cyst nematodes not only the perineal pattern and cone top structures are important but juveniles and males morphometrics are also equally important. De Man (1880) is the founder of the use of measurements and ratios derived therefrom. He had proposed the following formula for various measurements and ratios (originally designated by α, β, γ etc., but later changed to a, b, c in the roman version).

L: Total length of the body (head to tail tip)

a: Body length divided by the maximum body width

b: Body length divided by esophageal length (anterior end to esophago-intestinal junction)

b′: Body length divided by length from anterior end to the posterior end of esophageal glands

c: Body length divided by tail length

c′: Tail length divided by body width at anus

V: Percentage of length from anterior end to vulva to body length

V′: Percentage of length from anterior end to anus

T: Percentage of distance from cloacal aperture to anterior end of testis to body length

m: Percentage of length of conus to total stylet length

O: Percentage of distance of DEGO from stylet base to stylet length

MB: Percentage of length of median bulb from anterior end to esophageal length

R: total number of body annules

Ran: Number of annules on tai

Chapter 6

Interaction of Cyst and Root Knot Nematodes with Other Microorganisms

"Nature does not work with pure cultures. I suspect that many plant diseases are influenced by associated organisms to a much more profound degree than we have yet realized, not only as to inhibition but also as to acceleration of the process. It may be that a number of diseases require an association of organisms for their occurrence and cannot be produced by infection of one organism alone. Pure cultures we must, however, continue to use as a basis for the known mixtures and as controls on the activity of mixtures. Research with mixtures of organisms will not furnish an excuse for any less care than with pure cultures in excluding organisms foreign to the mixtures. Work with mixtures, however will not make the already complex problems of plant pathology, as a whole, an easier or less complex but it may throw much light on certain relationships which will never be discovered by the use of pure cultures of single organism"-Fawcett (1932). These statements of him are still true and invite our attention with the same vigor as they did about 80 years ago.

Soil is the habitat of an extremely complex microbiological community. There are billions to hundreds of billions of soil microorganisms in a mere handful of a typical field soil. That single handful might well contain thousands of different species of bacteria, hundreds of different species of fungi and protozoa, dozens of different species of nematodes plus a good assortment of various mites and other microarthropods. Almost all of these countless soil organisms are not only beneficial, but essential to the life giving properties of soil. They struggle for existence among themselves mostly in the rhizosphere. This struggle make certain groups of microorganisms as dominant forms in the community and others are relegated to a subdominant level. The presence of large numbers of organisms in the rhizosphere is evidence that roots supply considerable quantities of organic matter.

We have gone far ahead from the time of Cobb (1915) when he rightly quoted that our world will still be dimly recognizable if all the matter is removed apart from the nematodes as we have now huge diversity of nematodes and the niche they occupy. Under certain circumstances the nematodes can built themselves to large numbers but their number is kept under control by the presence of extensive population of organisms that prey and parasitize them. These organisms include fungi, bacteria, insects, mites, tardigrades, turbellarians and other predaceous nematodes. It is essential to be aware of the array of the organisms and their interactive nature influencing the crop quality and economic production. All the stages of the nematodes, sedentary and active have been found to be preyed upon by a number of other invertebrates significantly changing the community structure of nematodes although overall nematode biomass may remain unchanged (Bradford *et al.*, 2002).

Diseases and pests have the potential to cause significant reduction on biomass production of the crops and quality of the yield. Of the many pests and diseases, plant-parasitic nematodes are of great economic importance because they can directly influence plant biomass and predispose plants to attack by other soil-borne pathogens (Mekete, 2010). Root-knot and cyst nematodes are obligate, biotrophic pathogens of numerous plant species. These organisms cause dramatic changes in the morphology and physiology of their hosts either alone or in combination with other disease causing organisms. Plants resistant against fungal and bacterial pathogens may become susceptible in the presence of nematode infestation. In some instan-ces, root injuries caused by nematode larval penetration or feeding may provide an easy entry avenue for fungal and bacterial colonization, which otherwise they might not have been able to do so independently. In such cases the disease expression may be quite severe.

Many aspects of nematode involvement in plant disease complexes have been described. In essence plant-parasitic nematodes may serve as (i) vectors of viruses, fungi, or bacteria; (ii) wounding agents; (iii) necrotic wound causing agents; (iv) host substrate modifiers; (v) rhizosphere modifying agents; (vi) resistance breakers; (vii) pathogenicity inducers; (viii) predisposing agents; (ix) deterrents to other pathogens; and (x) disrupters of the symbiont activity (Swarup, 1990; Khan, 1993). Since this chapter will only focus on the interrelationships involving root-knot and cyst nematodes, certain type of interactions will not be discussed herein, *e.g.* nematode vectors of plant viruses limited to members of the Dorylaimida (*Xiphinema, Longidorus, Paralongidorus, Trichodorus*, and *Paratrichodorus* species) (Weischer, 1993).

After hatching in the soil, root-knot nematodes must locate and penetrate a root, migrate into the vascular cylinder, and establish a permanent feeding site. Striking parallels can be drawn between the nematode–plant interaction and plant symbioses with other microorganisms, and evidence is emerging to suggest that nematodes acquired components of their parasitic armory from these microbes (Bird, 2004).

6.1 Nematode-Fungal Interaction

Hundreds of studies have focused on documented and potential relationships between nematodes and fungi since Atkinson (1892) when he first observed that the severity of *Fusarium* wilt of cotton was enhanced in the presence of root-knot nematodes. Plant parasitic nematodes interact with fungi in a variety of ways to cause plant disease complexes. Interaction may be additive (damage caused by combined effect is equal to the damage caused by sum of individual effect) or, synergistic (combined effect damage is more than the sum of individual effect). Interaction among *M. incognita* and root-rot pathogen, *Rhizoctonia solani* on tomato roots (Abawi and Barker, 1984), *H. schachtii* and damping off pathogen, *Pythium aphanidermatum* on sugarbeet (Whitney, 1974), *Pratylenchus thornei* and *Bipolaris sorokiniana* on Wheat (Doyle *et al.*, 1987) are some of the examples of additive effect. Hassan (1985) studied synergistic interaction between *M. incognita, P. aphanidermatum* and *R. solani* on chilli and found that combined effect is more harmful than their individual action. On chickpea, *M. incognita* and *Fusarium oxysporum* caused synergistic effect. Majority of the interactions studied so far have been with either wilt/ root rot/ seedling blight fungi.

6.1.1 Nematode-Fungal Wilt Complexes

The influence of infection by *Meloidogyne* spp. on vascular wilt pathogen, *Fusarium oxysporum* has been most often studied on tomato and cotton, although the effects are probably similar on nearly every plant susceptible to these pathogens. In the field, wilt symptoms are more severe; develop more rapidly and at greater frequency when the plants are also infected by the nematode than in the absence of nematode infection (Manzanilla-Lopez and Starr, 2010).

Jenkin and Cursen (1957) has suggested that when two separate species of root knot are affecting the host, their infection has marked difference on the physiology of the plant as *M. hapla* was able to induce 60 per cent wilt symptoms in a wilt resistant plant whereas *M. incognita* was able to produce 100 per cent wilt in the same plant. Porter and Powell (1967) showed that both susceptible and resistant varieties of tobacco had increased incidence of wilt when *M. incognita, M. javanica* or *M. arenaria* were present. Wilt was more severe when nematodes were already present in the roots before fungus. The root-knot nematodes may predispose the plants to fungal infection within two to four weeks prior infection. *Meloidogyne incognita* maximally predisposed tobacco plants to Fusarium wilt (Porter and Powell, 1969) and to root decay caused by *Pythium ultimum* (Meléndez and Powell, 1970) when the nematode was inoculated four weeks prior to fungal inoculations. Griffin and Thyr (1986, 1988) reported that the severity of Fusarium wilt of alfalfa was increased when *M. hapla* inoculations preceded fungal inoculation. This phenomenon has been used to argue the basis of interaction must be physiological rather than due to a simple response to wounds made by the invading nematodes. It is well known that physiological activity in galls and nematode induced giant cells is entirely different from that of non infested host roots although, it has been difficult to correlate these changes in roots to the infection caused by Fusarium wilt. Wang and Bergeson (1974) suggested that changes in total sugar and amino acid concentration in the xylem sap which reach maximum concentration four weeks after nematode inoculation exuded as root leachates may contribute to the enhancement of Fusarium wilt in tomato infected with *M. incognita.*

A nematode species may interact differently with two fungi on the same plant. In a greenhouse experiment, *Verticillium* wilt was more severe in tomato cv. Bonny Best infected with *Heterodera (Globodera) tabacum* than in uninfected plants, while *Fusarium* wilt was less severe in the presence of the nematode than in its absence (Miller, 1975). The alterations induced by one species of nematode may predispose its host to infection by certain species of fungus only.

According to Johnson and Littrell (1969) all root-knot nematode species (*M. incognita, M. javanica* and *M. hapla*) infected roots of Chrysanthemum cultivars and caused characteristic root-knot symptoms but did not appreciably affect the plant growth characters measured by changes in plant weight. Nematodes did not break *Fusarium* wilt resistance of 'White Iceberg'; however, wilt symptoms appeared earlier and were more severe among 'Yellow Delaware' plants inoculated with *Meloidogyne javanica* and *F. oxysporum* than with similar combinations of the fungus and *M. incognita* or *M. hapla* or with the fungus alone. While studying the interaction of *Meloidogyne javanica* and *Fusarium oxysporum* f. sp. *ciceri* on Fusarium wilt-susceptible (JG 62 and K 850) and resistant (JG 74 and Avrodhi) chickpea cultivars Maheshwari *et al.* (1997) found that inoculation of *M. javanica* juveniles prior to *F. oxysporum* f. sp. *ciceri* caused greater wilt incidence in susceptible cultivars and induced vascular discoloration in roots of resistant cultivars. The nematode enhanced the wilt incidence in wilt-susceptible cultivars only at 25°C. Interaction between the two pathogens on shoot and root weights was significant only at 20° C, and *F. oxysporum* f. sp. *ciceri* suppressed the nematode density at this temperature.

Biochemical modifications of root leachates induced by root-knot nematodes appear to enhance the colonization of the rhizosphere by pathogenic fungi, to attract them to galled tissues, and favour their growth at the surface. Moreover, they also appear to lower the numbers of actinomycetes, antagonistic to *Fusarium oxysporum* f. sp. *lycopersici,* in the rhizosphere. Bergeson *et al.* (1970) observed a highly significant reduction in the number of actinomycetes and a significant increase in number of *Fusarium* propagules in the rhizospheric soil surrounding roots inoculated simultaneously with *Meloidogyne javanica* and *F. oxysporium* f. sp. *lycopersici* compared to those observed when the fungus was inoculated alone. A similar observation was made by Noguera and Smits (1982), who suggested that the reduction in number of actinomycetes antagonistic to Fusarium in the rhizosphere of *M. incognita* infected plants may partly be responsible for the enhancement of the pathogenic effect of the fungus.

According to McKenry and Roberts (1985) root knot nematode infection on both cotton and tomato cultivars normally resistant to fungal wilt can predispose these plants to severe Fusarium wilt infection. In the southern San Joaquin Valley, especially in Kern County, concomitant infection of *Meloidogyne incognita* and *Fusarium oxysporum* f. sp. *vasinfectum* can severely damage cotton.

The soybean cyst nematode, *Heterodera glycines,* and the fungus that causes sudden death syndrome (SDS) of soybean, *Fusarium solani* f. sp. *glycines,* frequently co-infest soybean (*Glycine max*) fields. Both *H. glycines* and *F. solani* f. sp. *glycines* reduced the growth of soybeans. Reproduction of *H. glycines* was suppressed by high inoculum levels but not by low levels of *F. solani* f. sp. *glycines*. The infection of soybean roots by *H. glycines* did not affect root colonization by the fungus, as determined by real-time polymerase chain reaction. Although both pathogens reduced the growth of soybeans, *H. glycines* did not increase SDS foliar symptoms, and statistical interactions between the two pathogens were seldom significant (Gao *et al.*, 2006).

Thuy (2010) observed that there was a negative relationship between the inoculation with *M. incognita* alone or in combination with *F. solani* and percentage of black pepper plants with yellow leaves and plant growth. No effect of inoculation with *F. solani* before, at the same time, or 2 week after inoculation with *M. incognita* was observed.

In natural conditions, banana roots are always infested with numerous fungi which are early colonizers of the lesions caused by nematodes (Brun and Laville, 1965; Stover, 1966); such associations and/or interactions should not be underestimated. The expression of Fusarium wilt was increased (etiological interaction) in the presence of *R. similis* (Loos, 1959). Pinochet and Stover (1980) reported that both fungi *Cylindrocarpon musae* and *Fusariuin moniliforme* significantly suppressed the multiplication of *Radopholus similis* (ecological interaction). In Martinique, Lorida (1989) demonstrated the own pathogenicity of a fungus *Cylindrocladium* sp., often associated with *R. similis* in banana root necrosis. Therefore, additional studies are also needed to verify the complex relationships among other microorganisms and nematodes on bananas.

Yang *et al.* (1976) have noted that the plants inoculated with *Fusarium* alone or in combination with *Trichoderma* did not develop wilt but simultaneous inoculation of 7-day-old seedlings with all three organisms (NTF) [Nematode-*Trichoderma-Fusarium*] produced earliest wilt. However, plants receiving nematodes at 7 days and *Fusarium* and *Trichoderma* at 2 or 4 weeks later (N-T-F, N-TF) developed the greatest wilt between 49-84 days after initial nematode inoculation. During the same period, *Fusarium* added 4 weeks after initial nematode inoculation (N-F) and *Fusarium* added 4 weeks after initial simultaneous inoculation of nematode and *Trichoderma* (NT-F) produced the least wilt. The addition of *Fusarium* inhibited nematode reproduction. Simultaneous inoculation with nematodes and *Trichoderma* (NT-) resulted in the greatest root gall development, whereas nematodes alone produced the greatest number of larvae. In comparison with non-inoculated controls (CK), treatments involving all three organisms inhibited plant growth, plants inoculated with the nematode alone (N-) or with nematodes and *Trichoderma* (NT-) simultaneously had greatest root weight. Any treatment involving the nematode resulted in fewer bolls per plant and greater necrosis on roots than the non-inoculated checks.

6.1.2 Nematode-Root-Rot Complexes

Disease complexes involving root knot nematodes and other root-rot pathogens may be among the most important disease complexes in term of total economic impact. Nearly every soil is infested with many root rot pathogens, but comparatively it is less infested with vascular wilt pathogens. Thus complexes involving root rot pathogens are much more to complexes associated with wilt pathogens.

When *M. incognita* inoculation preceded *Rhizoctonia solani* inoculation by at least ten days, *M. incognita* susceptible tobacco cultivars exhibited more severe root-rot than when nematode and fungus were inoculated simultaneously (Batten and Powell, 1971). *R. solani* penetration and development in roots of radish was increased with prior *M. hapla* root infection (Khan and Müller, 1982). The incubation period of *Cylindrocladium* black rot of peanut was shortened in the presence of *M. hapla* and was the same whether the fungus was inoculated simultaneously with the nematode or two weeks later (Diomande and Beute, 1981). Van Gundy *et al.* (1977) reported that the development of infection of tomato plants by *R. solani* was delayed by three to four weeks when the fungus was inoculated simultaneously with *M. incognita*. In Brazil, guava decline was associated with interaction of *M. myaguensis* and *Fusarium solani* where it was observed that the nematode infection predisposes the plants to fungal infection causing a synergistic effect on the severity of the disease (Gomes *et al.*, 2011).

Nematodes have to induce drastic structural, biochemical and/or physiological changes in the roots to render them suitable to fungal penetration and development. In the case of root-rot caused by *Rhizoctonia solani* on tomato, Van Gundy *et al.* (1977) attributed this delay to a modification of the root exudates which occurs three to four weeks after nematode infection when a high concentration of nitrogenous compounds in nematode infected root leachates is favourable for maximum virulence of the fungi. Christian *et al.* (2011) observed significant interaction between *H. schachtii* and *R. solani* on sugar beet. The susceptible cultivar showed synergistic levels of damage when simultaneously inoculated with *H. schachtii* and *R. solani*. Synergistic effects were also detected in the concomitant treatments on the *R. solani* tolerant cultivar. Conversely, damage was lower on the *H. schachtii* resistant cultivar in the concomitant treatments. They also reported that leaf reflectance gave reproduced results in the detection of *Rhizoctonia* crown and root rot disease development. Normalised Difference Vegetation Index (NDVI) values of leaf reflectance showed high correlations to plant and visual disease symptom rating variables. The NDVI allowed disease severity detection without damage to the plant. Further, the results demonstrated that hyperspectral reflectance (leaf reflectance) can be used effectively to monitor aetiopathology of *R. solani* and may be an effective tool for early detection of *Rhizoctonia* crown root rot symptoms in the field.

Golden and Van Gundy (1975) reported that on tomato and okra infected with *Meloidogyne incognita*, sclerotium of *Rhizoctonia solani* were formed only on galled tissues. While studying the interactive effect of *M. incognita* and *F. solani* on tomato various workers (Ahmad and Khan, 2011, Anwar and Khan, 2002 and Jonathan and Rajendran, 1998) have found that growth parameters and yield were severely and synergistically affected when both the organisms were present. Van Gundy *et al.* (1977) observed that 28 days after *M. incognita* inoculation of tomato roots, galled surface segments were abundantly colonized by *R. solani* sclerotia, which began to germinate. A similar observation was made by Khan and Müller (1982), who reported that following *M. hapla* infection mycelial growth of *R. solani* was more abundant on galled regions of radish roots than on non-galled areas. *Rhizoctonia solani* was specifically attracted to *M. incognita* galled tissues responding to stimuli which originated from the galled roots and passed through semi-permeable cellophane membranes (Golden and Van Gundy, 1975). Starr and Aist (1977) observed that *Pythium phytophthora* colonized preferentially the earlier galled areas of celery roots infected with *M. hapla* than non-galled root segments, suggesting that factors attractive to the fungus originated from galls caused by the root-knot nematode.

Galled okra and tomato roots infected with *M. incognita* are highly susceptible to infection by *Rhizoctonia solani*, a disease complex that results in root decay caused by the fungus within about one month after nematode infection. This interaction is common in the warm irrigated soils (McKenry and Roberts, 1985).

Van Gundy *et al.* (1977) demonstrated that, severe root-rot caused by *Rhizoctonia solaní* on Pixie Hybrid tomato may be associated with nutrient mobilization in root leachates induced by *M. incognita*. When root leachates of the plants were removed permanently, and inoculated simultaneously with the nematode and the fungus, no root-rot occurred. In contrast, when root leachates were not removed a severe root-rot developed. Moreover, when root leachates produced by *M. incognita* infected plants were applied to roots of plants inoculated with *R. solani* alone, severe root-rot occurred, whereas roots inoculated with *R. solani* receiving root leachates from control plants were free of decay. During the first 14 days after nematode infection, when carbohydrates were abundant and C/N ratio was high in *M. incognita* infected root leachates, *R. solani* growth was, stimulated in the rhizosphere and the fungus was attracted to the root.

Dong *et al.* (2009) observed that in the greenhouse, inoculation of 500–3000 eggs per plant of *M. arenaria* did not affect the level of root rot induced by 1·0 to 5·0 microsclerotia of *Cylindrocladium parasiticum* per g soil. In microplots, the root rot ratings from Georgia-02C and C724-19-25 were higher in plots infested with *M. arenaria* (0·4–2·0 eggs per cm^3 soil) and *C. parasiticum* than in plots with *C. parasiticum* alone; however, *M. arenaria* did not increase the root rot ratings on the nematode resistant C724-19-15.

6.1.3 Interaction with Mycorrhizal Fungi

Plant parasitic nematodes and vesicular arbuscular mycorrhizal (VAM) fungi are obligatory parasites which share plant roots as a resource for food and space for reproduction. The interest in VAM-nematode interactions lies in the possibility of enhanced resistance or tolerance of VAM-infected plants to nematodes. VAM fungi form a beneficial symbiotic association with roots that increases the plants' ability to absorb phosphorus, minor elements and water. Nematodes and VAM fungi often occur in vicinity in the rhizosphere and plant roots and therefore frequently encounter each other. Mycorrhizal fungi benefit the plant by increasing the adsorption of nutrients and water and by protecting the root from soil-borne diseases. Nematodes can interfere with any of these functions.

VAM fungi do not colonize regions infected by endoparasitic nematodes, and nematodes rarely infect regions colonized by VAM fungi. Nematode-mycorrhizal interactions appear to be very specific and highly dependent on the particular association of plant cultivar, fungal and nematode species, the order of colonization by the symbionts and the soil nutrient level. Plant-parasitic nematodes generally reduce the amount of benefit gained by the association with the fungus. However, VA mycorrhizae usually compensates more for the amount of nematode damage sustained by non-mycorrhizal plants, since their function is generally not affected except in close proximity to the nematode feeding site (Hol and Cook, 2005; Ingham, 1988).

Mycorrhizal fungi and endoparasitic nematodes are often mutually inhibitory, each reducing the population of the other. In other cases, the stimulation of root growth by the fungus provides greater habitat for nematodes with increase in its population. Some studies have also observed increased spore production and higher percent root colonization by the fungus when nematodes were present. Fungal-feeding nematodes may feed on the fungal symbiont to such an extent that *'in vitro'* cultures may be killed and *'in vivo'* infections significantly reduced. In some studies, plants were severely deprived of growth due to grazing pressure of the fungus, while in others plant growth was not affected. Under heavy grazing pressure by fungal-feeding nematodes, sufficient hyphae may be destroyed to limit phosphorus uptake to a level inadequate for nodulation in legumes.

Several review articles have presented the subject of vesicular arbuscular mycorrhizae and nematode interactions (Francl, 1993; Hussey and Roncadori, 1982; Smith, 1987). Many workers have reported a reduction in nematode population or disease incidence (Jain and Hasan, 1988; Cooper and Grandison, 1986). But, cases where nematode population remains unaffected (Hasan and Jain, 1987) or even increased (Atilano *et al.*, 1981; Kasab and Taha, 1990) under the influence of VAM are not uncommon. Studies conducted by Pandey *et al.* (1997) on interaction between *M incognita* and V A fungi indicated that root-knot nematode, *M. incognita* multiplies well in absence of VAM fungi and significantly reduced plant growth and yield.

The beneficial effect of VA mycorrhizae on a nematode-susceptible cotton cultivar offsets damage caused by *Meloidogyne incognita* (Roncadori and Hussey, 1977). Soybeans inoculated with *Glomus macrocarpum* and *M. incognita* had fewer galls per gram of root, increased root weights, and increased yields, compared with plants inoculated with the nematode alone (Kellam and Schenck,1980). When tomato plants were pre-inoculated with a VAM, penetration of second stage juveniles of *M. incognita* was less but they developed to maturity in roots of mycorrhizal tomato (Sikora, 1979). Tomato roots colonized by *Glomus fasciculatum* and infected by *M. incognita* or *M. hapla* had fewer and smaller galls than did nematode-infected nonmycorrhizal plants (Bagyaraja *et al.*, 1979).

Small giant cells with few nuclei and retarded nematode development was also observed in mycorrhizal tomato roots infected with *M. incognita* (Sikora, 1979). In addition to the effects on nematode penetration and development, there are indications that nematode reproduction may also be suppressed on mycorrhizal plants (Hussey and Roncadori, 1978; Sikora, 1979). Inoculation with either *Gigaspora margarita* or *Glomus mosseae* prior to nematode inoculation did not alter infection by *M. incognita* compared with nonmycorrhizal plants, regardless of Soil Phosphorus (P) level. However, plants grown in soil containing high level of P had greater root weights, increased nematode penetration and egg production per plant, and decreased colonization by mycorrhizal fungi, compared to plants grown in soil containing low level of P (Cason *et al.*, 1983). Such studies suggest VAM fungi have potential for reducing plant diseases caused by certain plant-parasitic nematodes. However, other reports suggest that the beneficial effects of and that the reproduction of *M. incognita* increased on mycorrhizal soybeans (Schenck *et al.*, 1975).

Ryan *et al.* (2000) have found that in the presence of potato plants, *Globodera pallida* exhibited delayed in-soil hatch compared to that of *G. rostochiensis*, with significantly fewer *G. pallida* second-stage juvenile nematodes hatching in the first two weeks, though the difference disappeared after four weeks. Inoculation of potato plants with arbuscular mycorrhizal fungi eliminated this delay in *G. pallida* hatch, so that both the species of potato cyst nematode (PCN) exhibited similar in-soil hatch rates. When the corresponding *'in vitro'* hatching activities of root leachate from uninoculated and mycorrhiza-inoculated plants were compared, similar effects were revealed. *G. pallida* hatch in root leachates from uninoculated plants increased significantly from one-week-old to two-week-old plants, but this increase was not significant in the mycorrhizal-inoculated plants. When the in-soil experiment was repeated using the PCN non-host plant strawberry, mycorrhizal inoculation induced no significant increase in *G. pallida* hatch. The results indicated that mycorrhizal inoculation of potato plants stimulated production of *G. pallida*-selective hatching chemicals, either hatching factors or hatching factor stimulants.

Inoculation of *G. fasciculatum* has resulted in maximum P uptake in presence of root-knot nematodes (Suresh and Bagyaraj, 1984; Jothi and Babu, 2000; Labeena *et al.*, 2002; Shreenivasa *et al.*, 2007). Combined application of vermicompost, *Glomus aggregatum* and *Bacillus coagulans* increased plant growth characters and reduced Root-knot nematode reproduction rate, number of galls and egg masses on tomato cv. Pusa Ruby in sandy loam acidic soils (Serfoji *et al.*, 2010).

A number of mechanisms can be proposed to explain the antagonistic effect of VAM on nematode parasitism and their activities. These may be either physical or physiological in nature. VAM can alter physiology of the root, including the exudates responsible for chemotactic attraction of nematode (Mac Guidwin *et al.*, 1985). VAM offsets yield loss normally caused by nematode by enhancing the uptake of phosphorous and other nutrients leading to improvement of plant vigor and growth (Hussey and Roncadori, 1982). According to Heald *et al.* (1989) the improvement of growth of the plant (muskmelon) infected with *M. incognita* was attributed to improved nutrient status and not to reduction in nematode infection. Nematode development and reproduction might have retarded in mycorrhizal plants due to decrease in available food (Salen and Sikora, 1984). Physiological changes in roots inoculated with VAM may result in development of resistance due to production of antagonistic substances (Suresh *et al.*, 1985). Mycorrhizal fungi may alter the microbial activity in the rhizosphere affecting the survival of nematodes and penetration of roots by nematodes (Timothy and Robert, 1992). The above mechanisms operate singly or in combination to make mycorrhizal plants resistant against invasion of plant pathogens. VAM will colonize much faster than nematode resulting in alteration of root physiology of host plant like increasing wall thickness and changing the chemical composition of root exudates (Sivaprasad *et al.*, 1990).

6.2 Role of Nematodes in Fungal Interaction

6.2.1 As Wound Causing Agents

All root-parasitic nematodes cause mechanical injuries as they penetrate within or feed on root tissues, providing ready avenues of entry for other pathogens. For many years, the role of nematodes in fungal disease interaction was limited to wounding agents. However, mechanical wounding does not always promote fungal penetration within the root tissues. Hart and Endo (1981) reported that wounding roots of celery prior to inoculation of *Fusarium oxysporum* f. sp. *apii* had no effect on the development of Fusarium yellows. Sumner and Minton (1987) reported that wounding roots with a knife did not increase wilt caused by *Fusarium oxysporum* f. sp. *tracheiphilum* race 1 in soya bean cv. Cobb, while the disease was more severe in the presence of *Belonolaimus longicaudatus* and *Pratylenchus brachyurus*. There seems to be something special about the role of some nematode wounds in fungal disease interaction (Pitcher, 1965).

Several mechanisms have been proposed to explain the increased susceptibility of many nematode-infected plants to certain fungal pathogens (Back *et al.*, 2002). Wounding by the nematode (providing an entrance route for the fungus) was long considered important in increasing susceptibility to various fungi (Bergeson, 1972, Storey and Evans, 1987). Powell (1971), however, proposed that the increased capacity of certain *Meloidogyne* spp. to enhance fusarium wilt on tobacco when the nematode infection preceded the fungus by a few weeks is an indication of more elaborate mechanisms is being involved. Most artificial wounding in these types of tests does not realistically mimic nematode injury.

6.2.2 As Predisposers

Statements such as 'Nematodes predispose plants to fungal diseases' or 'Nematode infestations enhance fungal development and fungal disease symptoms in plants' are based on many research papers on interactions involving nematodes and fungi. Other researchers have indicated that nematodes may affect fungal resistance in plants and that they may render plants susceptible to fungi which are innocuous in the absence of nematodes (Batten and Powell, 1971; Powell, 1971). These results and statements imply that nematodes induce or produce factor(s) which can alter the susceptibility of their hosts to fungal diseases or enhance ability of the fungi to penetrate and develop within the plant tissues. Most would agree that plant-parasitic nematodes induce physiological, biochemical and structural changes in their hosts. On the other hand, many abiotic and biotic factors can predispose plants to diseases that would other- wise occur to a lesser extent (Lockwood, 1988). In addition, almost any stress or stimulus seems to influence fungal disease development in plants. For example, Shawish and Baker (1982) reported that a gentle shaking of stems and leaves for 1 minute each day increases *Fusarium* wilt symptom expression in flax, pea and tomato.

Observations indicated that plant predisposition to fungal diseases by plant parasitic nematodes require a minimum level of nematode infestation. Addition of 1000-2000 *Meloidogyne incognita* juveniles per plant significantly increased the number of infected cotton cvs. Deltapine Smooth Leaf and Pima S-2 with *Verticillium alboatrum*. However, addition of 250 and 500 juveniles per plant did not give a similar result (Khoury and Alcorn, 1973). A minimum initial population density of *M. incognita* was required to increase *Rhizoctonia solani* infection on cotton seedlings (Carter, 1975). A significant interaction occurred between *M. incognita* and *Fusarium oxysporum* f. sp. *vasinfectum* on cotton at a higher nematode population, while no interaction was observed at a lower nematode population level (Starr *et al.*, 1989).

Predisposition factor induced by the nematode can be translocated to considerable distances from the nematode infection site. Using split root technique it has been shown that *Meloidogyne* spp. can result in a systemic change in the host so that plant tissues remote from the site of nematode infection are rendered more susceptible to nematode infection (Prot, 1993). Bowman and Bloom (1966) exposed one part of root system of a wilt resistant variety of tomato to *Fusarium oxysporum* f. sp. *lycopersici* and other part to *M. incognita*, demonstrated that plant even become diseased even though the nematode and fungus were on a separate root system.

The shortening of the incubation period and an increase in severity of a fungal disease when the number of nematodes infecting the plant increase may also be related to a wounding action of the nematodes; however, it is also suggested that the nematodes have to induce a certain level of physiological changes to modify the reaction of their host to the fungus.

6.2.3 As Host Substrate Modifier

The delay in predisposition of plants to fungal diseases by root-knot nematodes suggested that these nematodes are not just wounding agents facilitating the penetration of the fungi within the roots. If their role in disease complexes with fungi was limited to wounding agents, the fungal invasion of the plant tissues would begin almost immediately after nematode infection.

Nematodes modify plant tissue in such a way that it becomes a better substrate for the fungus and thus increases their growth and reproduction to the detriment of the host. Quantitative and qualitative changes in root exudates which are induced by certain nematodes stimulate the germination, growth, and reproduction of fungal propagules in the rhizosphere. These exudates may also indirectly inhibit components of the rhizosphere microflora (*e.g.* actinomycetes) which are antagonistic to some plant pathogens. Depending on the interacting species of nematode and fungus, concomitant infections may stimulate nematode reproduction (*Pratylenchus-Verticillium*) or inhibit reproduction (*Heterodera-Fusarium*) (Bergeson, 1972).

Nematodes seem to favour all stages of fungal infection and development, by modifying the composition of the root leachates they can promote the growth of fungi in the rhizosphere and favour their pathogenic development. Moreover, these modifications of the rhizospheric environment may limit the development of organisms antagonistic to the pathogenic fungi. Their feeding sites and the cells they modify, especially the giant cells induced by root-knot nematodes, may serve as a favourable substrate which helps the fungi to establish within the plant and promote their development. Nematode-induced or produced factors appear to be translocated from the nematode feeding sites to other parts of their host, especially in the above ground parts. These factors seem to modify the resistance of the host tissues to the fungi and/or directly stimulate fungal growth.

The feeding sites of sedentary endoparasitic nematodes (giant cells or syncytia) are zones of high metabolic activity. These nutrient-rich cells could be the substrate for fungal colonization (McLean and Lawrence, 1993; Abdel-Momen and Starr, 1998). A 3-4 week nematode pre-inoculation has been found to be critical in investigations of some nematode-fungus disease complexes (Golden and van Gundy, 1975; Negron and Acosta, 1989). Taylor (1990) suggested this could be linked to syncytial development which will take 3-4 weeks to reach peak activity in a susceptible host.

Exudates from galled roots may provide stimuli for germination of chlamydospores, accounting for the increase in numbers of propagules (Bergeson *et al.*, 1970) and colonies (Noguera and Smits, 1982) of *F. oxysporum* f. sp. *lycopersici* in the rhizosphere of tomato plants infected with *M. incognita* and *M. javanica* compared to those observed around non-galled roots.

6.2.4 As Pathogenicity Inducing Agents

There are some evidences that root-knot nematodes interact with saprophytic soil fungi and induce pathogenicity in them. Inhabiting fungi like *Aspergillus ochraceus, Curvularia trifolli, Penicillium martensii, Botrytis cinerea, Pythium ultimum* and *Trichoderma harzianum*, which were considered to be non-pathogenic to tobacco, caused severe root necrosis in tobacco when inoculated four weeks after *M. incognita* infection (Powell *et al.*, 1971).

6.2.5 As Resistance Breakers

Many authors have concluded that monogenic resistance (I gene) of tomato to *F. oxysporum* f. sp. *lycopersici* was rendered ineffective by infection with root-knot nematodes. Jenkins and Coursen (1957) reported that both *Meloidogyne hapla* and *M. incognita* have broken the resistance in tomato against Fusarium wilt. Sidhu and Webster (1977) indicated that monogenic resistance to *F. oxysporum*. f. sp. *lycopersici* race 1 in tomato cultivar possessing resistance gene I-2 was ineffective in the presence of *M. incognita*. Studying interaction between *M. hapla* and *F. oxysporum* f. sp. *medicaginis* on alfalfa, Griffin and Thyr (1988) reported that in cv. Synthetic XX, resistant to both organisms, the nematode promoted the fungal infestation only at 30°C, while resistance to *M. hapla* was lost. Noguera (1982) suggested that the loss of *Fusarium* resistance in tomato when infected with *M. incognita* was associated with the lack of rishitin, an antifungal substance present in healthy plants.

Contradictory to previous reports Jones *et al.* (1976) observed that the monogenic resistance of tomato cvs. Florida MH-1 and Manapal to *F. oxysporum* f. sp. *lycopersici* race 1 and race 2 were not altered by either the modification of the rhizospheric environment or simultaneous or prior inoculation with *M. incognita*. Similar results were obtained by Abawi and Barker (1984) with tomato cvs. Nematex, Manapal and

Floradel against the *Fusarium* resistance gene I-1 and cv. Florida MH-1 with the *Fusarium* resistance gene I-2. In addition, they observed that *Fusarium* resistance of Nematex was not reduced by the loss of *M. incognita* resistance at 35°C. Inconsistencies between the observations may correspond to differences in experimental conditions, cultivar, nematode species associations, environmental and especially edaphic factors and nonetheless population levels of pathogens.

6.3 Nematode-Bacterial Interaction

Bacteria and nematode constitute two of the most numerous groups of organisms in the biosphere. Most of the nematodes at one time during their life history share their environment with bacteria. Nematode-bacteria interactions can be conveniently classified on the basis of their effect or role on the hosts concerned as: (i) nematode acting as a vector (ii) bacteria as parasite of nematode (iii) bacteria as symbionts of nematodes (iv) synergistic effect on the hosts (v) nematodes as wound causing agents and (vi) nematodes causing physiological changes in the host for bacterial invasion.

The complexes/interactions involving nematodes and bacteria are relatively few compared to the number involving nematodes and fungi. For this less interaction between bacteria and nematodes the reasons may be: (i) number of pathogenic bacteria known is very less as compared to fungus or virus. Thus, there is less probability of a nematode to come in contact with bacteria (ii) host specificity of bacteria leads to drastic reduction or death of bacteria in absence of the specific host (iii) most of the bacterial diseases are foliar in nature while most of the nematodes (except few) are in soil thus reduced chances of interaction (iv) unlike fungi, there is no resting or dormant stages in bacteria, thus, they are very much sensitive to environmental stresses (v) bacteria reproduce only through binary fission while fungi have various means of reproduction (vi) bacteria are thermophilic and grow better in alkaline media. Most of the bacteria are non-motile except few.

It has been experimentally proved that the plant parasitic nematodes like root knot nematodes acquired parasitism through horizontal gene transfer from bacteria as both organisms establish intimate developmental interactions with host plants and thus facilitating genetic material transfer from bacteria (Jaubert *et al.*, 2002). All plant-parasitic nematodes of major economic importance have been observed to be parasitised by either *P. penetrans* or closely related species (Sayre and Starr, 1988; Sturhan, 1988; Chen and Dickson, 1998). Based on host range, life cycle and morphology, three nominal species of *Pasteuria* able to parasitise plant parasitic nematodes have been proscribed, namely: (1) *P. penetrans,* which is a parasite of *Meloidogyne* spp. (2) *Pasteuria thornei*, which is a parasite of *Pratylenchus* spp., and (3) *Pasteuria nishizawae*, which is a parasite of *Heterodera* and *Globodera* spp. (Chen and Dickson, 1998).

6.3.1 Nematodes as Vectors

There are several instances where plant parasitic nematodes assist as carriers of the pathogenic bacteria. Nematode juveniles carry bacterial cells on their external surface and introduce them into the infection site. Gupta and Swarup (1966, 1968) demonstrated that wheat seed gall nematode, *Anguina tritici* had obligate relationship with *Clavibacter tritici* to cause 'tundu' disease in wheat. *Ditylenchus dipsaci* transports *Pseudomonas fluorescens* in garlic resulting in "Café au lait bacteriosis" (Caubel and Samson, 1984). The foliage and meristem disorder in Strawberry known as "cauliflower disease" involving *Aphelenchoides ritzemabossi* and *Corynebacterium fascians*, and stem nematode, *Ditylenchus destructor* and *Clavibacter michiganense* subsp. *insidiosum* causing crown rot are the two other example. *Globodera pallida* acts as a carrier for *P. solanacearum* in potatoes. In all the above associations the developing bacterial colonies are detrimental to the nematodes and the association is not mutualistic, but disease complex requires both nematode and the bacterium for expression of the complete disease syndrome.

6.3.2 As Wound Producers

For the development of crown gall caused by *Agrobacterium tumefaciens*, some sort of injury on the crown or roots is essential before the bacterium can invade the host. In raspberry, bacteria inoculated alone into roots could not cause crown gall, but together with *M. hapla*, they induced crown gall. Normal alafalfa had a low incidence of wilt caused by *Clavibacter michiganense* subsp. *insidiosum* but when roots were injured by *M. hapla*, the incidence of wilt was increased. Zutra and Orion (1982) reported that *M. incognita* enhanced the intensity of infestation of *A. radiobacter* var. *tumefaciens* in cotton. Sitaramaiah and Sinha (1985) had observed severe bacterial wilt in brinjal due to interaction of *P. solanacearum* biotype 3 and *M. javanica*.

6.3.3 As Resistance Breaker

Jatala *et al.* (1975) reported that resistance of potatoes to *P. solanacearum* was broken down due to infection of *M. incognita acrita* in the field. Napiere and Quimio (1980) found that the development of wilt occurred earlier with higher mortality rate and reduced yield in both wilt resistant and susceptible tomato varieties grown in *P. solanacearum* and *M. incognita* infested soil. Johnson and Powell (1969) reported that the bacterial wilt caused by *Pseudomonas solanacearum* is enhanced in presence of *M. incognita* when the roots are exposed 3-4 weeks to nematodes prior to bacterial infection. Routary *et al.* (1987) studied response of some nematode resistant tomato cultivars to mixed inoculation of *M. incognita* and *P. solanacearum*. No wilting was observed in resistant tomato plants when *M. incognita* and *P. solanacearum* were inoculated separately. But the simultaneous inoculation of the nematode and bacterium led to breakdown of resistance in all varieties except Ronita. Maximum wilting occurred in susceptible variety Pusa Ruby. Overall greatest wilt incidence occurred when *M. incognita* was inoculated 7 days prior to inoculation of the bacterium.

6.3.4 Synergistic Effect on Host

Jatala *et al.* (1975) reported that resistance of potatoes to *P. solanacearum* was broken down due to infection of *M. incognita acrita* in the field. Napiere and Quimio (1980) found that the development of wilt occurred earlier with higher mortality rate and reduced yield in both wilt resistant and susceptible tomato varieties grown in *P. solanacearum* and *M. incognita* infested soil. Johnson and Powell (1969) reported that the bacterial wilt caused by *Pseudomonas solanacearum* is enhanced in presence of *M. incognita* when the roots are exposed 3-4 weeks to nematodes prior to bacterial infection. Routary *et al.* (1987) studied response of some nematode resistant tomato cultivars to mixed inoculation of *M. incognita* and *P. solanacearum*. No wilting was observed in resistant tomato plants when *M. incognita* and *P. solanacearum* were inoculated separately. But the simultaneous inoculation of the nematode and bacterium led to breakdown of resistance in all varieties except Ronita. Maximum wilting occurred in susceptible variety Pusa Ruby. Overall greatest wilt incidence occurred when *M. incognita* was inoculated 7 days prior to inoculation of the bacterium.

Jatala *et al.* (1975) reported that resistance of potatoes to *P. solanacearum* was broken down due to infection of *M. incognita acrita* in the field. Napiere and Quimio (1980) found that the development of wilt occurred earlier with higher mortality rate and reduced yield in both wilt resistant and susceptible tomato varieties grown in *P. solanacearum* and *M. incognita* infested soil. Johnson and Powell (1969) reported that the bacterial wilt caused by *Pseudomonas solanacearum* is enhanced in presence of *M. incognita* when the roots are exposed 3-4 weeks to nematodes prior to bacterial infection. Routary *et al.* (1987) studied response of some nematode resistant tomato cultivars to mixed inoculation of *M. incognita* and *P. solanacearum*. No wilting was observed in resistant tomato plants when *M. incognita* and *P. solanacearum* were inoculated separately. But the simultaneous inoculation of the nematode and bacterium led to breakdown of resistance in all varieties except Ronita. Maximum wilting occurred in susceptible variety Pusa Ruby. Overall greatest wilt incidence occurred when *M. incognita* was inoculated 7 days prior to inoculation of the bacterium.

6.3.5 Interaction with Rhizobacteria

Based on genome-to-genome analyses of gene sequences obtained from plant-parasitic nematodes (*Meloidogyne* spp., *Heterodera* spp., *Globodera* spp.), it seems likely that certain genes have been derived from bacteria by horizontal gene transfer. Strikingly, a common theme underpinning the function of these genes is their apparent direct relationship to the nematodes' parasitic lifestyle. Phylogenetic analyses implicate rhizobacteria as the predominant group of 'gene donor' bacteria. Root-knot nematodes and rhizobia occupy similar niches in the soil and in roots, and thus the opportunity for genetic exchange may be omnipresent. Further, both organisms establish intimate developmental interactions with host plants, and mounting evidence suggests that the mechanisms for these interactions are shared too. It has been proposed that the origin of parasitism in plant parasitic nematodes may have been facilitated by acquisition of genetic material from soil bacteria through horizontal gene transfer, and that such events represented key steps in speciation of plant-parasitic nematodes (Bird *et al.*, 2003).

The impact of plant-parasitic nematodes on nodulation and N_2 fixation varies with nematode feeding habit, initial inoculum levels, host plants, and epidemiology (Huang, 1987). Parasitism of soybean by *Heterodera glycines*, especially race1, may have striking effects on the morphology and physiology of developing *Bradyrhizobium* nodules; fewer rhizobia cells were observed on root hairs of soybeans infected by *H. glycines*. The magnitude of the suppression of nodulation in Soybean by the bacterium, *Bradyrhizobium japonicum* varied with the soybean genotype, host races and sequence of inoculation of *H. glycines* (Ko *et al.*, 1985). On the contrary, *H. trifolli* as well as

Meloidogyne spp. may develop compatibly within nodule tissue (Taha and Raski, 1969; Barker and Hussey, 1976).

The *H. glycines-B. japonicum*-soybean interactions undoubtedly are very complex and impacted by a number of other plant physiological process. From the split root experiments, it was evident that *H. glycines* might have systemic impact on nodulation of soybean. On the split-root soybean, nodulation was suppressed on *H. glycines* infected root half compared to uninfected one. Thus, *H. glycines* may suppress nodulation by inducing or producing a translocated compound that inhibits nodulation, or nematode may impact nodulation by serving as a metabolic sink (Ko *et al.*, 1984). Hallman *et al.* (2001) has studied the colonization of *Rhizobium* in roots in the presence of *M. incognita* found that the bacterium mostly colonized interior of nematode galls.

6.4 Interaction among Nematodes

Interactions between two species of nematodes may be beneficial to one or both species or have no effect, or most commonly interfere with the wellbeing of one or both species. The antagonistic interaction (competition) can be caused by a spatial occupation or physical alteration or destruction of feeding sites, or by a physiological alteration of the host that decreases its suitability (Norton, 1978; Eisenback, 1985). As it is well known, the outcome of interspecific interactions may depend upon: i) initial population densities, ii) environmental conditions (edaphic factors, temperature, humidity, etc.) and iii) the genetic constitution of the strains of interactive species (Pianka, 1981).

Metakaberhan and Dey (2003) reported that the effect of increasing *H. glycines* proportions on the infection rate of *M. incognita* was generally adverse, the rate deviated significantly from a trend of linear decline at the 75 per cent *H. glycines* level in one of two experiments. All lack-of-fit F-tests for the *H. glycines* and *Pratylenchus penetrans* mix were significant, indicating that infection rates for both nematodes varied considerably across inocula. The infection rate of *H. glycines* decreased with increasing *P. penetrans* proportions. The rate of *P. penetrans* infection increased with increasing *H. glycines* proportions up to the 50 per cent level, but declined at the 75 per cent level. Competition had no effect on nematode development. The general adverse relationships between *M. incognita* and *H. glycines* and those between *P. penetrans* and *H. glycines* showed a linear trend. The relationship between *H. glycines* and *P. penetrans* indicated that the former may be competitive when present at higher proportions than the latter.

The attractiveness of *Radopholus similis* towards the banana roots seems to be governed by the rhizogenic activity of some parts of the rhizome rather than by the strict origin of the roots on the rhizome. In the absence (or reduced densities) of one or more competitors, *R. similis* is able to colonize, with high population levels, on the entire set of banana roots. In the presence of competitors, e. g. *Helicotylenchus multicinctus*, the arrangement and density of *R. similis,* in the soil as well as in the roots, is reduced and restricted to areas close to the rhizome. The higher the density of competitors, the more restricted was the dispersion area of *R. similis* around the rhizome (Quénéhervé, 1990).

Nematodes may predispose hosts for other nematodes which may lead to competition and facilitation effects. More complex the relationship between root feeding nematodes and their host plants, more competitions there are within and among species (Eisenback, 1993). Another issue of competition and facilitation is that between root feeding nematodes and other soil invertebrates. These interactions may lead to contrasting patterns such as induction of antagnostic effects by earthworms against cyst nematodes (Blouin *et al.*, 2005) and to facilitative effects of wireworms on root knot nematodes by direct promotion of root knot nematodes abundance in mixed plant communities (De Deyn *et al.*, 2007).

6.4.1 Interaction among Meloidogyne Species

Johnson and Nusbaum (1970) in a greenhouse pot experiment on the pathogenicity and interactions of *Meloidogyne incognita, M. hapla* and *Pratylenchus brachyurus* on four cultivars of tobacco, found that the cultivars 'Hicks' and 'NC 2326' were susceptible to each nematode and 'NC 95' and 'NC 2512' were resistant only to *M. incognita.* Mean heights of susceptible plants were depressed but fresh weight of tops did not differ significantly. *Meloidogyne* spp. increased fresh weight of susceptible (but not the resistant) roots. Reproduction of *M. incognita* was decreased in the presence of *P. brachyurus* in one case. *M. hapla* reproduction was less with either of the other nematodes in five out of eight cases. It showed the suppressive effect of *M. incognita* over *M. hapla* in tobacco.

It has been proved that different root-knot nematode species have different temperature requirements for various physiological processes. *M. incognita* dominance over *M. javanica* and *M. hapla* occurs at high temperature of 25-32°C. Populations of *M. javanica* and *M. hapla* matured slowly at 15 and 25° C while *M. hapla* maturing a few days sooner than *M. javanica,* which matured faster at 30° C (Kinloch and Allen, 1972). Khan and Haider (1991) found mutually inhibitory interaction between *M. incognita* and *M. javanica* but these negative interactions did not affect plant growth.

Much of the available information on *Meloidogyne* spp. and other nematodes have been generated from glasshouse and laboratory observations under controlled experimental conditions that usually include one species of nematode on one plant species, an uncommon situation in agricultural soils, where multi-species and polyphagous nematode communities occur (Johnson and Nusbaum, 1970; Norton, 1989). Only few studies of interactions with other microorganisms are supported with observations from field studies.

Nature does not work with pure cultures, most plant diseases particularly those of roots are influenced by the associated microorganisms. Under natural conditions, the roots of plants are constantly exposed to these soil microorganisms. It has been estimated that one square meter of high fertility field soil may contain as many as 3×10^{14} bacterial cells (300g), 5×10^{8} protozoa (39g), 1×10^{7} nematodes (12g) and 400g of fungi. But this may not be true for all soils; some may have a little different soil biota. However, the number of soil microorganisms present in the root zone of any crop plants is sufficient to cause innumerable and diverse interactions (Mai and Awabi, 1987).

Extensive studies of nematode effects on host gene expression is required to develop testable hypotheses that may eventually shed light on those complex biological systems. Biological basis of interaction and their statistical significance should be investigated thoroughly. Molecular approaches and understanding of the host-parasite relationship are advancing rapidly; genome sequence information of *Meloidogyne* spp. (Abad and Opperman, 2010), along with the genomes of other nematode species, microbes and host plants, will be of great use for the study of these biological systems, which will facilitate generation of more valuable information on the complexes involving plant parasitic nematodes and other microorganisms.

Chapter 7
The Cyst Nematodes

We, who are engaged in unraveling the science of nematodes, owe a lot to the discovery of sugar beet cyst nematode, *Heterodera schachtii* Schmidt, 1871 by Schacht (1859) in Germany. This nematode was found responsible for the 'beet-tiredness' symptoms in sugar beet and it made zoologists, entomologists and plant pathologists realize the crop losses due to plant parasitic nematodes at that time because many factories processing sugar beet for sugar were closed as they suffered huge losses due to the nematode infestation. The follow-up studies for finding this nematode in other parts of the country led to the discovery of so many other plant parasitic nematodes including many other cyst nematodes, most significant among them, however, were many root-knot nematodes though, Berkeley (1855) had reported the occurrence of root knot nematode in cucumber in green house for the first time and the presence of another plant parasitic nematode, *Ditylenchus dipsaci* on *Dipsacus fullonum* was reported by Kuhn (1857). But it was the destruction of sugar beet crop by the sugar beet nematode which made the scientific fraternity in most parts of the world, especially in Europe, aware of the economical importance of these tiny creatures. It was the time for the beginning of the growth of Plant Nematology. In the meantime, discovery of another devastating nematode, the golden nematode of potato was made in England and other parts of Europe in the early twentieth century although Kuhn (1881) had reported its presence and that of oat cyst nematode, *Heterodera avenae* (Kuhn,1874) in Germany. At that time, such was the impact of damage of these (sugar beet cyst nematode and golden nematode of potato) nematodes to the two agriculturally and economically important crops that funds provided for their research and management exceeded the funds allotted for the research and management of all other plant pathogens put together. According to Shah (1915) more than forty-five people were working on sugar beet cyst nematode alone in Europe, who had published papers by 1911. The losses caused by these and other nematodes were so enormous that they served as requisite catalysts for the growth and development of science of Plant Nematology in the world *per se.*

7.1 The Cyst

Many cyst nematodes like, oat cyst nematode, *Heterodera avenae* by Kuhn, 1874, golden nematode of potato, *Globodera (Heterodera) rostochiensis* by Kuhn, 1881, the pea cyst nematode, *H. goettingiana,* Liebscher, 1892, were discovered but the usage of the term 'cyst' for these nematodes was familiarized almost 30 years after the discovery of the sugar beet cyst nematode, *H. schachtii.* The words cyst and 'cyst-like' had earlier been used by M. J. Berkeley for the root knot female when he reported the presence of these nematodes on green house grown cucumbers. He observed, *'certain cyst-like bodies filled with a multitude of minute eggs'* and *'it is not conceivable that such a cyst could have been deposited by the 'Vibrio' itself'* and described it as 'an irritation caused by eggs'. As is clear the 'cyst' or 'cyst-like' was used for the adult root knot female not for the cyst nematodes. A French scientist, Chatin (1887) was working on *H. schachtii* and he sent a note to the French Academy of Sciences which said, "*When winter is approaching, changes of particular interest are observed in the female. The integument, previously very thin, thickens progressively; its glands provide an abundant secretion which, agglutinating organic and mineral substances, forms around the female a type of adventitious test of mixed composition. While developing this covering ends in obscuring the buccal, anal and vulval openings; the cephalic stylet that maintained the parasite fixed in the plant tissue can no more act and any contact is ruptured between the worm and the plant. It is no longer an animal we see before our eyes but a cyst, full of eggs similar to an oothecum, which falls into the soil mixed with the roots. One may see how such a cyst differs from the fecund female as she is observed with white color, thin and fragile integument that rupture following the smallest shock or under the smallest osmotic pressure*". Then, he reported that these cysts were of variable shapes and sizes, brown in color and protected by very thick and hard permeable walls, which can overcome unfavorable environmental conditions giving a powerful protection to the eggs they contain.

Till few years back, there was lot of confusion in defining the cyst as many workers defined it in their own suitable way. A majority of them defined the cyst as, 'part of the eggs remain within the female worm which now form a cyst, a sac like structure, consisting of the female cuticle in which egg-masses are contained' (Filipjev and Schuurmans Stekoven, 1941); 'transformation of the body wall of adult female into tough resistant cyst or sac containing the embryonated eggs' (Franklin, 1951); 'the female cuticle transform into a tough, brown cyst-like sac protecting the eggs which have been formed within the body (Thorne, 1961); the female develops a thick cuticle, its uterus becomes packed with fertilized eggs that mature to second stage juveniles but not further and when the adult female dies its cuticle turns into a brown leathery sac, the cyst (Dropkin, 1980); the cyst is tanned cuticle of the mature female body in which eggs are retained.' (Maggenti, 1981); and Siddiqi (1986) referred to it as 'the female body turns into a hard walled protective cyst'. In Chatin's description the cyst is equated with an oothecum and he emphasized that the cyst differs from the fecund female which is white and has a thin fragile body wall and that the cyst contained eggs that are protected against the adverse environmental conditions.

Chatin (1891) confirmed the usage of the term cyst when he studied the life cycle of *H. schachtii* from cyst to cyst. He, like Franklin (1951) was of the opinion that the subcrystalline layer was fundamental to the formation of cyst but the later investigators (Brown *et al.*, 1971 and Zunke, 1986) showed its distinct origin and that it occurs only in some species of *Heterodera* while it is also highly developed in genus like *Atalodera* which is not a cyst but cystoid nematode.

Along this prevailing controversy, Luc, Weischer, Stone and Baldwin (1986) proposed that the cyst wall is supposed to be derived from the female body wall but it was not clear if all the components of the body wall take part in the formation of the cyst wall and also that the thickness of the cuticle was not correlated with the capacity to form a 'cyst'. After thorough deliberations, the definition of the cyst was suggested as, '*a heteroderid cyst is a persistent tanned sac which retains eggs and is derived from some or all components of the mature female body wall*'. The persistence of the cyst, its tanning and retention of eggs inside are the characters that are best considered for its qualification and in the occurrence of this combination of the characters that conforms to be called as a cyst.

7.1.1 Characters of a Cyst

The cyst is a key character in the systematics of the family, Heteroderidae along with second stage juveniles because these are the stages generally encountered in the soil samples. Like the root knot nematodes whose identification highly depends upon the perineal pattern near the vulval region in the posterior most part of the female, in cyst nematodes also the structure in the above area is equally reliable for diagnosis of characters of a species. It was Franklin (1939) who observed the presence of circular shaped hyaline areas marked vulval regions and inside the cone, knob-like dark brown structures in *Heterodera* cysts. Taylor (1952) and Oostenbrink and den Ouden (1954) reported that these knob-like bodies were different in many of the species like, *H. schachtii, H. avenae, H. glycines, H. trifolii* and *H. galeopsidis.* They were first to differentiate the cyst forming nematode species using characters like bullae, underbridge and fenestrae. Cooper (1955) was the first to introduce a variety of terminology for various structures in the vulval cone of the cyst as he found that certain morphological features were of diagnostic value. He proposed a system of nomenclature for the structures associated with the cone and also divided the genus into two groups, Bullata and Abullata. The various terminologies used were:

Vulval Cone

It is the tapered terminal end or the protuberance in the posterior end of the cyst. The species without a cone are classified as having a terminal area as in *Globodera* and *Punctodera*. In some of the species like *H. mothi* and *H. cajani*, the cone is quite prominent and deep while in *H. avenae*, it is very low *i.e.* the protuberance is small. In *H. sorghi*, the protuberance is more sharp and conical whereas *H. spinicauda* is apparently without any posterior protuberance.

Fenestra

It is the translucent region or a thin cuticular area (mostly circular or a half circle) at the posterior-most region of the vulval-cone, present on either side of the vulva. As the cyst matures, the thin area disappears and only a hole or a window is left for the juveniles to emerge from the cysts. All species have vulval fenestra but *Punctodera* spp. have both vulval and anal fenestrae whereas *Afenestrata* is without any fenestrae. There are mainly three types of fenestrations present in the cysts: circumfenestration, ambifenestration and bifenestration.

Circumfenestration

It is the type of fenestration where there is only a single fenestra present in the cyst. It is present in round and globose cysts like *Globodera, Punctodera, Dolichodera, Betulodera, Vittatidera* and *Cactodera*. In *Punctodera*, in addition to vulval fenestra, anal fenestra is also present. In white females of these cysts, the terminal portion has a very thin walled cuticular area having a slender vulval slit. As the female matures into a cyst the vulval slit disintegrates along with the thin cuticular area giving rise to a window like single opening, the fenestra and thus the condition, circumfenestrate.

Ambifenestration

A type of fenestration in which the vulval bridge divides the fenestra into two 'D' shaped semifenestrae. The ambifenestration is observed in majority of the cyst forming nematodes belonging to the genus, *Heterodera* like *H. cajani, H. zeae, H. sorghi, H schachtii* and *H. graminis*.

Bifenestration

It is a type of fenestration where two fenestrae are widely separated by a wider vulval bridge having short vulval slit. They are not 'D' shaped but almost circular. Such type of fenestration is present in the hop cyst nematode, *H. humuli*. Rarely this type of fenestration can also be observed in some specimens of *H. avenae* complex species.

Underbridge

This structure is composed of muscles bands, gives supports to the whole vulval cone. It is in reality the remnants of two arms of the didelphic ovary which form these muscular bands. The bands harden with the formation of cyst and varies in size and shape among various species. *H. sorghi* and other members of Group 4 (Mulvey, 1972) have very strong underbridge while in the members of Group 5, it is not strong but week. It is usually present 10-70μ below the fenestra and vulval bridge and runs parallel with the vulval bridge. It is a reliable character in the identification of species. There may be another underbridge present in early stages of cyst formation running perpendicularly to the first as in case of *H. trifolii* and *H. glycines*. This second underbridge is called as Mulvey's bridge.

The general perception is that the species having strong underbridge have no or less number of bullae and vice-versa, *e.g.* in *H. avenae* or its related group species, it is either absent or a delicate thin underbrigde is present but all these species have too many knob-like dark brown bullae which sometime even overlap the fenestrae. But in *H.sorghi*, there is a very strong and long underbrigde that ends with bifurcating arms but has few, weak circular shaped bullae around the semifenestrae. If the underbridge is moderately present, moderate occurrence of bullae will also be there (*e.g. H. cajani* and *H. zeae*). In *H. graminis*, the presence of vulval knobs on both sides of vulval bridge suitably makes for the presence of slender underbridge and absence of bullae. In *H. spinicauda*, the 'bow-tie' type or 'dumble' shaped structure (call it an underbridge) supports the vulval region and keeps it intact although it arises from the parapets the vulval slit. In some cysts like *H.trifolii*, *H. sorghi* and *H. skohensis*, a 'sheaf'-like organ may be present in the middle of the underbridge.

Vulval Bridge

A thick cuticular band bearing the vulval slit and divides the fenestra into two semifenestrae. The width of the vulval bridge differs significantly in the ambifenestrate and bifenestrate cone tops.

Bullae

The knob-like, dark brown cuticular projections around perianal region in the vulval cone, situated below the fenestral level or at the level of and below the underbridge. They vary in size, shape and numbers. Cysts with bullae are termed bullate and without them as abullate. Bullae in *Heterodera zeae* and *H. cajani* are mostly finger shaped. In *H.* zeae, there is another set of finger-like bullae which run diagonally on both sides and are present below the normal bullae and termed as 'cross bullae'. Presence or absence of bullae and their positional occurrence is also a reliable and prominent character in identification of species of cyst nematodes.

Basin

The thick cuticular area in a band surrounding the fenestrae, distinctly present in some of the species.

It was only in the late 1950's that cyst-forming species having round shape with circumfenestration were included in the separate genus, *Globodera* (Skarvilovich, 1959) Behrens, 1975; those with additional fenestration around the anus in the genus, *Punctodera* (Mulvey and Stone, 1976); those with slight but prominent protuberance in the genus, *Cactodera* Krall and Krall, 1978; those with vulva in the contour of the body in *Dolichodera* Mulvey and Ebsary, 1980. The terminal area in all these genera is circumfenesrate and mostly without bullae. The rest lemon shaped ambifenestrate cysts remained in the genus, *Heterodera*. The genus *Afenestrata* included the cysts without fenestration and a having prominent sunken vulva and trophied vulval lips in the cone tops (Baldwin and Bell, 1985).

7.2 Identification of Cyst Nematodes

The cyst nematodes have a variety of morphological characters which differentiate them into different genera *viz., Heterodera, Globodera, Punctodera, Dolichodera, Paradolichodera, Cactodera, Betulodera, Vittatidera* and *Afenestrata. Heterodera* (Schmidt, 1871) was proposed for the nematode having sexual dimorphism, the males being slender thread-like while the females become swollen at maturity. Taking this into account initially the root knot nematodes and other such nematodes were included in this genus, *Heterodera radicicola* or *H. marioni*. Stone (1973) described a new species of potato cyst nematode, *H. pallida* diffentiating it from *H. rostochiensis. Globodera* was erected as a subgenus by Skarbilovich (1959) for round cyst nematodes.Later on Beherns (1975) elevated the subgenus to generic rank and almost at the same time Mulvey and Stone (1976) also proposed *Globodera* to the generic level as they were ignorant of the Beherns (1975) work. At the time of proposing generic rank to *Globodera*, Mulvey and Stone (1976) described a new genus, *Punctodera* (*punctum* (Latin) =dot) for species having punctuation around the anal region. The anal opening was also found to be quite big and called it as anal fenestra. Soon *H. rostochiensis* and *H. pallida* were known as *G rostochiensis* and *G. pallida* after *Globodera* was raised to generic rank.Golden (1986) provided a comprehensive identification key for the species of *Punctodera*. Mulvey and Ebsary (1980) reported the presence of a cyst nematode in an acquatic habitat in Quebec, Canada and described it as a new genus, *Dolichodera* (*dolichos* (Greek) = long and *deros* means skin) on the basis of cyst being oblong, circumfenestrate and without any anal fenestration (Siddiqi, 2000). *Paradolichodera* (Sturhan *et al.*, 2007) was described as a new genus (*para* (Greek) =near and *Dolichodera*) from plants growing along a lagoon south of Christchurch, New Zealand on the basis of presence of a very slender J2 different from all other known heteroderids. Group 2, of Mulvey's (1972) grouping of *Heterodera* species into five groups, included species like *H. cacti*, *H. betulae*, *H. estonica* and *H. weissi* which were having posterior protuberance and circumfenestrate vulval region. Krall and Krall (1978) erected a genus, *Cactodera* from the species epithet of the type species name (*kaktos* (Greek) = prickly plant) for these species with the addition of two more species, *H. amaranthi* and *H. acquatica*. After this eight more species were added in this genus.The description of the genus *Cactodera* was amended by Mulvey and Golden (1983). In 2002, Sturhan (2002) shifted *C. betulae* to a new genus, *Betulodera* on the basis of differential host range and cyst morphology. The genus, *Afenestrata* was erected by Baldwin and Bell (1985) while redefining the genus *Sarisodera* Wouts and Sher, 1971 to accommodate *S. africana* Luc, Germani and Netscher, 1973 on the basis of it being a cyst nematode without fenestration. Wouts (1985) also erected a new genus, *Afrodera* for the same species but *Afenestrata* had priority because it was published earlier than *Afrodera*. Recently on the basis of molecular phylogenetic analysis of ribosomal RNA and protein-coding genes, *Afenestrata* was synonymised with *Heterodera* (Mundo-Ocampo *et al.*, 2008). But here it is considered as a separate genus as in cyst nematodes of *Afenestrata* apart from absence of fenestration in the vulval region and presence of sunken vulva along with hypertrophied vulval lips are character present only in this genus. A new genus, *Vittatidera* was described by Bernard *et al.* (2010) infesting maize in U.S.A. and characterized by presence of lateral fields in females and cysts.

During the course of time, many species of cyst nematodes (> 110 species) have been described around the world since 1871, however, initially, most of the species were recorded as the races of *H. schachtii*, except the description of *H. goettingiana* infesting pea by Liebscher (1892). Woolenweber (1923, 1924) distinguished *H. schachtii* from the golden nematode of potato and the oat cyst nematode on certain morphological and physiological basis naming them as *Heterodera* (*Globodera*) *rostochiensis* and *H. avenae*, respectively. Soon Goffart (1932) described *H. trifolii*, the clover cyst nematode and Filipjev (1934) described *H. humuli*, the hop cyst nematode as new species of cyst nematodes. Meanwhile in Canada, Thorne (1928) described *H. punctata* a spherical shaped cyst infesting wheat. Filipjev (1934) reviewed the genus *Heterodera* and reported the presence of seven species, *H. avenae*, *H. punctata*, *H. goetingiana*, *H. trifolii*, *H. rostochiensis*, *H. schachtii* and *H. humuli*. Goffart (1936) described *H. galeopsidis*, the hemp cyst nematode and Filipjev and Schuurman Stekhoven (1941) named another cyst nematode from cactus as *H. cacti* which was described by Adams (1932) but not named. Franklin (1951) published a monograph on cyst forming species of *Heterodera* adding *H. cruciferae* (Franklin, 1945) to the above list.

It is only after mid-1950s (Oostenbrink and den Ouden, 1954 and Cooper, 1955) that morphological characters were used for the identification of the cyst nematodes when the significance of the characters of the vulval region was realized. And again after 1970, the stability of the many of the characters of the second stage juveniles was established and used for the identification purposes. A number of cyst nematodes have been described and named after their type hosts though they were differentiated on the basis of morphological characters. Some of such species are *Heterodera amaranthi* Stoynov, 1972; *H. avenae* Wollenweber, 1924; *H. cajani*, Koshy, 1967; *H. carotae* Jones, 1950; *H. cynodontis* Shahina and Maqbool, 1989; *H. cyperi* Golden, Rau and Cobb, 1962; *H. glycines* Ichinohe, 1952; *H. graminis* Stynes, 1971; *H. hordecalis* Andresson, 1975; *H. lespedezae* Golden and Cobb, 1963; *H. medicaginis* Kirjanova, 1971; *H. millefolii* Kirjanova and Krall, 1965; *H. mothi* Khan and Hussain, 1965; *H. oryzae* Luc and Brizuela, 1961; *H oryzicola* Rao and Jayaprakash; *H. rumicis*, Pogosyan, 1961, *H. sacchari*, Luc and Merny, 1973; *H.salixophila* Kirjanova; *H. sancophila* Kirjanova, Krall and Krall,1976; *H. scleranthii* Kaktina, 1975; *H. trifolii*, Goffart,1932; *H.sorghi*, Jain *et al.*, 1982 and *H. zeae* Koshy, Swarup and Sethi,1971. There are other species which are named after the place of their origin (*Globodera rostochiensis*) and then there are also species which have been given names on the basis of a specific character in them (*G. pallida, Punctodera, Afenestrata*). Because of this enhanced knowledge, there has really been a big addition to the existing number of described species of cyst nematodes. The two species, *H. australis* Subbotin, Sturhan, Rumpenhorst and Moens (2002) and *H. pratensis* Gabler, Sturhan, Subbotin and Rumpenhorst (2000) which though morphologically resemble *H. avenae*, have been described separate species based on analyses of IEF, RFLP and sequences of the ITS of rDNA obtained from these populations.

Various diagnostic characters used in identification of cyst nematodes include the shape and size of the cyst, presence or absence of subcrystalline layer, cone top structures and various morphometrics of second stage juveniles and males.

Shape

There are two basic shapes of the cysts *i.e.* round and lemon. Cysts belonging to the genera *Globodera, Cactodera, Punctodera, Dolichodera, Betulodera* and *Vittatidera* are mostly round or globose and the genera, *Heterodera* and *Afenestrata* have mostly lemon shaped cysts. Certain species of *Cactodera* may have lemon shaped cysts. Overall cyst shape allows for the initial separation of the *Heterodera* groups from the other groups in which the lemon shape feature is absent.

Size

It is an extremely variable character in the populations of the cyst nematodes. Generally, cysts of *Globodera rostochiensis, G. pallida, H. avenae, H. mothi* and *H. sorghi* are large whereas *H. zeae, H. cajani, H. graminis* and *H. skohensis* are smaller sized cysts. But in these both types variations are there *i.e.* a large sized population may have smaller cysts and vice-verse.

Colour

It is again a character of minor taxonomic importance as majority of cyst are dark brown in colour. However, *H. zeae* and *H. mothi* are light (yellow or brown) coloured cysts with thin cuticle whereas cysts like *H. avenae, H. cajani* and *H. sorghi* are dark coloured which may even turn black as they mature.

Subcrystalline Layer

Presence or absence of sub-crystalline layer and patterns on cyst wall are other characters that can be used in the identification of the cysts. In many cyst nematodes but more prominently in the cereal cyst nematode, *H. avenae,* a thick chalky coloured, sub crystalline layer is quite prominent which is formed almost on the whole body barring the vulval region.

All these characters described above can only be used at the specific or generic level and are of not much use in population studies. Characters that can be used include vulval slit length, fenestral shape, presence or absence of bullae and underbridge, stylet length and total length of J2 shape and shape and size of stylet knobs, tail length and hyaline portion of the tail (Mulvey, 1972; Cook, 1982; Mulvey and Golden, 1983).

Cone Top Structures

Extensive use of cone top structures to differentiate species and genera of cyst nematodes was reported by Mulvey (1972). He studied the posterior ends of the cysts of 39 species of *Heterodera* and arranged them into five major groups based on variations in cone top structure and cysts. The round or pear shaped cysts, circumfenestrate without cone-top, rear presence of underbridge and bullae and with reduced vulval slit (<15μ long) comprised the Group1 (*Globodera, Punctodera, Dolichodera*); the round to lemon cyst, circumfenestrate with little but strong protuberance, underbridge absent, bullae present or absent and short vulval slit (<20μ long) in Group 2 (*Cactodera* and *Betulodera*); lemon shaped cysts with cone-top, very nearly bifenestrate (still ambifenestrate) cysts, bullae and underbridge present or absent with small vulval slit (<16μ long) in Group3 (*Heterodera avenae* complex group); cysts lemon shaped to globose with posterior protuberance, ambifenestrate, bullae and underbridge prominently developed with a long vulval slit (> 35μ long) in Group 4 (*H. schachtii* group); and lemon or spherical shaped cysts, underbridge slender or absent, bullae in a row around fenestrae or scattered and having a long vulval slit (>30u long) in Group 5 (*H. goettingiana* group). Based on the morphological characters, this system formed the model for the cyst nematodes groupings (Evans and Rowe, 1998). The progress made in the molecular diagnostics has resulted in a modification of this grouping. The phylogenetic analysis of ITS rRNA and D2-D3 expansion segment of the 28S-rRNA gene sequences confirmed the monophyly of the tribe Punctoderini with the genera *Globodera, Cactodera, Punctodea, Dolichodera* and *Betulodera* and the monophyly of the tribe Heteroderini with the genus *Heterodera*. Subbotin *et al.* (2001) have reported that the combination of molecular data with the morphology of the vulval structure and number of lateral lines of second stage juveniles, six groups within the genus *Heterodera* can be recognized: (i) *Avenae* (ii) *Cyperi* (iii) *Goettingiana* (iv) *Humuli* (v) *Sacchari* and (vi) *Schachtii* groups. Close relationships have been revealed between the *Avenae* and *Sacchari* groups and in some other groups. Some inconsistencies between molecular phylogeny and earlier proposed morphological groupings may be attributed to homoplastic evolution, *e.g.* a bifenestral vulval cone developed independently at least three times during the evolution of cyst nematodes and also the presence of three incisors in the lateral field of J2 seems to have arisen twice (Subbotin *et al.*, 2001).

The structure of the vulval cone and associated internal and external characters has been used reliably for the precise identification of cyst nematodes species and in studying intra-specific variations. Vulval cone of cyst nematodes comprises the vulval slit, vulval bridge, underbridge, fenestrae and anus and surrounded by folds of primary and secondary bullae. In case of *Heterodera zeae*, the bullae are finger-shaped and almost similar is the case with *H. cajani*. At the terminus of the cone is the vulval slit that is bordered by the vulval lips which forms parapets of the vulval bridge, which crosses a thin walled area, fenestra, on the cone top. The vulval bridge thus divides the fenestra into two semifenestrae. The semifenestrae show no useful characters but their shapes and sizes are important in species diagnosis. In the older cysts, the semifenestrae break down leaving holes with more or less well defined margins and provide an exit to the hatching juveniles. The type of fenestration, length and width of semifenestrae, distance from anus to the nearest edge of fenestra and even the position of the anus are important diagnostics for cyst nematodes. The other character on the vulval cone is the basin, a band of thick cuticle around the fenestra, is quite prominent in *H. zeae*, *H. sorghi* and *H. skohensis*. Length and breadth of underbridge and presence or absence of bullae and their number are also good taxonomic characters for cyst nematodes. Commonly, the species with a strong underbridge have few (*H. sorghi*, *H. mothi*) or no bullae (*H. graminis*) whereas species with prominent and more numbers of bullae, have no underbridge (*H. avenae*) or a weak underbridge (*H. filipjevi*). In species like *H. zeae* and *H. cajani* that have strong and finger-like bullae but few in numbers, are provided with a medium sized underbridge, which is equal to the length of vulval bridge. The vulval bridge may be short and wide (*H. avenae*) or it may be long and slender (*H. zeae* and *H. sorghi*). The fenestrae are mostly 'D' shaped in these cysts. In *H. mothi*, the fenestral length is, most of the times, less than the fenestral width and identifies as a good taxonomic character.

Mulvey (1959) recorded another bridge-like structure crossing the vulval cone in newly formed cysts of *H. trifolii*. This bridge, known as Mulvey's bridge, is possibly a hardened muscle tissue, runs at right angle to the underbridge and found just below the level of fenestra. In *H. glycines* also, a similar bridge is encountered with one arm being bifurcated. But this bridge gets dislodged in the older cysts.

Although no two vulval patterns are exactly alike *i.e.* the cone top structure of individual and populations within a species may vary and is exploited for population studies. But the basic species character does not change significantly. In *Punctodera* spp., the anal opening is bigger enough in size taking the shape of a fenestra and also has subcuticular punctuations around it. In *Cactodera* spp. there is a prominent but short posterior protuberance but the fenestration is circumfenestrate. Even in some of the older cysts, after the thin cuticular area disintegrates, the vulval slit remnants can clearly be seen in many species of this genus.

On the basis of similarities of characters in some species and differentiating them from other species in these nematodes, new genera were erected. Krall and Krall (1978) proposed the genus, *Bidera* for *H. avenae* and other such species, which were having short vulval slit and well-separated fenestrae. But Mulvey and Golden (1983) and Luc *et al.* (1988) considered it to be a junior synonym of *Heterodera.* Then, Shagalina and Krall (1981) erected the genus, *Ephippiodera* for those species in *Bidera*, which had very wide vulval bridge and fenestrae had saddle-like depressions. This character was also considered as variation in the fenestrae (Wouts, 1985) and again returned to *Heterodera* (Luc *et al.*, 1988).

In cyst nematodes, there are many species that produce egg masses of various sizes and lay eggs in different numbers. This varies even in different species belonging to a single genus. In *Heterodera avenae* and the species encompasses in this complex either produce a rudimentary egg sac without any eggs in it or the egg sac is altogether absent. The smaller sized species of *H. zeae*, *H. sorghi* and *H. graminis* secrete a small egg sac that contain a few or one third of the eggs produced by these species but *H. cajani* has a egg sac which is two to three times bigger than the size of the cyst itself and may contains two-third of the eggs that are laid out by this species. Then there is another cyst nematode, *H. assamiensis* (n.sp.) which has an egg sac seven to eight or more times bigger than the body of the cyst. No egg sac is found in the genera *Globodera, Punctodera, Dolichodera, Paradolichodera, Betulodera* or *Afenestrata* but very large egg masses are produced by *Vittatidera*.

In addition to the various cone top structures, J2 characters like length and width of the body, length of the stylet, hyaline portion and tail length and position of DEGO are also observed. Shape and size of stylet knobs, position and size of the phasmids and sharpness of the tail terminus are some other characters of diagnostic value. Some characters in males also may be of diagnostic value like shape of the spicular tip which may be flat or rounded, bifid or tridentate. Lip pattern may also be of diagnostic value if observed under SEM.

Granek (1955) has used the anus-vulva distance relative to the diameter of vulval opening to differentiate *H. tobaccun* from *H. rostochiensis*. This ratio was expressed as the distance from anus to the nearest edge of vulval opening divided by diameter of the vulval opening. This ratio was modified by Granek (1968) and Hesling (1973) where vulval opening was replaced with fenestra and diameter of the fenestra was changed with length of the fenestra.The modified ratio is, the distance from anus to the nearest edge of fenestra divided by the lengh of the fenestra and mostly used to differentiate *Globodera* species.

7.2.1 Molecular Level Identification of Cyst Nematodes

The identification of *Heterodera* species using morphological and morphometrical characteristics is time consuming and requires great skill and training by the observer. However, there are many other characteristics allowing discriminating between different species of nematodes *e.g.* biotechnological tools (Romero *et al.*, 1996;

Rumpenhorst *et al.*, 1996; Rivoal *et al.*, 2003; Subbotin *et al.*, 2003). As a consequence, the development and use of new tools to identify nematodes using molecular technologies increases exponentially (Rivoal *et al.*, 2003). The analysis of coding and non-coding regions of ribosomal DNA (rDNA) became a favorite way for nematode identification (Vrain *et al.*, 1992; Wendt *et al.*, 1993; Zijlstra *et al.*, 1995). The internal transcribed spacer region (ITS) is variable and therefore useful for nematode identification and phylogenetic studies at species level. Polymerase Chain Reaction-Restriction Fragment Length Polymorphism (PCR-RFLP) based on ITS-regions of the rDNA repeat units has provided a reliable tool for quick and precise identification of cyst nematode species and 3 subspecies (Bekal *et al.*, 1997; Subbotin *et al.*, 1999; Subbotin *et al.*, 2000; Rivoal *et al.*, 2003; Madani *et al.*, 2004; Abidou *et al.*, 2005; Smiley *et al.*, 2008). This was a start of many studies leading to an explosion of RFLP-patterns and sequences. Comparisons of sequences of the ITSrDNA of unknown nematodes with those published or deposited in GenBank (Ferris *et al.*, 1994; Orui, 1997; Szalanski *et al.*, 1997; Subbotin *et al.*, 1999, 2000, 2001; Sabo *et al.*, 2001; Tanha Maaû *et al.*, 2003) facilitated the quick identication of most species of cyst nematodes. By a simple PCR-reaction, it was then possible to detect the nematode species for which the primers were designed. Currently, most of the agriculturally important cyst-forming nematodes of the genus *Heterodera* are identified by using PCR-ITS-RFLP and sequencing of the ITS-rRNA genes, and PCR using species-specific primers developed for *H. schachtii* and *H. glycines*.

Recent progress in sequencing nematode genomes, including the recent sequencing of the *H. glycines* genome, supports the search for more reliable markers for use in diagnostics. DNA sequencing costs have decreased more than 100-fold over the past decade, fuelled in large part by tools, technology and process improvements developed as part of the successful effort to sequence the human genome. New technology opens the door to the next generation of sequencing methods, which include, pyrosequencing, sequencing-by-synthesis and sequencing using nanopores. There are many opportunities to reduce the cost and increase the throughput of DNA sequencing, which are likely to lead to very different and novel approaches to diagnostics (Perry *et al.*, 2007).

7.3 Cyst Nematode Biology

There are various species of cyst nematodes, all follow a similar life cycle *i.e.* going through the stages of egg, juvenile and adult with the entire process completing in 24 to 30 days if environmental conditions are favorable. The second stage juveniles emerge from the cyst under suitable conditions of moisture and temperature, enter roots of susceptible plant preferably in the region just posterior to the root tip and move to the vascular cylinder, piercing cell walls with their stylets and disrupting cells as they move. The orientation of infective juveniles in the root has the head near vascular cylinder and the tail towards the root tip. Nematodes secrete cellulase in the saliva from the secretory granules of the subventral glands and use mechanical force to migrate through the root. The cellular damage usually causes necrosis along the migration path of the nematode. Nematode invasion of roots and their migration to their feeding sites results in changed root architecture and significant reductions in nutrient and water uptake and consequent crop yields.

7.3.1 Syncytium Formation

The host–parasite relationship is governed by a complex network of interactions and in susceptible interactions there is a subtle interplay between parasite survival strategies and host defense mechanisms. Upon reaching the vascular cylinder, they establish a feeding site, apparently by injecting stylet secretions. The nematode completely alters its behavior when it arrives at suitable cells for feeding, where it gently pierces the initial syncytial cell and injects saliva into the cytoplasm. The nematode secretes esophageal gland-derived effector proteins into the host cell. These effector proteins developmentally reprogram the host cell into a specialized feeding site, called a syncytium (Davis *et al.*, 2008). The nematode also induces an up-regulation of cell wall-modifying proteins (Goellner *et al.*, 2001; Wieczorek *et al.*, 2006, 2008; Karczmarek *et al.*, 2008), causing the cell wall to dissolve. This allows the syncytium to expand in size by incorporating neighboring cells into the growing feeding site, a multinucleate syncytium which has the manifestation of dissolution of the cell wall, breakdown of the vacuole, formation of granular cytoplasm and fusion of neighbouring protoplasts with walls of irregular thickness (Endo, 1964).

The phytohormone indole-3-acetic acid (IAA), or auxin, has been implicated in syncytium development (Goverse *et al.*, 2000; Grunewald *et al.*, 2009). Auxin is involved in a variety of plant cellular and developmental processes. Multinucleate syncytia are produced from protoplasts of procambial cells after partial dissolution of the cell wall; the nuclei are hypertrophied in two days after infection. Genes for putative cell wall degrading enzymes were also upregulated, and some of them were among the most strongly induced such as those coding for pectate lyases and for a β-glucanase (Wieczorek *et al.*, 2008). The parasitic process from host root penetration through formation and maintenance of the syncytium is regulated, directly or indirectly, by a suite of effector proteins secreted from the nematode stylet that are developmentally- regulated to alter host cell physiology to promote parasitism (Baum *et al.*, 2007; Mitchum *et al.*, 2007; Davis *et al.*, 2008, 2009). Many genes are involved in adapting cyst nematodes to a parasitic life style. However, only genes whose products are expressed in the nematode secretory glands and are injected into the host cells through the stylet (*i.e.*, parasitism proteins) are called parasitism genes, which in turn have been termed parasitome in their entirety (Baum *et al.*, 2007). Thus, it is possible that the degradation of cell walls within the syncytium is solely achieved through endogenous plant proteins, and that the nematode-specific enzymes are only responsible for cell wall degradation and loosening during the passage of the nematodes through the plant root towards the central cylinder. A leading hypothesis is that cyst nematodes secrete effector proteins into the host cell and that these proteins function, among other things, to manipulate auxin flow through the feeding site by modulating auxin transport proteins or auxin signaling pathway components.

Syncytia have high metabolic activity, with hypertrophied nuclei, increased cytoplasmic density and increased cell organelles. Smooth and rough endoplasmic reticulum accumulates in the thick-walled initial syncytial cell. Nematode secretions are surrounded by a narrow region of endoplasmic reticulum that integrates with the rough endoplasmic reticulum of the host cytoplasm. Initial syncytial cell and adjacent cells show slight enlargement at 18 hrs and have definite disruptions in their connecting cell walls. The cell walls surrounding the syncytia thicken and develop ingrowths to increase plasmalemma surface area and to enhance solute uptake by the parasite (Hussey and Grundler, 1998; Lilley *et al.,* 2005). The secreted effector proteins originate in secretory granules in three elaborate secretory gland cells within the nematode oesophagus and are released through complex valves via exocytosis for secretion through the nematode stylet into the host tissues (Wyss and Zunke, 1986; Hussey, 1989; Davis *et al.,* 2008). An electron dense "feeding plug" is formed around the stylet to form a seal when the stylet is removed. The nematode also makes a feeding tube (after 24 hrs. of feeding site selection) which allows molecules of about 40kDa and less pass into the nematode. After the nematode completes feeding, the tubes are disassociated from the stylet orifice and are dispersed in the syncytial cytoplasm. The feeding tube lumen is filled with cytoplasm and the wall gradually degrades. No difference has been observed between feeding tubes formed by male and female juveniles.

The nematodes become sedentary and the life cycle differs between males and females.The development of the male through the third and fourth stage juvenile proceeds inside the cuticle of the second ctage juvenile without feeding like that of root knot nematodes. At the completion of the fourth molt the vermiform male emerges. The cell wall degrading enzymes aid in the exit from the root tissue. Females of some cyst nematodes emit sex pheromones that attract the emerging male but little is known about the chemical nature of these pheromones. Vanillic acid was identified as attractant for mals of *H. glycines* (Jaffe *et al.,* 1989) and partial purification of components of *H. schachtii* sex pheromones has been achieved (Jonz *et al.,* 2001). In case of females there is continuous development and growth which become saccate apparently feeding at each stage. From second stage to fourth stage flask-shaped they took about 15 days but they took only three days prior to molting to mature adult females. Females become lemon shaped, and can be seen as small white-yellow dots, protrude from the root surface crushing the plant cells, only the neck and head remain embedded in the

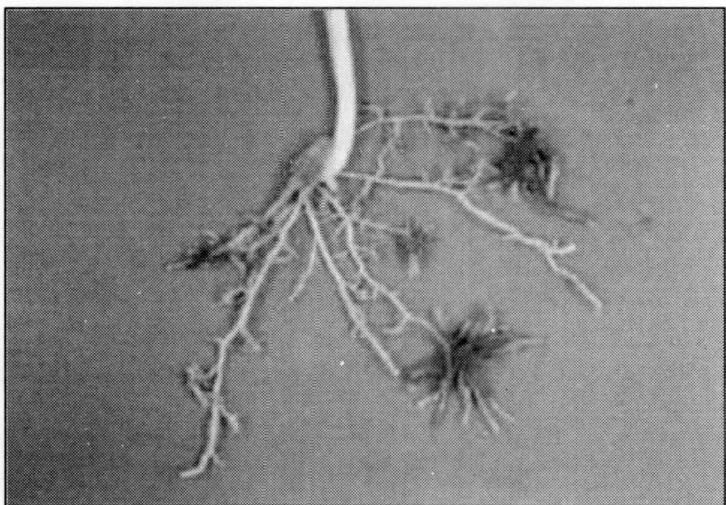

Figure 2: 'Bushy' or Coral Roots Symptoms with Cyst Nematodes Infection

vascular tissue surrounded by syncytial cells. They generally reproduce sexually, and once fertilized, the female becomes full of eggs. When all the eggs are laid the female dies and due to the action of polyphenol oxidase on polyphenols in the cuticle (2-3 per cent polyphenol), its body becomes a protective cyst for the eggs. During the chromogenesis process, the females turn yellow, light brown to dark brown or black in colour.

In some species, the cyst is coverd by an additional deposit called *subcrystalline layer*, a thick chalky crystallized double layer protecting the cyst from adverse environmental conditions. The inner zone of the sub-crystalline layer of cysts of certain *Heterodera* spp. is comprised largely of n-tetracosanoic acid, whereas the outer layer consisting of a mixture of this acid and its calcium salts (calcium tetracosanate) and is fractured into pyramidal blocks presenting an irregular polygonal appearance. Some species like *H. avenae* and *H. trifolii* may have additionally hexacosanoic acid in its composition. It is produced by the exudates released by the nematode itself (Zunke and Eisenback, 1998).

A few eggs may be deposited outside in the gelatin matrix but most of them remain in the cyst itself. Eggs can persist in the cyst for many years and in some species are stimulated to hatch in response to the exudates from the host plant but some eggs do hatch immediately.In the absence of the host the decline within the cyst eggs and larval content may be around 50 per cent annually.

7.4 Hatching in Cyst Nematodes

Nematodes and other organisms are influenced by host signals as regards hatching, host location and selection, feeding-site location, sex determination and intra-host migration. Signals contain information, which react with a receptor and elicit a response. Most signals between organisms and their host plant are based on phytochemicals from the host plant, but the organism can also release chemicals to the host plant. The host plant then starts to defend itself (Hirsch *et al.*, 2003). Nematodes lay eggs and the egg in the egg shell enables the juvenile to develop while protected against environmental stresses. In general, under the favourable conditions devoid of any stresses, hatching of juveniles in most species occurs without any hinderance but in some species this is synchronized with the cultivation of the host plants. The eggs in cyst nematode species are either laid in the egg sac or remain in the cyst or a certain portion of egg content is laid in both in the egg sac as also in the cyst itself. In cereal cyst nematode, *Heterodera avenae* and potato cyst nematodes, *Globodera rostochiensis* and *G, pallida*, the eggs are retained in the cysts whereas in species like *H. cajani* and *H. trifolli*, majority of the eggs are laid in gelatin matrix (egg sac) and in species like *H. zeae* and *H.sorghi* only few eggs are deposited in the egg sac while rest are retained in the cysts.The hatching of eggs and emergence of juveniles from the cyst or from the egg sac is greatly influenced by many environmental factors like aeration, moisture, temperature and availability of root exudates. After the physiological maturity, the some eggs of *H.avenae* may hatch (eclosion from the egg) in the cyst by June ot July but will not emerge out of the cyst till proper moisture and temperature conditions are condusive with the onset of winter in late October or early November.This is diapause which allows nematode to arrest development before existing the cyst where the length of dormancy period is dictated by the seasonal changes, enabling the the eggs or juveniles to withstand the harshness of the external environment. But it has also been observed that if these cysts are exposed to differential temperature regimes of 10 and 20° C, juveniles do emerge from the cysts even in the months of June and July. In this species, about 50-60 per cent of eggs hatch in one season freely in water and 50 per cent will hatch in the next year and so on. Juveniles of *H. oryzicola* are always dependent on host root diffusate for substantial hatch, irrespective of generation, and can remain dormant until the subsequent growing season (Ibrahim *el al.*, 1993). An analysis of the hatching from cysts and eggsacs of six successive generations of *H. cajani* produced on cowpea during a single growing season in glasshouse pot cultures demonstrated that the majority of eggs in egg sacs hatched within seven days in distilled water or host root diffusates irrespective of generation (Gaur *el al.*, 1992). It has also been observed that the eggs in egg sacs hatched more readily than those in cysts and probably provides J2 a chance of secondary infection and multiplication; this is also the situation with *H. oryzae* (Merny, 1966), *H. glycines* (Ishibashi *el al.*, 1973), *H. carotae* (Greco, 1981) and *H. cruciferae* (Koshy and Evans, 1986). *H. sacchari* rarely has an egg sac and encysted eggs hatch readily in water without the need for hatch stimulation by rice root diffusate (Bridge *et al.*, 1990).

Aeration plays an important role in the movement of juveniles and their hatching. It has also been reported that the eggs on the periphery of cyst hatch first and those in the centre hatch later on (Onion, 1955). This may be due to the oxygen tension gradient in the cyst. The cysts of *H. schachtii* when exposed to a small oxygen concentration, hatching in root diffusates was decreased (Wallace, 1955) and when the cysts were placed in water under various mixtures of oxygen, the hatch was closely correlated with the amount of oxygen. Temperature also influences the hatching rate. The emergence of juveniles from the eggs from the cyst or egg sac in case of *H. cajani* takes place only at particular temperature of 28± 1°C. The juveniles emerge readly from the eggsacs than from the cysts.

Baunacke (1922) observed that the emergence of juveniles of *H. schachtii* got stimulated by the presence of root diffusates from the host whereas the potato cyst nematodes *G. rostochiensis* and *G. pallida* require a proper stimulant in the form of root exudates for hatching and emrgence (Hominick. 1986). O'Brien and Prentice (1930) reported the hatch-inducing activity of potato root exudates on the potato cyst nematode whereas Triffith (1930) was the first to investigate in detail the excretions of hatching factors by potato roots. The dependence of species like *Globodera rostochiensis* on host root diffusates to stimulate hatch is well documented and the sequence of events in the hatching process has been the subject of extensive research (Perry, 1987, 1989). Hatching is dependent on host root diffusates, ensuring synchrony between emergence of infective juveniles and the presence of roots at the initial stages of host growth (Perry, 1987). Just before hatching the juveniles of *G. rostochiensis* move quite vigorously inside the eggshell with rhythmic movements of the stylet against the shell making perforations until a slit appears through which it escapes outside. Hatching is stimulated by root diffusates of host plants but in some cases it may be stimulated by non hosts. *H. schachtii* is stimulated by diffusates of *Hesperis matronalis* and *Coronopus squamatus* which are not hosts of this species whereas *H. cruciferae* does not get stimulated by its host *C. squamatus* when all brassicas stimulate hatching in this species. So there is no correlation between resistance to attack by a given species and the production of root diffusates active to that species. The cyst nematodes can be classified in three broad categories on the bases of their hatching response to root diffusates: 1. Few J2s hatch in water, very large number hatch in root diffusates (*e.g. G rostochiensis, G. pallida, H.cruciferae, H. carotae, H.goettingiana* and *H.humuli.*) 2. Moderate hatch of J2s in water, very large hatch in response to root diffusates (*e.g. H. trifolii, H.galeopsidis, H.glycines*). 3. Large hatch of J2s in water, very large hatch in root diffusates (*H.schachtii, H.avenae*).

In plant parasitic nematodes, where host root diffusate are required for hatching, considerable information is available on characterization of diffusate activity, the chemicals involved and their effects. Many factors including plant age, development, cultivar and growing conditions modify the hatching activity of a root diffusate towards a particular nematode. It is also evident that the activities of the hatching factors in root diffusates are not constant throughout the life of some host plants but get reduced as the plant matures. This is apparent from the results of *in vitro* hatching tests with *Heterodera goettingiana*, for example, diffusates from 4- and 6-week-old pea gave good hatch of juveniles but diffusates from 2- or 10-week-old plants has reduced hatching of less than 10 per cent (Perry *et al.*, 1980). Hatching factors in potato root diffusate are produced along the length of the root, but particularly at or just behind the root tip (Rawsthorne and Brodie, 1986). The persistence and mobility of hatching factors in the soil depend on several factors like soil type, moisture content, temperature and microbial activity. Hatching activity towards *G. rostochiensis* was detected 100 days after removal of the plants (Tsutsumi, 1976). Hatching factors are highly mobile in the soil as their activity towards *G.rostochiensis* and *G. pallida* was detected in soil up to 80 cms away from the potato root zone (Rawsthorne and Brodie, 1987 and Malinowska, 1996). Fractionation of root diffusates has revealed the presence of many hatching factors. Glycinoeclepin A has been isolated from kidney beans roots rather than the root diffusates which induces hatching in *H. glycines*. Almost a similar fraction (solanoeclepin A) has been isolated from potato roots inducing hatch in potato cyst nematode, *G. rostochiensis*. This has recently been synthesized artificially (Tanino *et al.*, 2011) so that it can be used to stimulate early hatching and let the nematode die in the absence of the host. Solanoeclepin A is derived from potatoes and, as a natural product, is suitable for an environmentally friendly nematode control method (Blaauw *et al.*, 2001).

There are many artificial hatching factors consisting of organic acids *e.g.* anhydrotetronic acid, picrolonic acid, fumaric acid stimulating hatching in *G. rostochiensis;* flavionic acid stimulating hatching in *H.glycines* and *H. schachtii.* These acids mostly increase the rate of metabolism of stimulated juveniles (Perry, 1986). There are also a number of inorganic chemicals like sodium metavanadate, zinc chloride, zinc sulphate, calcium chloride, calcium sulphate that stimulate hatch in many cyst nematodes. An isolate of the rhizobacterium, *Pseudomonas fluorescens* which is used as a biocontrol agent has been shown to stimulate hatch in *G. rostochiensis* by exporting a hatch stimulatory metabolite, an antibiotic, 2.4-diacetylphloroglucinol (DAPG) but it decreased the mobility of the hatched juveniles (Cronin *et al.*, 1997). It is a species specific chemical that does not affect all the nematodes.

There are many chemicals (including many nematicides) and plant exudates which have been exhibiting hatching inhibitory (HIs) reactions.The inhibitors are produced at an earlier stage than the stimulants, protecting the plant during its most sensitive stage of growth (Byrne *et al.*, 1998). Several plant based hatching inhibitors like DMDP(2,5-dihydroxymethyl-3,4-dihydroxypyrrolidine), a sugar analog from *Derris* sp. (Birch *et al.*, 1993) and the lignans bursehernin and metairesinol from the umbellifer, *Bupleurum salicifolium* inhibited hatch of *G. rostochiensis* and *G. pallida* (Gonzalez *et al.*, 1994). Forest (1989) reported hatch inhibitory effects of white mustard towards potato cyst nematodes. HIs appear to be widespread throughout the Solanaceae. *Petunia* spp. are the member of the nonhost subfamily, Cestroideae, and produce HIs but not hatching factors in their root diffusates, leading to a net inhibition of *G. rostochiensis* in the soil in the presence of *Petunia* plant compared with the hatch in the absence of plant (Twomey, 1995). Neem and its various derivatives have shown promise in the reduction of haching in cyst nematodes. Dhawan and Kaushal (1988) showed reduction in hatching incidence of *H avenae* larvae by the use of neem emulsion.

7.5 Classification

The cyst nematodes with *Heterodera* (including root knot nematodes), *Tylenchulus*, and *Paratylenchus* were proposed as a new subfamily Heteroderinae by Filipjev and Schuurmans Stekhven (1941). The subfamily was raised to familial rank Heteroderidae by Skarbilovich (1947) to superfamily rank Heteroderoidea by Golden (1971) and Stone (1975) and even a suborder Heteroderata (Heteroderina) was proposed by Skarbilovich (1959). Such higher rankings have mainly not been accepted and Maggenti *et al.* (1987) has considered Heteroderidae to be valid at family level and root knot nematodes in a separate family Meloidogynidae even by other authors such as Stone (1978) and Siddiqi (1986). Several subfamilies have been split from Heteroderidae (like Meloiderinae Golden, 1971; Ataloderinae Wouts, 1973; Sarisoderinae Hussain, 1970; Punctoderinae Krall and Krall, 1978; Verutinae Essar, 1981) and from Meloidogynidae (Nacobboderinae Golden and Jensen, 1974; Meloinematinae Hussain, 1976; Meloidoderellinae Hussain, 1976). Luc *et al.* (1988) have also accepted Heteroderidae as a family with three subfamilies, Heteroderinae, Meloidogyninae and Nacabboderinae.

Baldwin and Schouest (1990) gave a classification of Heteroderinae with several tribes: Heteroderini, Verutini, Cryphoderini, Meloidoderini, Sarisoderini, and Ataloderini, based on phylogenetic analysis of morphological characters. Siddiqi (2000) accepted the division of the family Heteroderidae into three subfamilies Heteroderinae, Meloidoderinae and Ataloderinae as proposed by Wouts (1973) after considering many options. So the systematic position of *Heterodera* is:

Kingdom	Animalia (Bilateralia)
Phylum	Nematoda

Class	Secernentea
Subclass	Diplogasteria
Order	Tylenchida
Suborder	Hoplolaimina
Superfamily	Hoplolaimoidea
Family	Heteroderidae Filipjev and Schuurmans Stekhoven, 1941
Subfamily	Heteroderinae Filipjev and Schuurmans Stekhoven, 1941
	= Heteroderini Filipjev and Schuurmans Stekhoven, 1941 (Coomans, 1979)
	= Punctoderinae Krall and Krall, 1978
	= Sarisoderinae, Husain, 1976
Genus	*Heterodera* Schmidt, 1871

The molecular diagnosis has facilitated a better understanding of phylogeny of nematodes. As a result the morphology based systematic schemes are being changed by hierarchies based on molecular phylogeny. There is a proposal for infraorders and bringing together groups which have been regarded as distinctly related (De Ley and Blaxter, 2002). So if Siddiqi (2000) and De Ley and Blaxter (2002) are to be followed simultaneously, the systematic position of the genus *Heterodera* will be upgraded to-

Phylum	Nematoda Potts, 1932
Class	Chromodorea Inglis, 1983
Subclass	Chromodoria Pearse, 1932
Order	Rhabditida Chitwood, 1933
Suborder	Tylenchina Thorne, 1949
Infraorder	Tylenchomorpha De Ley and Blaxter, 2002
Superfamily	Tylenchoidea Orley, 1880
Family	Heteroderidae Filipjev and Schuurmans Stekhoven, 1941
Subfamily	Heteroderinae Filipjev and Schuurmans Stekhoven, 1941
Genus	*Heterodera* Schmidt, 1871

7.6 Diagonosis of Subfamily Heteroderinae (modified after Subbotin *et al.*, 2010)

Heteroderidae

Mature female- spherical, pear or lemon shapedwith a short neck, turning into a tough, hard -walled, yellowish, light to dark brown or black cyst containing eggs and juveniles,in some specie eggs may be laid in large gelatinous matrix. Cuticle surface with zig-zag or lace-like pattern, or with elongated fusiform ridges, overlaying fine pattern of annulations, or annulations absent on mature cysts and females. Vulva and anus usually close together, almost terminal on raised vulval cone or in flat to concave vulval basin. Clear hyaline single or double translucent vulval fenestrae present, but absent in the genus *Afenestrata*, anal fenestra present only in *Punctodera*.Vulval lips rarely persistent (*Vittatidera*). Esophageal gland nuclei contained in a single lobe or subventral gland nulei of the female contained in separate small lobe (*Vittatidera*). *Male*-developed through metamorphosis, labial region annulated, lateral field with three or four incisures, tail short, hemispherical without bursa. J2- with three or four incisures in lateral field and stylet over 14µ long.

Type Genus

Heterodera Schmidt, 1871

Other genera

Afenestrata, Baldwin and Bell, 1985

Betulodera Sturhan, 2002

Cactodera Krall and Krall, 1978

Dolichodera Mulvey and Ebsary, 1980

Globodera Skarbilovich, 1959

Paradolichodera Sturhan, Wouts and Subbotin, 2007

Punctodera Mulvey and Stone, 1976

Vittatidera Bernard, E. C., Z. A. Handoo, T. O. Powers, P. A. Donald, and R. D. Heinz, 2010

7.6.1 Identification Key to Genera of Heteroderinae (after Subbotin et al., 2010) (With modifications)

1. Cysts without vulval fenestration .... *Afenestarta*
   Cysts with vulval fenestration 2
   Cysts and females with persistent lateral field .... *Vittatidera*
2. Cysts circumfenestrate 3
   Cysts ambifenestrate or bifenestrate .... *Heterodera*
3. Cysts with terminal cone 4
   Cysts without terminal cone 5
4. Vulval slit in cysts 12-28µm long, J2 with four incisures .... *Cactodera*
   Vulval slit in cysts 5-8µm long.J2 withthree incisures .... *Betulodera*
5. Anal region with fenestration .... *Punctodera*
   Anal region without fenestration 6
6. Mature female and cyst spheroidal, perineal tubercles usually present .... *Globodera*
   Mature female and cyst elongate-oval shape, perineal tubercles absent 7

7. Bullae absent in cysts, J2 with DGO=11-15µm, c=9-11, stylet<20µm .... *Paradolichodera*

   Bullae present in cysts, J2 with DGO=7-8µm, c=5-6, stylet=22-24µm .... *Dolichodera*

7.7 The Characters of Various Genera

7.7.1 *Heterodera* Schmidt, 1871

In 1850, Herman Schacht in Germany observed exhaustion in many sugar beet fields as the plants were stunted with hairy roots and they were giving very low yields. This condition of plants in the beet fields was termed as 'beet weariness' or 'beet tiredness' (*Rubenmudigkeit*). On examination, he found that the plant roots had small, white lemon shaped bodies, which were described as females of some species of nematodes and held them responsible for the losses in the sugar beet production. Schmidt (1871) named the nematode as *H. schachtii* as a mark of honor to its discoverer.

Heterodera Schmidt, 1871

Syn. *Tylenchus* (*Heterodera*) Schmidt, 1871

Heterodera Schmidt, 1871

Heterobolbus Railliet, 1896

Bidera Krall and Krall, 1978

Ephippiodera Shagalina and Krall, 1981

The females or cysts are lemon-shaped or globose, light to dark brown in color when fully developed and bear a short neck. The posterior protuberance may be slight or quite prominent. The cuticle is provided with a lace-like or zigzag pattern. 'D' layer is absent and subcrystalline layer is present or absent. Vulva is terminal and vulval lips are generally amalgamated into a vulval cone bearing vulval slit 10-60 µm long, which may be too short for passage of eggs in some species. The anus is sub-terminal on dorsal side of vulval cone, without fenestra-tion; vaginal remnants, underbridge and bullae are often present. Cysts are ambifenestrate or bifenestrate. Most eggs are retained in body, but in some species, the eggs are also laid in gelatinous matrix.

Type Species

Heterodera schachtii Schmidt, 1871

Other Species

H. amygdali Kirjanova and Ivanova, 1975

H. artemisiae Eroshenko and Kazachenko, 1972

H. aucklandica Wouts and Sturhan, 1995

H. australis Subbotin, Sturhan, Rumpenhorst and Moens, 2002

H. avenae Wollenweber, 1924

syn. *H. ustinovi* Kirjanova, 1969

H. major Schmidt, 1930

H. bergeniae Maqbool and Shahina, 1988

http://www.russjnematology.com/rjnvol91.htm - Wouts*H. betae* Wouts, Rumpenhorst and Sturhan, 2001

H. bifenestra Cooper, 1955

syn. *H. longicaudata* Seidel, 1972

H. cajani Koshy, 1967

syn. *H. vigni* Edward and Misra, 1968

H. canadensis Mulvey, 1979

H. cardiolata Kirjanova and Ivanova, 1969

H. carotae Jones, 1950

H. circeae Subbotin and Sturhan, 2004

H. ciceri Vovlas, Greco and diVito, 1985

H. cruciferae Franklin, 1945

H. cynodontis Shahina and Maqbool, 1989

H. cyperi Golden, Rau and Cobb, 1962

H. daverti Wouts and Sturhan, 1978

H. delvii Jairajpuri, Khan, Setty and Govindu, 1979

H. elachista Oshima, 1974

H. fici Kirjanova, 1954

H. filipjevi (Madzhidov, 1981) Stelter, 1984

syn. *Bidera filipjevi* Madzhidov, 1981

H. gambiensis Merny and Netscher, 1976

H. glycines Ichinohe, 1952

H. glycyrrhizae Narbaev, 1987

H. goettingiana Liebscher, 1892

H. goldeni Handoo and Ibrahim, 2002

H. graminis Stynes, 1971

H. graminophila Golden and Birchfield, 1972

H. hordecalis Andersson, 1975

H. humuli Filipjev, 1934

H. iri Mathews, 1971

H. kirjanovae Narbaev, 1988

H. latipons Franklin, 1969

H. lespedezae Golden and Cobb, 1963

H. leuceilyma Di Edwardo and Perry, 1964

H. limonii Cooper, 1955

H. litoralis Wouts and Sturhan, 1996

H. longicolla Golden and Dickerson, 1973

H. mani Mathews, 1971

H. medicaginis Kirjanova, 1971

H. mediterranea Vovlas, Inserra and Stone 1981

H. menthae Kirjanova and Narbaev, 1977

H. mothi Khan and Husain, 1965

H. oryzae Luc and Berdon-Brizuela, 1961

H. oryzicola Rao and Jayaprakash, 1978

H. oxiana Kirjanova, 1962

H. pakistanensis Maqbool and Shahina, 1986

H. phragmitidis Kazachenko, 1986

H. plantaginis Narbaev and Sidikov, 1987

H. pratensis Gabler, Sturhan, Subbotin and Rumpenhorst, 2000

H. raskii Basnet and Jayaprakash, 1984

H. riparia Subbotin, Sturhan, Waeyenberge and Moens, 1997

H. rosii Duggan and Brennan, 1966

H. sacchari Luc and Merny, 1963

H. salixophila Kirjanova, 1969

H. sinensis Chen and Zheng, 1994

H. skohensis Kaushal, Sharma and Singh, 2000

H. sonchophila Kirjanova, Krall and Krall, 1976

H. sorghi Jain, Sethi, Swarup and Srivastava, 1982

H. spinicauda Wouts, Schoemaker, Sturhan and Burrows, 1995

H. tadshikistanica Kirjanova and Ivanova, 1966

H. trifolii Goffart, 1932

syn. *H. galeopsidis* Goffart, 1936

H. paratrifolii Kirjanova, 1963

H. rumicis Pogosyan, 1961

H. scleranthii Kaktina, 1975

H. scutellariae Subbotin and Sturhan, 2004

H. turangae Narbaev, 1988

H. turcomanica Kirjanova and Shagalina, 1965

H. urticae Cooper, 1955

H. uzbekistanica Narbaev, 1980

H. vallicola Eroshenko, Subbotin and Kazachenko, 2001

H. zeae Koshy, Swarup and Sethi, 1971

Species Inquirendae

H. aquatica Kirjanova, 1971

H. arenaria Cooper, 1955

H. graduni Kirjanova (in Kirjanova and Krall), 1971

H. methwoldensis Cooper, 1955

H. polygoni Cooper, 1955

7.7.2 *Globodera* (Skarbilovich, 1959) Behrens, 1975

syn. Heterodera (Globodera) Skarbilovich, 1959

Skarbilovich (1959) erected a subgenus *Globodera* (Latin *globus*=globe and Greek *dera* =skin) to accommodate round cyst nematodes without a terminal cone originally included in the genus *Heterodera*. Behrens (1975) elevated this subgenus to the generic rank. Even, Stone (1973) while describing *Heterodera pallida* reported that it belonged to the subgenus *Globodera*. Almost at the same time, Mulvey and Stone (1976) based on differences in the biological and morphological characters in the females, also proposed *Globodera* at the generic level. The potato cyst nematode is a menace to potato growers in Europe and rest of the world. Balwin and Mundo-Ocampo (1991) emphasised the importance of the species causing economic losses to potato though included other lesser important species while reviewing the *Globodera* species.

Diagnosis

Female and Cyst

The females are nearly spherical when fully grown without terminal cone; cuticle with a lace-like pattern; vulval lips absent; vulva located in a cavity below outline of body, with vulval slit less than 15 µm long, too short for passage of eggs; in mature females vulva with tubercles on dorsal and ventral wall of vulval cavity; anus terminal or sub terminal, without fenestration; vaginal remnants, underbridge and bullae rarely present; no fenestration around anus. Cyst's vulval area is circumfenestrate. All eggs are retained in the body.Males have lateral field with four lines, spicules > 30µm, distally pointed. J2s are with four incisures. Tail is conical, pointed, phasmids punctiform. En face pattern is typically with six separate lips, sometimes with fusion of adjacent submedial lips.

Type Species

Globodera rostochiensis (Wollenweber, 1923) Behrens, 1975

syn. *H. rostochiensis* Wollenweber, 1923

H. pseudorostochiensis Kirjanova, 1963

H. schactii solani Zimmerman, 1927

H. solani Zimmerman, 1927

G. arenaria Chizhov, Udalova and Nasonova, 2008

Other Species

G. artemisiae (Eroshenko and Kasachenko, 1972) Behrens, 1975

syn. *H. artemisiae* Eroshenko and Kasachenko, 1972

G. hypolysi Ogawa, Ohshima and Ichinohe, 1983

G. bravoae Franco, Cid del Prado and Lamothe-Argumedo. 2000

G. chaubattia (Gupta and Edward, 1973) Wouts, 1984

syn. *H. chaubattia* Gupta and Edward, 1973

G. mali (Kirjanova and Borisenko, 1975) Beherns, 1975

G. leptonepia (Cobb and Taylor, 1953) Behrens, 1975

syn. *H. leptonepia* Cobb and Taylor, 1953

G. maxicana (Campos-Vela, 1967) Subbotin, Mundo-Ocampo and Baldwin, 2010

syn. *H. mazicana* Campos-Vela, 1967

G. millefolii (Kirjanova and Krall, 1965) Beherns, 1975

syn. *H. millefolii* Kirjanova and Krall, 1965

G. achilleae (Golden and Klindic, 1973) Beherns, 1975

H.achilleae Golden and Klindic, 1973

G. pallida (Stone, 1973) Behrens, 1975

syn. *H. pallida* Stone, 1973

G. tabacum (Lownsbery and Lownsbery, 1954) Behrens, 1975

syn. *H. tabacum* Lownsbery and Lownsbery, 1954

H. solanacearum Miller and Gray, 1972,

H. virginiae Miller and Gray, 1968

G. zelandica Wouts, 1984

7.7.3 *Cactodera* Krall and Krall, 1978

Krall and Krall (1978) erected the genus *Cactodera* (kaktos =prickly plant and dera =skin) for the species of *Heterodera* having posterior protuberance in the cyst with circumfenestrate vulval region.The shape of cysts of *Cactodera* is generally between the *Heterodera* and *Globodera* as a full grown cyst will almost be globose with a very small posterior protuberance which sometimes may not be easily visible. Some of the younger cysts look exactly like cysts of the genus *Heterodera* with pronounced posterior protuberance.

Diagnosis

Female almost spherical, with slight vulval cone when fully developed. Cuticular has D-layer and the surface pattern is provided with roughly parallel ridges. Vulval lips are absent. Vulval region protruding slightly out of body contour; vulval slit short, less than 30 μm, slit not associated with a vulval cavity, without tubercles on dorsal and ventral sides adja-cent to slit; denticles present around vulva. Underbridge and bullae are lacking. Anus is subterminal with no fenestration around it. Cysts are circum-fenestrate. All eggs are retained in body.

Type Species

Cactodera cacti (Filipjev and Schuurmans Stekhoven, 1941) Krall and Krall, 1978

syn. *H. cacti* Filipjev and Schuurmans Stekhoven, 1941

Other Species

C. acnidae (Schuster and Brezina, 1979) Wouts, 1985

syn. *H. acnidae* Schuster and Brezina, 1979

C. amaranthi (Stoyanov, 1972) Krall and Krall, 1978

syn. *H. amaranthi* Stoyanov, 1972

C. betulae (Hirschmann and Riggs, 1969) Krall and Krall, 1978

syn. *H. betulae* Hirschmann and Riggs, 1969

C. eremica Baldwin and Bell, 1985

C. estonica (Kirjanova and Krall, 1963) Krall and Krall, 1978

syn. *H. estonica* Kirjanova and Krall, 1963

C. johanseni Sharma *et al.*, 2000

C. milleri Graney and Bird, 1990

C. thornei (Golden and Raski, 1977) Krall and Krall, 1978

syn. *H. thornei* Golden and Raski, 1977

C. weissi (Steiner, 1949) Krall and Krall, 1978

syn. *H. weissi* Steiner, 1949

Cactodera johanseni (Sharma *et al.*, 2000) has very erroneously been shifted to the genus *Heterodera* by Sturhan (2002) by just looking at the cone top structure which appears as fenestrated with a vulval bridge but that is not true. It is a *Cactodera* cyst species and at younger stage vulval slit (not vulval bridge) is always there whether it belongs to *Globodera*, *Punctodera* or *Cactodera* species and that is what, is visible in the published paper. On maturity, this vulval slit disintegrates making it single fenestral window- circumfenestrate.

7.7.4 *Betulodera* (Hirschmann and Riggs, 1969) Sturhan, 2002

This genus was erected by Sturhan (2002) on the basis of work done by many workers (Golden and Raski, 1977; Cliff and Baldwin, 1985; Baldwin and Schoeust, 1990; Othman *et al.*, 1998 and Sturhan, 2000) who considered it to be morphologically different citing various peculiarities and found it to be an intermediate betw en *Heterodera* and *Cactodera*. It was first described by Hirschmann and Riggs (1969) as *H. betulae* and subsequently moved to *Cactodera betulae* as the species belonging to *H. cacti* group were put in the genus *Cactodera*.

Cyst is spherical to pear shaped. The cuticle has a lace-like diagnostic pattern having a thick subcrystalline layer. Vulval cone is low and button-like. Cuticular punctuations are absent. The vulval slit is short (less than 10 µm) and has circumfenestration without underbridge. Bullae are sometimes present. Male body is twisted with prominent phasmids. Spicules have bifid tips. Second stage juveniles have three to four head annules and the lateral field has three incisures.

Type and Only Species

Betulodera betulae

7.7.5 *Afenestrata* (Luc, Germani and Netscher, 1973) Baldwin and Bell, 1985

Baldwin and Bell (1985) observed that *Sarisodera hydrophila* (Wouts and Sher, 1971) does not possess cysts but *S. africana* (Luc, Germani and Netscher, 1973) does possess a cyst and thus justified erecting a new genus *Afenestrata* for it. Few other differences like formatiom of single giant cell by *S. hydrophila* but a syncytium by *A. africana*, absence of D-layer in *Afenestrata* but present in former genus and presence of punctiform phasmids in *Afenestrata* but lens-like in *S. hydrophila*, were also highlighted. On the basis of molecular phylogenetic analysis of ribosomal RNA and protein-coding genes, *Afenestrata* was synonymised with *Heterodera* (Mundo-Ocampo *et al.*, 2008) and considered as '*Afenestrata* group' under the genus *Heterodera* by Subbotin *et al.* (2010). Here, it is retained as a separate genus on the basis some stable characters like the presence of ornamented cuticular vulval flaps (genus *Heterodera* lacks these vulvular cuticular extensions), no fenestration and deep sunken vulva (*Cactodera* has also sunken vulva but less deep).

Diagnosis

The cuticle of cyst is provided with a zig-zag or lace-like pattern of ridges. Cysts are tanned to brown, lemon shaped. Vulval slit is small without tubercles. Vulva sunken into terminal cone, lacking fenestration or bullae, underbridge may be present in some species, anus dorsal to vulval lip. Fenestration is mainly undone by the extended vulval cuticular flaps which are provided with differential annular markings. *Male:* Spicules are terminal enclosed by sheath; tail is absent. *Second-stage juvenile:* Stylet is less than 38µ long; phasmid opening is pore-like, punctiform.

Type Species

Afenestrata africana. (Luc, Germani and Netscher, 1973) Baldwin and Bell, 1985

syn. *Sarisodera africana* Luc, Germani and Netscher, 1973

Afrodera africana (Luc, Germani and Netscher, 1973) Wouts, 1985

Other Species

A. axonopi Sousa, 1996

A.koreana Volvas, Lamberti and Chao, 1992

A.orientalis Kazachenko and Lebedev, 1989

A. sacchari Kaushal and Swarup, 1989

7.7.6 *Punctodera* Mulvey and Stone, 1976

Mulvey and Stone (1976) raised the subgenus *Globodera* to the generic rank, they also described a new genus *Punctodera* (*Punctum*=dot) for the species having punctations (small subcuticular dots) around the anal fenestra which incidentally was large enogh to be called so like the vulval fenestra in this genus.

Diagnosis

Cyst/female is ovoid or spherical, with short projecting neck and heavy subcrystalline layer, without any posterior protuberance. Cuticle is provided with a reticulate pattern or with roughly parallel ridges. Subcuticle has punctuations. D-layer is present. Vulval lips are absent, vulval slit is short, less than 5 µm long, vulva not in a cavity, without tubercles on dorsal and ventral side. Underbridge absent. Anus is sub-terminal and has a fenestra of similar size and shape as vulval fenestra. Cysts are circumfenestrate. Bullae may be present or absent. All eggs retained in the body. **Males** are vermiform less than 1.5 mm long. DGO 2.6-4.6 µm. Spicules 31-33µm long and are distally pointed. Cloaca is without a cloacal tubus. **J2** body 0.35-0.49 mm. Stylet is 24-26µm having a conical tail, 63-78µm long. Lateral field with four incisures. Phasmids are punctiform without lens like structure.

Type Species

Punctodera punctata (Thorne, 1928) Mulvey and Stone, 1976

syn. *Heterodera punctata* Thorne, 1928

Other Species

P. chalcoensis Stone, Sosa Moss and Mulvey, 1976

P. matadorensis Mulvey and Stone, 1976.

P. stonei Brzeski, 1998

7.7.7 *Dolichodera* Mulvey and Ebsary, 1980

Female spherical and posteriorly rounded when fully developed. Cuticle is provided with roughly parallel fine ridges. Vulval lips are absent; vulval slit too short for passage of eggs, located more or less in contour of body. Vulva area is without tubercles. Underbridge and vulval denticles are absent. Bullae are present. Anus is subterminal. Cyst is circumfenestrate. All eggs retained in the body. Subcrystalline layer is present.

Second Stage Juveniles

Length= 570± 40 µm. Stylet < 25 µm.Tail=107± 15µm. Body is slender (a=27-31). Lateral fields are provided with four lines. Oesophageal glands fill the body cavity. Tail is conical, elongated with hyaline terminal half. Phasmids are punctiform.

Type and Only Species

Dolichodera fluvialis Mulvey and Ebsary, 1980.

7.7.7 Paradolichodera Sturhan, Wouts and Subbotin, 2007

A new cyst nematode was found parasitising the rush *Eleocharis gracilis* (Cyperaceae) on the banks of a lagoon in Selwyn, Canterbury near Christchurch, New Zealand. The slender body of the second-stage juveniles (a = 61-79) makes the type species, *P. tenuissima* unique among known Heteroderidae species. The females are weakly tanned after death and retain eggs. The cysts are elongate-ovoid with rounded posterior end. A circumfenestral vulva area is in terminal position with an indistinct anus and a cuticle with faint striation anteriorly and punctations posteriorly. Bullae are absent. The male body is not twisted, a cloacal tube is present and phasmids are lacking. Morphologically, the new genus is closest to *Dolichodera* in the subfamily Punctoderinae.

Type and Only Species

Paradolichodorus teneussima

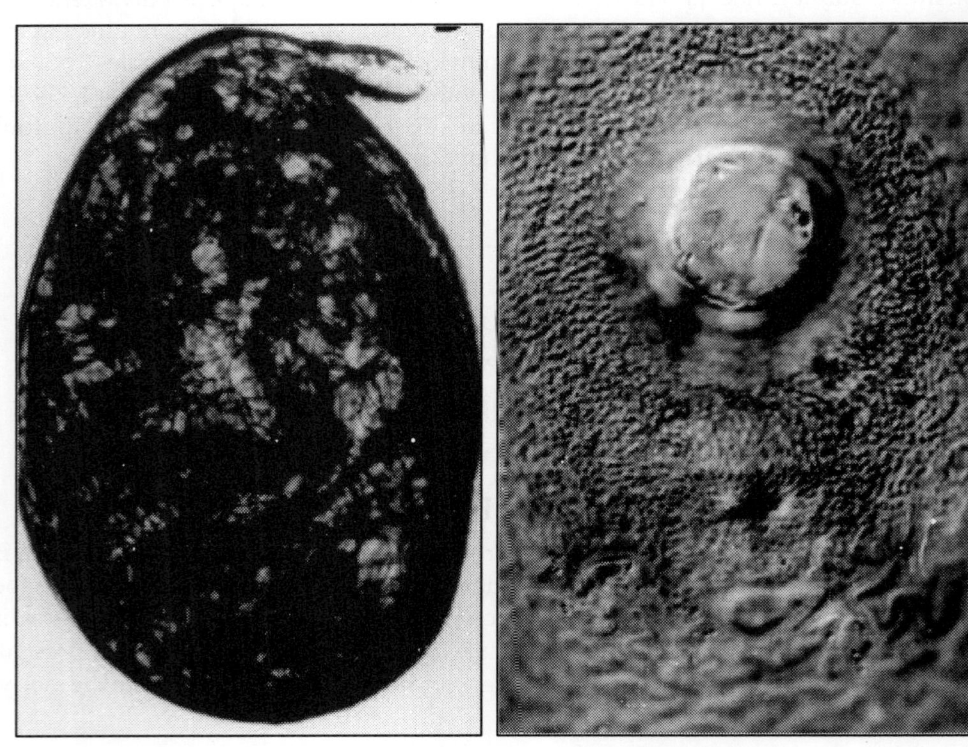

Figure 3: *Paradolichodera* Cyst and Vulval Region

7.7.8 Vittatidera Bernard, Handoo, Powers, Donald and Heinz, 2010

A new genus and species of cyst nematode, *Vittatidera zeaphila*, is described from Tennessee, U.S.A (Latin *Vittatus* means striped and Greek *dera* means skin). The new genus is superficially similar to *Cactodera* but is distinguished from other cyst-forming taxa in having a persistent lateral field in females and cysts, persistent vulval lips covering a circumfenestrate vulva, and subventral gland nuclei of the female contained in a separate small lobe. Infective juveniles (J2) are distinguished from all previously described *Cactodera* species by the short stylet in the second-stage juvenile 14-17μ; J2 of *Cactodera* species have stylets at least 18μ long. The new species also is unusual in that the females produce large egg masses. Known hosts are corn (*Zea mays*) and goosegrass (*Eleusine indica*).

Type and Only Species

Vittatidera zeaphila Bernard, Handoo, Powers, Donald and Heinz, 2010

7.8 Cystoid Nematodes (No Cyst Stage Present)

7.8.1 *Verutus* Esser, 1981

Body is generally saccate, sausage-shaped or kidney-shaped with thick cuticle which is annulated over total body length. D-layer is absent but multiple B-layers is present. Subcrystalline layer is present. Vulva is sub-median, large, with protuberant vulval lips. No phasmids. Remnant of tail very short to absent. Eggs not retained in body, but deposited individually, without gelatinous matrix.

Type Species

V. volvingentis Esser, 1981

Other Species

V. mesoangustus Minagawa, 1986

7.8.2 *Meloidodera* Chitwood, Hannon and Esser, 1956

Body is globose with short neck. Cuticle is of medium thickness, whole body annulated, with modified pattern at anal-vulval region. D-layer is absent. Subcrystalline layer is present. Vulva median; vulval lips not protruding. Eggs are either deposited or retained in body.

Type Species

M. floridensis Chitwood, Hannon and Esser, 1956

Other Species

M. alni Turkina and Chizkov, 1986

M. belli Wouts, 1973

M. charis Hopper, 1960

M. eurytyla Bernard, 1981

M. tianshanica lvanova and Krall, 1985

Species Inquirendae

M. armeniaca Poghossian, 1960

M. sikhotealiniensis Eroshenko, 1978

M. tadzhikistanica Kir'yanova and Ivanova, 1966

7.8.3 *Atalodera* Wouts and Sher, 1971

syn. *Sherodera* Wouts, 1974

Body is globose with projecting neck and terminal cone. Cuticle is thic annulated on fore part of body, with lace-like pattern on the posterior part. D-layer present. No subcrystalline layer. Vulva on prominent terminal cone, with thick cuticl constituted by protruding and developed vulval lips. Anus situated on the side posterior vulval lip. Eggs are retained in body.

Type Species

Atalodera ucri Wouts and Sher,. 1971

Other Species

A. lonicerae (Wouts, 1974) Luc, Taylor and Cadet, 1978

syn. *Sherodera lonicerae* Wouts, 1974

7.8.4 *Cryphodera* Colbran, 1966

syn. *Zelandodera* Wouts, 1973

This cystoid nematode has a globose body with projecting neck and has n terminal cone. Cuticle is thick, annulated on whole body, except the vulval-anal are Vulva is terminal; vulval lips nearly flush with body contour to slightly protrude Anus is located at some distance from vulva. Eggs are retained in body.

Type Species

Cryphodera eucalypti Colbran, 1966

Other Species

C. *coxi* (Wouts, 1973) Luc, Taylor and Cadet, 1978

syn. *Zelandodera coxi* Wouts, 1973

C. *nothophagi* (Wouts, 1973) Luc, Taylor and Cadet, I 978

syn. *Z. nothophagi* Wouts, 1973

C. *podocarpi* (Wouts, 1973) Luc, Taylor and Cadet, 1978

syn. *Z. podocarpi* Wouts, 1973

7.8.5 *Bellodera* Wouts, 1985

Body globose with projecting neck; terminal cone well developed, but posteri ex-tremity somewhat flattened. Cuticle is thick with superficial irregular transver striae and minute pits between striae; D-layer absent; subcrystalline layer presen Excretory pore forwardly situated 27-56 µm from an-terior end. Vulva terminal; vulv slit large, not deeply sunken between hypertrophied vulval lips constituting t terminal cone; no fenestrae; no underbridge; no bullae. Anus situated at some distan from vulva, on posterior side of the cone. Eggs retained in body.

Type and Only Species

Belodera utahensis (Baldwin, Mundo-Ocampo and Othman 1983) Wouts, 1985

syn. *Cryphodera utahensis* Baldwin, Mundo-Ocampo and Othman, 1983

.8.6 *Thecavermiculatus* Robbins, 1978

Body is globose, with prominent neck and no terminal cone. Cuticle is thick, ınulated in fore part or on the major part of body, with lace-like pattern on posterior art; D-layer present. Subcrystalline layer present (very thick in *T. crassicrustatus*). ulva sub-terminal, close to anus; vulval lips non-protruding; anal-vulval region ush with body contour. Eggs retained in female body, together with hatched second-age juveniles.

ype Species

T. gracililancea Robbins, 1978

ther Species

T. andinus Golden, Franco, Jatala and Astogaza, 1983

syn. *Dolichodera andina* (Golden, Franco, Jatala and Astogaza, 1983) Wouts, 985

T. carolynae Robbins, 1986.

T. crassicrustallls Bernard, 1981

.8.7 *Rhizonema Cid del Prado Vera*, Lownsbery and Maggenti, 983

Body is globose with prominent neck and terminal cone. Cuticle is thick, wavy, ntire body annulated, except terminal cone. D-layer is absent. No subcrystalline ıyer. Vulva and anus situated on terminal cone having a thickened cuticle. Phasmids ot observed. Eggs retained in female body together with hatched second stage ıveniles.

ype and Only Species

R. sequoiae Cid del Prado Vera, Lownsbery and Maggenti, 1983

syn. *Sarisodera sequoiae* (Cid del Prado Vera, Lownsbery and Maggenti, 1983) Vouts, 1983.

Thecavermiculatus sequoiae (Cid del Prado Vera, Lownsbery and Maggenti, 1983) iddiqi, 1986

.8.8 *Hylonema* Luc, Taylor and Cadet, 1978

Body is long, ovoid to globose with prominent neck and no terminal cone. Cuticle s thin and annulated only in anterior fore end, posteriorly thick rugose with non riented striae. No subcrystalline layer. Vulva is subterminal, on flattened posterior ody end, vulva lips not protruding. Underbridge present. No phasmids. No remnant f tail. Eggs are not retained in body but deposited individually in a gelatinous natrix.

Type and Only Species

Hylonema ivorense Luc, Taylor and Cadet, 1978

7.8.9 *Sarisodera* Wouts and Sher, 1971

Body is globose, with neck of medium length, posterior end conical-truncated with deep vulval cleft, but with no defined terminal cone. Cuticle is thick, with superficial lace-like pattern, D-layer present. Vulva is terminal, with prominent hypertrophied lips, separated by a deep depression. Anus is in the inner side of the posterior lip. Eggs are retained in body.

Type and Only Species

S. hydrophila Wouts and Sher, 1971

7.9 Cysts Nematodes Species in India

There are 22 cyst nematodes species that are either reported or described from this country. The species which have been described are *H. cajani* Koshy, 1967; *H. delvii* Jairajpuri *et al.*, 1979; *H. mothi* Khan and Hussain, 1965; *H.oryzicola* Rao and Jaiprakash, 1978; *H.raskii* Basnet and Jaiprakash, 1984; *H.skohensis* Kaushal *et al.*, 2000; *H.sorghi* Jain *et al.*,1982; *H. swarupi* Sharma and Siddqui,1992 and *H.zeae* Koshy *et al.*,1970.

7.9.1 *Heterodera* Species in India

1. *Heterodera avenae*
2. *H. cajani*** Koshy, 1967
3. *H. carotae**
4. *H. cyperi*
5. *H. delvii*** Jairajpuri *et al.*, 1979
6. *H. filipjevi*
7. *H. galeopsidis**
8. *H. gambiensis*
9. *H. glycines*
10. *H. graminis*
11. *H. iri*
12. *H. indica* (under publication)
13. *H. mothi*** Khan and Hussain, 1965
14. *H. oryzicola*** Rao and Jayaprakash, 1978
15. *H. raskii*** Basnet and Jayaprakash, 1984
16. *H. sacchari*
17. *H. skohensis*** Kaushal *et al.*, 2000
18. *H. sorghi* ** Jain *et al.*, 1982
19. *H. spinicauda*
20. *H. swarupi* ** Sharma and Siddiqi, 1992
21. *H. trifolii*

22. *H. zeae* ** Koshy *et al.*, 1971
   **: Species described from India
   *: Species of doubtful presence

7.9.1.1 The Cowpea Cyst Nematode, *Heterodera cajani* Koshy, 1967

Swarup *et al.* (1964) reported the presence of *H. trifolii* from the IARI fields infesting pulse crops but detailed studies on cyst and larval characters by Koshy (1967) revealed it to be a new species, *Heterodera cajani* from the same locality on pigeon pea. Almost at the same time, Edward and Misra (1968) recorded the occurrence of another species, *H.vigni* infesting cowpea in Allahabad (U.P.). This species was, however, later on, synonymized with *H.cajani* (Kalha and Edward, 1979). It differs from *H. trifolii* in not having a Mulvey's bridge and the 'sheaf' like organ in the underbridge. The presence of a perpendicular second underbridge (with bifurcating one side) in *H.glycines* differentiates it from *H. cajani*. The species has also been reported on cowpea in Egypt (Aboul-Eid and Ghorab, 1974). Its presence in Egypt shows that it could be established in tropical, subtropical and even Mediterranean countries.

The pigeon pea cyst nematode, *H.cajani* is now known to be present in seventeen states of India namely, J&K, H.P., Punjab, Haryana, Rajasthan, Gujarat, Delhi, U.P., Bihar, Jharkhand, West Bengal, M.P., Chhatisgarh, Maharashtra, Tamil Nadu, Karnataka and Andhra Pradesh (Kaushal *et al.*, 2007). Among cyst nematode species, it causes enormous losses in pulse crops almost throughout the country due to its widespread distribution. It is specially an important constraint in the production of chickpea in the state of Andhra Pradesh. In north and western India, it is more severe on cowpea and pigeon pea. It is wide spread in sandy-loam soils in northern India and in black cotton soils in western and southern India (Reddy *et al.*, 1993; Varaprasad *et al.*, 1997).

On the basis of field observations and experiments by various authors (Koshy, 1967; Koshy and Swarup, 1972; Bhatti and Gupta, 1973; Janarthanan, 1972; Verma and Yadav, 1975; Gaur and Singh, 1977) the following host plants have been established: *Cajanus cajan, Cicer arietinum, Cyamopsis tetra- gonoloba, Dolichos biflorus, D. lablab, Glycine max, Phaseolus atropurpureus, P. aureus, P. calcaratus, P. lathyroides, P. lunatus, P. mungo, P. vulgaris, Pisum sativum, Sesamum indicum, S. orientale, Vivia narabenesis* cv. *sativa*, Vigna *sinensis* and *Zea mays*. With the exception of the last plant species, all these host plants belong to Leguminosae or Pedaliaceae.

Usually, root diffusates are not required for hatching of juveniles from the old cysts but their exposure to almost a constant temperature of 28°C facilitates good hatching. Hatching from the egg-masses is profuse and immediate but from older cysts, it is very temperature specific and takes place at 28°C only. The life cycle of this nematode is completed in about 20-22 days at 25-30°C. More than nine generations per year have been reported (Koshy and Swarup, 1971). The cyst usually has 2-3 times bigger eggmass in which majority of the eggs are laid. Only one-third of the eggs are retained in the cysts. A single female may lay about 250- 300 eggs.

Walia and Bajaj (1988) reported the presence of two races, pigeonpea (PPR race A) and cluster bean races (CBR race B) of *H. cajani* on the basis of their reproduction

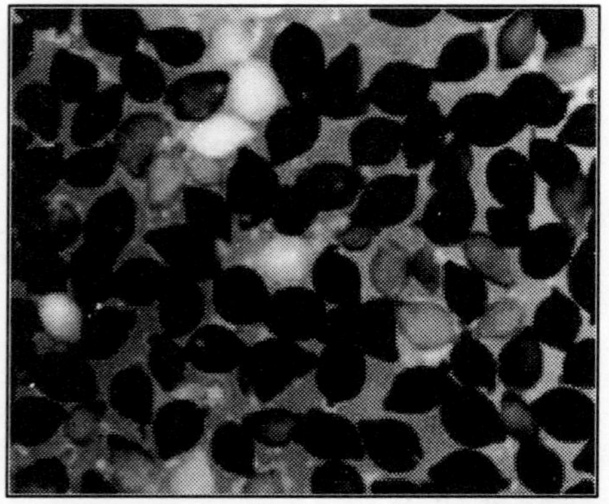

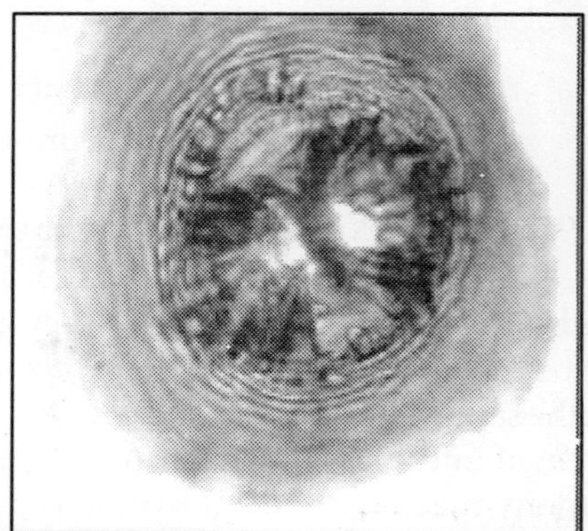

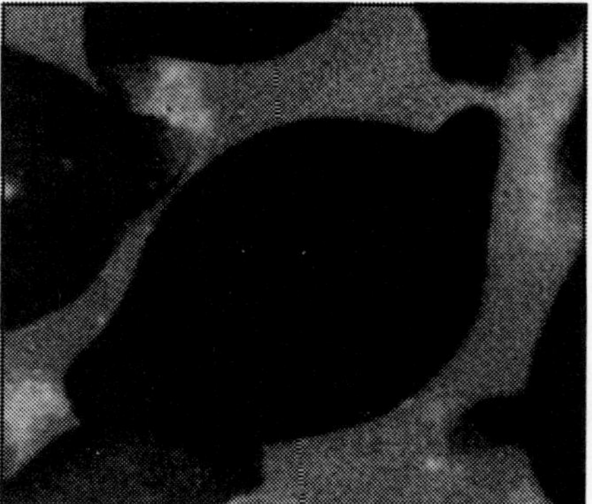

Figure 4: The Original Description of the Species has been Supplemented by Koshy *et al.* (1971)

on these hosts for seven populations originating from Ambala (Punjab), Faridabad (Haryana), Bhiwani (Haryana), Ludhaina (Punjab), Delhi (Delhi), Coimbatore (Tamil Nadu), and Hyderabad (Andhra Pradesh). These two races were differentiated in vulval cone structure and male morphology also. However, the mean values for these characters were overlapping and well within the range of the species. Biochemically these two races showed variation in expressing different MDH and esterase isozymes mobility patterns. Two distinctive bands with Rf 0.58 and 0.633 were found associated with race B but not with race A. Six bands of MDH activity were detected in the two races with relative migration rates of 0.25, 0.269.and 0.294 and 0.117, 0.294, 0.318 and 0.353 for race A and B respectively, with one band at 0.294 common with both the races. Use of the host differentials of cowpea, mungbean, soybean and pigeonpea accessions confirmed the presence of races of *H. cajani* (Mehta and Bajaj, 2005). A second study discriminated three races among 14 populations from seven districts of Uttar Pradesh (Siddiqi and Mahmood 1995).

Diagnostic Features

Cysts are lemon- shaped dark brown to black in colour, narrow towards posterior half and have ambifenestrate type of fenestration. Vulval protuberance is prominent, strong underbridge and scattered finger-like bullae around basin but clustered on

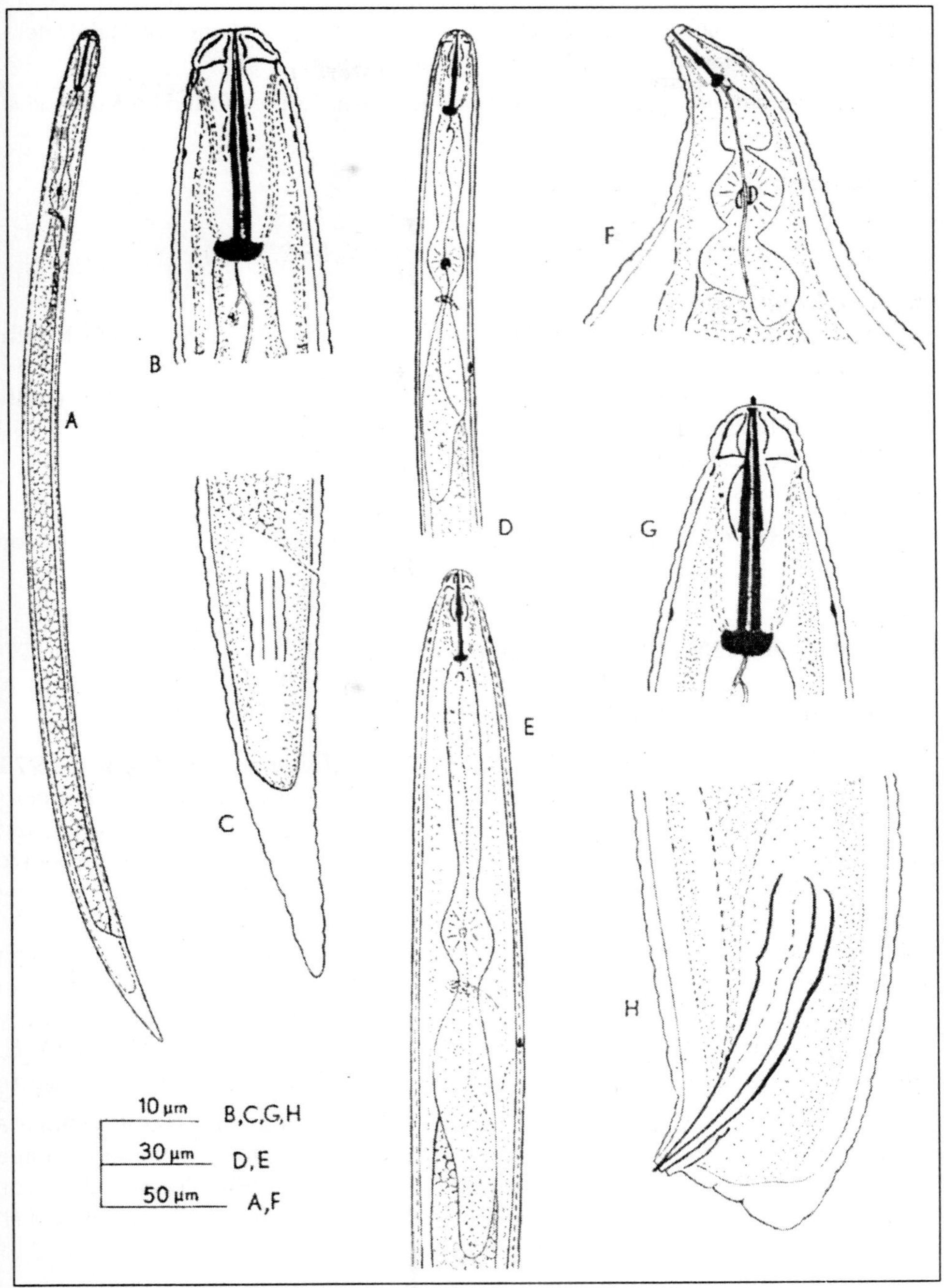

Figure 5: ***Heterodera cajani***
A-D: J2. A: Entire body; B: Anterior region; C: Tail; D: Pharyngeal region; E, G, H: Male. E: Pharyngeal region; G: Anterior region; H: Tail; F: Anterior Region. After Koshi *et al.,* 1971

fenestral sides, is present. Second stage juveniles (400µ) have well developed stylet with anteriorly directed knobs. DEGO is 3-5 µ behind the stylet knobs and the lateral field is with four lateral lines. Males (1050µ) are present.

The original description of the species has been supplemented by Koshy *et al.* (1971).

Measurements

Eggs

Length: 84-109 (95) µ; Width: 34-50 (43) µ; L/W: 1.8-2.7 (2.2).

Second Stage Larvae

Body length: 345-515 (435)µ; Maximum body width- 15-20 (18)µ ; Stylet length: 23-27 (26)u; Anterior end –medium bulb: 56-68 (64) µ ; AE- excretory pore: 73-110 (86) µ ; AE- esophageal glands: 110-154 (135) µ ; Anal body width: 7-11 (9.5) µ; Tail length: 32-52 (45)µ; Hyaline tail length: 17-30 (24)µ; 'a' value: 18.3-28. (23.9); 'b' value: 3.2-5.3 (4.1); 'b2' value: 2.5-3.7 (3.1); 'c' value: 8.0-11.7 (9.6): 'c2': 2.9-5.1 (4.1).

Females

Length: 480-640 (569) µ; Width: 285-480 (378) µ; L/W; 1.3-1.9 (1.5); Cone top height: 43-88 (63) µ; Stylet length: 25-28 (26) µ.

Males

Length: 780-1280 (1090) µ; Maximum body width: 20-26 (24) µ; Stylet length: 27-30 (28) µ; AE-esophageal glands: 142-163 (157) µ; Spicule length: 31-41 (34) µ; 'a' value: 31.6-51.4 (41.1); 'b'value; 5.8-11.8 (7.6); 'b2' value: 5.0-9.0 (6.2).

7.9.1.2 The Turf Grass Cyst Nematode, *Heterodera graminis* Stynes, 1971

The turf grass cyst nematode was described by Stynes (1971) from Hunter's valley at New Castle in N.S.W., Australia. In India, it was reported from Delhi and Jaipur by Sharma *et al.* (1984) infesting lawn grass. Now, it is known to be present in Delhi, H.P., Rajasthan and Haryana. It is really a problematic nematode causing great damage in the lawn grass especially in the 'greens' area in golf grounds. The big, dry water soaked patches can easily be seen as the nematode hampers the growth of the grass which not only spoils the beauty of the 'greens' but also interfere in smooth play of the game. It has been found in soil at Vanua Levu, Fiji Islands (Kirby *et al.*, 1978) and on *Cynodon dactylon* in a sugar cane plantation in Trinidad. *H. graminis* is of the 'goettingiana group', and appears to be closely related to *H. cyperi.*

H. cardiolata (Kiryanova and Ivanona, 1969) cysts are reported to have a 'heart'shape. Some cysts of *H. graminis* were also observed to have the same shape but with the characters of the genus. Similarly, *H.cynodontis* (Shahina and Maqbool, 1989) described from Pakistan has the similar morphological characters as that of *H. graminis*. They may all be synonym species.

Second stage juveniles from the cyst hatch continuously throughout the year. But the nematode reproduces well in the months of March-April and September-October under Indian conditions. The life cycle is completed in about 22-24 days at 25-30°C. Sometime a small egg-sac may be seen attached with the female or young cyst having few eggs. Males are present.

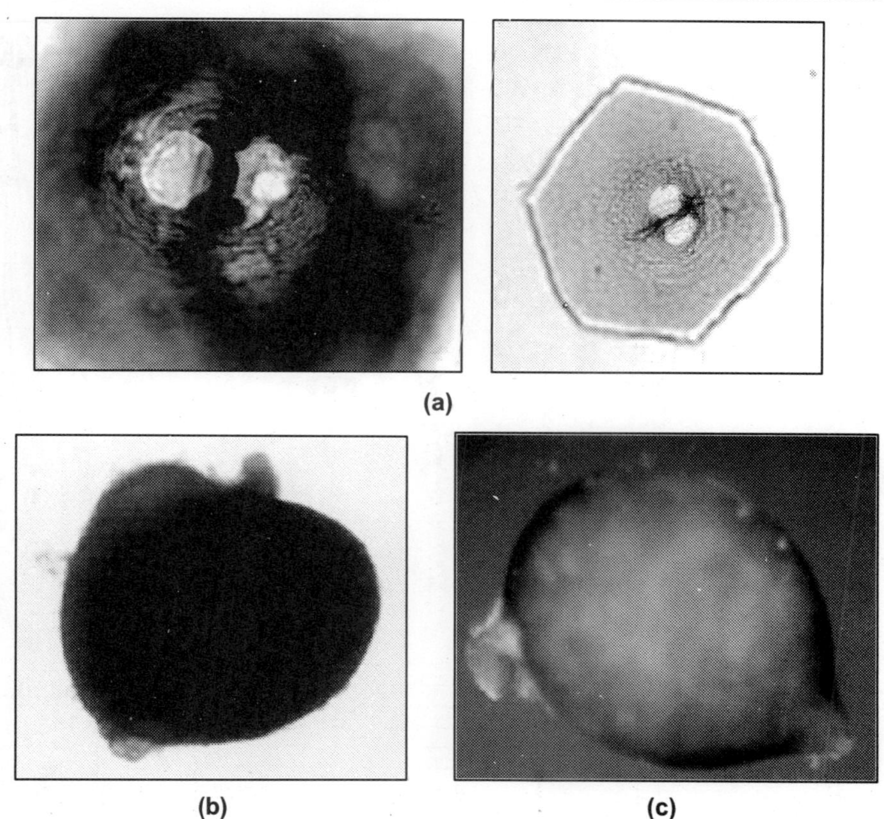

Figure 6: *H. graminis*
(a) Cone tops, (b) and (c) Cysts

Diagnostic Characters

The small, dark brown cysts are characterized by moderate vulval-cone protuberance, ambifenestration, moderately weak underbridge and are provided with diagnostic character of having swollen vulval ends (vulval knobs) at both the sides. Many cysts are found having such a 'heart'shape. The second stage juveniles (380µ) have well sclerotized head. DEGO is 4-6µ behind the stylet knobs. Lateral field is four lateral lines. Tail terminus is bluntly pointed. Males (850µ) are present.

Measurements

Second Stage Juveniles

Length: 343-444 (391) µ; Maximum body width; 14-21 (17) µ; AE- excretory pore: 76-120 (96) µ; AE- esophageal glands: 117-172 (143) µ; Anal body width: 8-12 (11.5)µ; Tail length: 44-65 (56) µ; Hyaline tail length: 23-36 (29) µ; a: 18.27 (22.4) µ; b : 2.3-3.3 (2.7) µ; b2: 2.5-3.4 (2.6);'c': 6.1-9.9 (7.1), c2: 3.8-5.9 (4.8).

Eggs

Length: 88-116 (99) µ; Width: 34-51(40) µ; L/W: 1.9-3.2 (2.6).

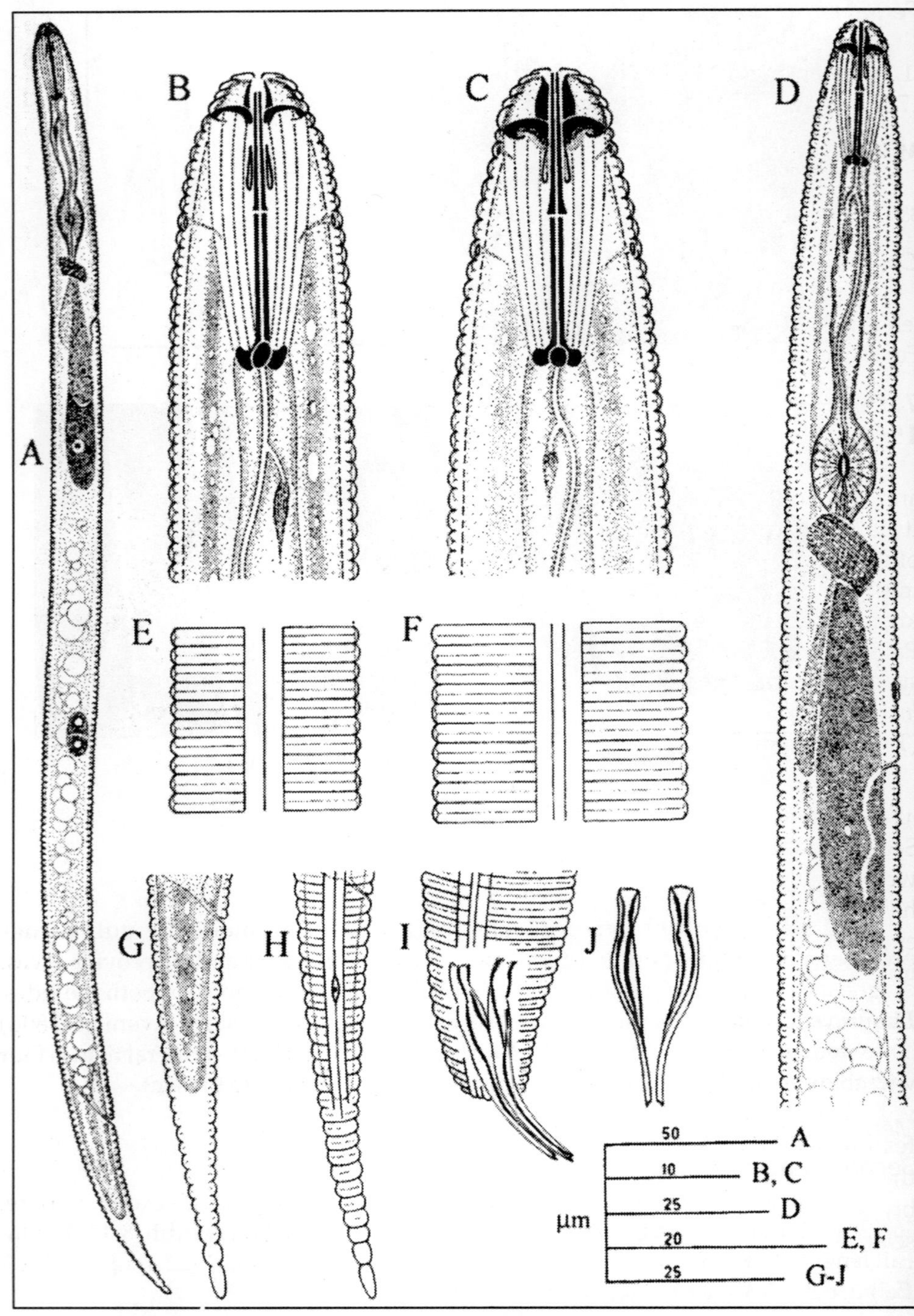

Figure 7: ***Heterodera graminis*.**

A. N. E. G, H: J2. A: Entire body; B: Anterior region; E: Lateral Field. G, H: Tails, C, D, F, I J: Male. C: Anterior region, D: Pharyngeal region; F: Lateral field; I: Tail, lateral, J: Spicules, ventral. After Stynes, 1971.

Females

Length (excluding neck): 312-636 (466) µ; Width: 1995-545 (356) µ; L/W: 1.1-1.7 (1.3); Stylet length; 19-24 (21) µ; Median bulb length: 23-46 (33) µ; Median bulb width: 17.39 (28) µ.

Males

Length: 571-924 (779) µ; Maximum body width: 17-29 (24) µ; Spicule length: 31-35 (33) µ; Gubernaculum length: 9-12 (11) µ; a value: 27-40 (33); b¢ value: 4.5-6.3 (5.2).

Cyst

Length: 381-927(561) µ; Width: 297-701 (441) µ; L/W: 1.1-1.7 (1.3); Underbridge length: 70-107(84) µ; Underbridge width: 4-10(6) µ; Fenestral length: 32-63 (50) µ; Fenestral width: 24-43 (32) µ; Vulval slit length: 27-46(38) µ.

7.9.1.3 The 'Nut Grass' Cyst Nematode, *Heterodera mothi* Khan and Husain, 1965

This is the first species of cyst nematodes described from the country by Khan and Husain (1965) infesting nut-grass, *Cyperus rotundus* in Aligarh (Uttar Pradesh). It is known to be present in almost whole of northwestern India including the states of J&K, H.P., Punjab, Haryana, Delhi and Uttaranchal. It is not of much economic importance except that it can be utilized in managing this weed in smaller plots and kitchen gardens where this is a perennial menace. It has also been reported from Iran, Pakistan and U.S.A. Other host records are from *Chrysanthemum carinatum* and *Hibiscus syriacus* in India (Yaqoob and Khan, 1969); *Cyperus esculentus, Glycine max, Gossypium hirsutum*, in Georgia, USA (Minton *et al.*, 1973); in soil from sugar-beet fields in Iran (Talatschian and Achyani, 1976) and sugarcane (Maqbool, 1981).

The life cycle of this nematode is completed in about 24 days at 25-30°C. The juvenile penetrated the rootwithin 48 hrs without any preference for the root tissue. Second moult could be seen afer 96 hrs and third after 6 days of inoculation. Males were out after 8th day and females were seen only after 14th day. The young cysts may have egg sacs containing 100 eggs sometime and the cyst may have about 150-200 eggs but they get detached soon afterwards. Emergence of juveniles from the cysts is immediate.

Diagnostic Characters

Cysts are oblong having larger protuberance of the vulval cone; light brown with thin cuticle and characterized by the presence of a moderate underbridge, slender vulval bridge and few peripheral, round bullae around basin. Another characteristic feature of this species is that in majority of the cysts, the fenestral length is less than the fenestral width. The vulval slit, vulval bridge and underbridge are often found broken in the older cysts. The second stage juveniles are about 450µ long with a well sclerotized head. Stylet knobs are round. Tail is sharper in this species.

Females

Length (excluding neck): 460-690 (565) µ; Width: 110-350 (270) µ; Cone top height: 45-84 (56) µ; Stylet length 17-21 (19) µ; Female L/W: 1.5-2.4 (1.9).

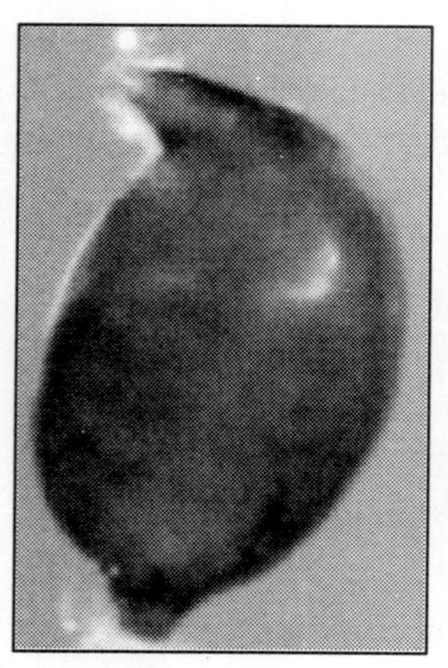

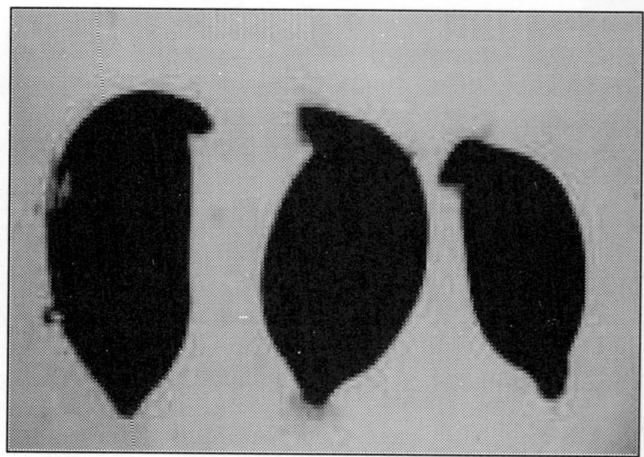

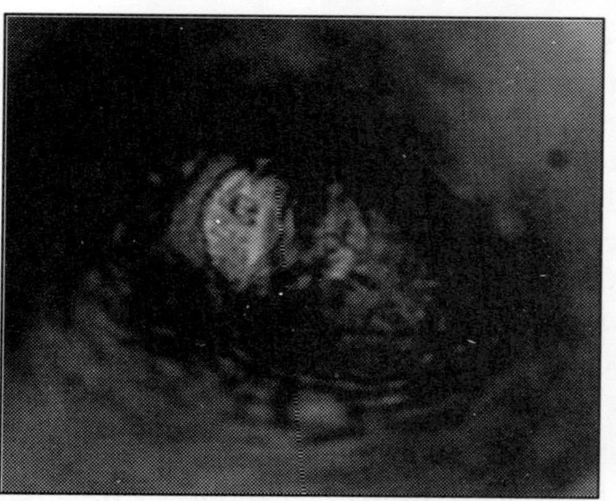

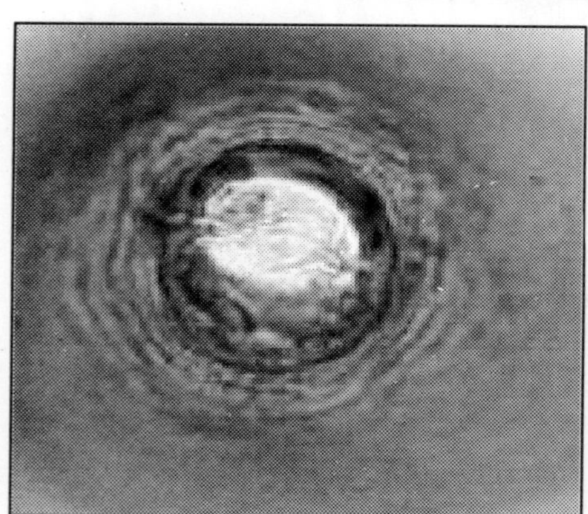

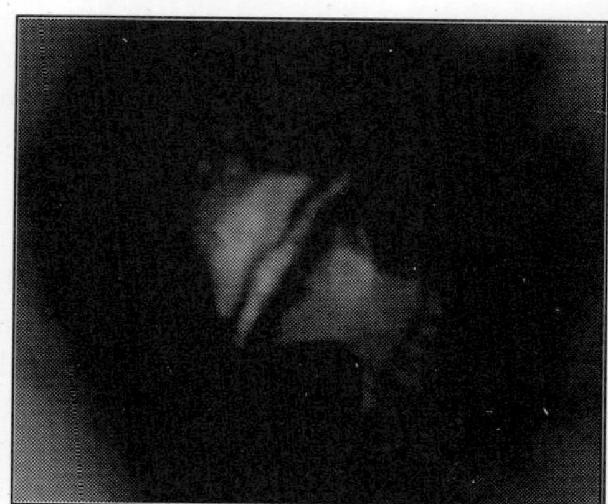

Figure 8: Cysts and Cone Tops of *H. mothi*

Cysts

Length: 470-790 (630) μ; Width: 230-280(292) μ; Cyst L/W: 1.5-2.6 (2.1); Fenestral length: 28-48 (37) μ; Fenestral width: 21-34 (27) μ; Underbridge length: 52-83(65) μ; Underbridge width: 11-18 (15) μ; Depth of underbridge: 16-25 (22) μ.

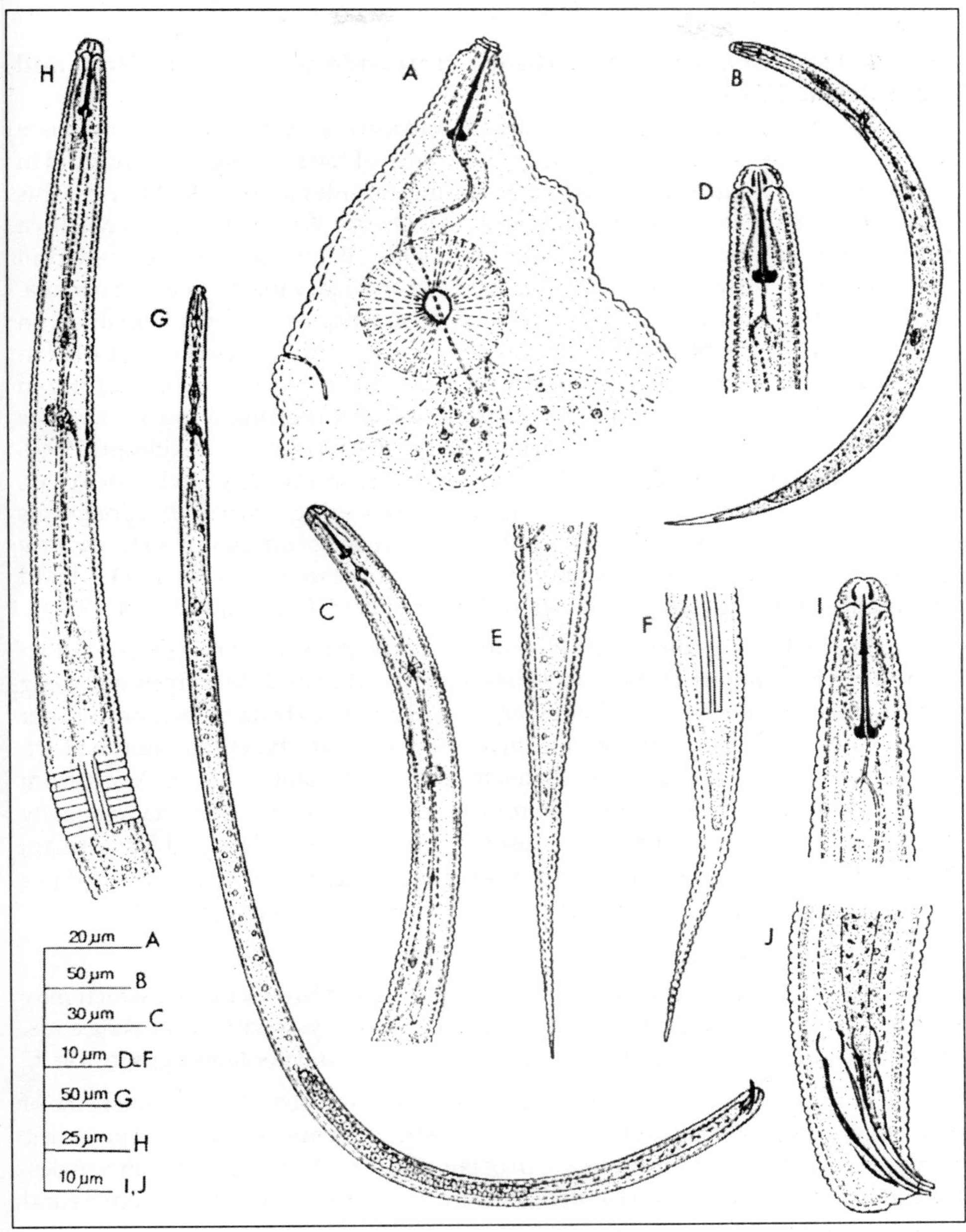

Figure 9: ***Heterodera mothi*. Female.**
A: Anterior region; J2. B: Entire body, C: Pharyngeal region, D: Anterior region, E. F: Tails, Male; G: Entire body, H: Pharyngeal region; I: Anterior region, J: I: Tail. After Shahina and Maqboo, 1991.

H. indocyperi

Husain and Khan in Khan *et al.*, 1964 (cited by Hesling, 1978) is considered to be a *nomen nudum* and a synonym of *H. mothi*.

7.9.1.4 The Rice Cyst Nematode, *Heterodera oryzicola* Rao and Jayaprakash, 1978

Rao and Jayaprakash (1978) described *Heterodera oryzicola* from the roots of rice in Pattambi area of Kerela state though appearance of the nematode was noticed in 1976 during kharif season and was found causing considerable loss to the crop. This nematode is a parasite of upland rice in Kerala, Orissa, West Bengal, and Madhya Pradesh states of India (Rao *et al.*, 1986). Subsequently, it has also been reported from Karnataka, Goa (Prasad, 2002), Madhya Pradesh and Haryana on rice and banana. Charles and Venkitesan (1984) reported the occurrence of this cyst nematode on banana (*Musa* sp.) cv. Nendran from the state of Kerala, India. Before its isolation from banana, this nematode was considered to be one of the major pests of rice in Kerala (Usha, 1980). Other hosts are *Cynodon dactylon* (Bermuda grass), *Kyllinga monocephala* and *Urocloa decumbens* (signal grass) (Charles and Venkitesan, 1990). The nematode completes 12 generations in a year on continuous growing rice crop. Browning and chlorosis of leaves, stunting, early flowering, partial filling of grains and seedling mortality in patches are commonly associated symptoms with *H. oryzicola* infestation (Rao and Jayaprakash, 1977; Jayaprakash and Rao, 1984). Field experiments have depicted yield losses of 21-42 per cent (Usha and Kurien, 1981).

The life cycle of this nematode is completed in 30 days in the rice season but on banana it has been reported to complete its life cycle earlier in 23 days from egg to egg (Charles and Venkitesan, 1995). The emergence of juveniles from cysts is only 50 per cent in water, which may show that the cysts do require hatching stimulant. *H. oryzicola* is dependent on root diffusates to induce substantial hatch. Apart from cysts produced after 30 days, diffusates from banana and rice stimulated hatch equally well; banana has been recorded as a host for *H. oryzicola* (Charles and Venkitesan, 1984). Males are present in abundance. The average number of eggs produced per individual adult female is 200 in the cyst body and about 100 in the egg sac.

Diagnostic Characters

Cysts are ambifenestrate with peripheral, small and brown bullae, which may be few in numbers. A weak underbridge is present in the cysts with bifurcating ends. Vulval slit is quite long (39µ). Gelatin matrix is present and contains eggs inside it.

Morphlogically it is very close to *H. oryzae* but differs from it by smaller size of cysts, length and width of fenestrae, short stylet length in females, males and juveniles, shorter spicular length and hyaline length of tail. Stylet knobs are spherical in females, males and juveniles while in *H. oryzae* they are anteriorly concave in juveniles and posteriorly sloping in females and males.

Measurements

Eggs

Length: 95-104 (101) µ; Width: 39-45 (42) µ; L/W: 1.7-2.6 (2.1).

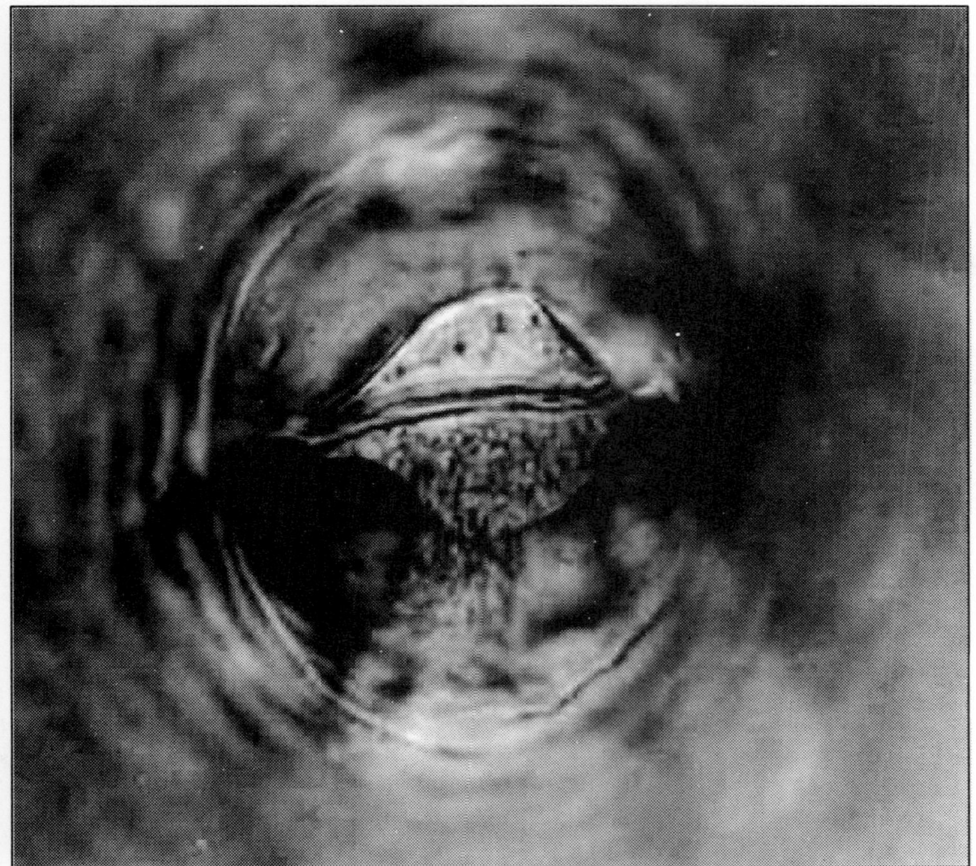

Figure 10: Vulval Cone Top of *H. oryzicola*

Second Stage Juveniles

Length: 370-428 (392) µ; Max. Body width: 18-22 (20) µ; Stylet length: 17-19 (18) µ; Ant. end- excretory pore: 76-102 (87) µ; Anal body width: 7-13 (10) µ; Tail length: 50-60 55) µ; Hyaline tail; 22-2 (28) µ; a-value: 19-20(19.4); b-value: 3.5-4.6(4.2); b2-value: 2.34-3.3 (2.9); c-value: 5.2-8.7 (6.8); c2-value: 4.2-8.7 (6.1)..

Females

Length: 414-520 (484) µ; Width: 280-348 (314) µ; Neck length: 50-96 (82) µ; Stylet length: 18-20 (20) µ; L/w ratio: 1.2-1.7 (1.3).

Males

Length: 896-980(934) µ; Width: 22-34 (27) µ; Stylet length: 20-30 (22) µ; Spicule length: 23-32 (27) µ; Ant. end-excretory pore: 104-120 (116) µ.

Cysts

Length: 420-530 (475) µ; Width: 280-484 (372) µ; Vulval slit length: 36-47 (39) µ; Fenestral length: 27-40 (32) µ; Fenestral width: 20-39 (30) µ; Vulval bridge width: 5-6

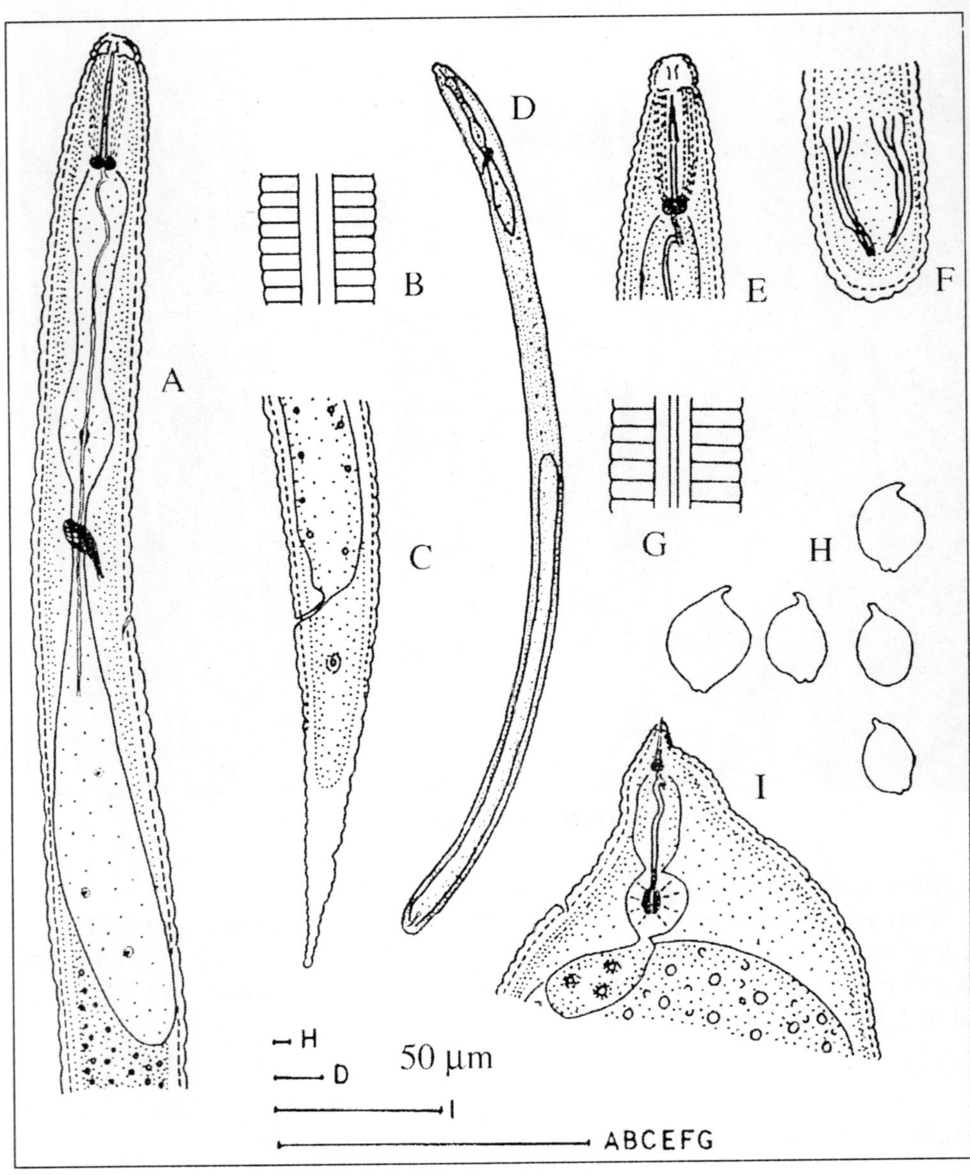

Figure 11: ***Heterodera oryzicola.***
J2. A: Anterior end, B: Lateral field, C: Tail, Male; D: Entire body, E: Anterior region, F: Tail, G: Laterial field, H. Cysts, I: Anterior region of female. After Rao and Jayaprakas, 1978.

(5.5) μ; Underbridge length: 84-112 (105) μ; Underbridge width: 6-20 (12) μ; L/B: 1.2-1.8 (1.5). AE-excretory pore: 104-125(114) μ; Spicule length-19-25 (22) μ; Head length/width: 1.0-1.5 (1.2); a-value: 32.0-35.0 (33.4); b-value: 7-9 (7.9). b′-value: 6-8 (6.5); c-value:7.3 (6.4-8.8).

7.9.1.5 The Sorghum Cyst Nematode, Heterodera sorghi Jain, Sethi, Swarup and Srivastava, 1982

The sorghum cyst nematode was described from Raispur village in Ghaziabad, U.P. (Jain *et al.*, 1982) from the roots of sorghum. This cyst species is very similar to *H. gambiensis* described by Merny and Netscher (1976) from Gambia, also infesting sorghum.

Although it is described from sorghum but maize is the preferred host for this nematode (Srivastava, 1990). It has also been reported from Pakistan. In India, it is known to be present in the states of J&K, H.P., Punjab, Haryana, Delhi, Rajasthan, Uttar Pradesh, M.P., A.P. and Maharastra (Kaushal *et al.*, 2007). It is quite severe in some parts of rain fed areas of H.P. and J&K. Few empty cysts were also isolated from soil samples collected from Orissa (Kaushal, 2004). Damage to maize or sorghum crop occurs when the initial inoculum in the field is quite high. This initial attack to the crop usually manifested in lower yield. The life cycle of this nematode is completed in 22-24 days with 4-5 generations during a crop season are common. The

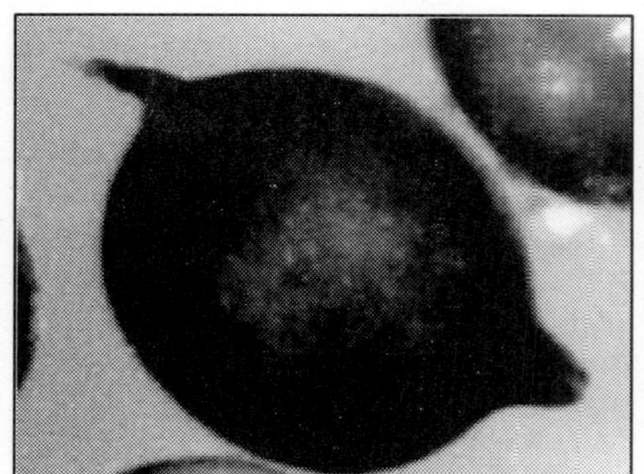

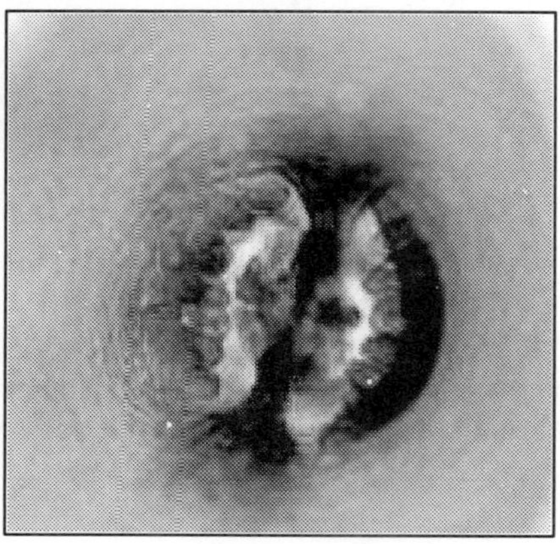

Figure 12: Cysts and Cone Tops of *H. sorghi*

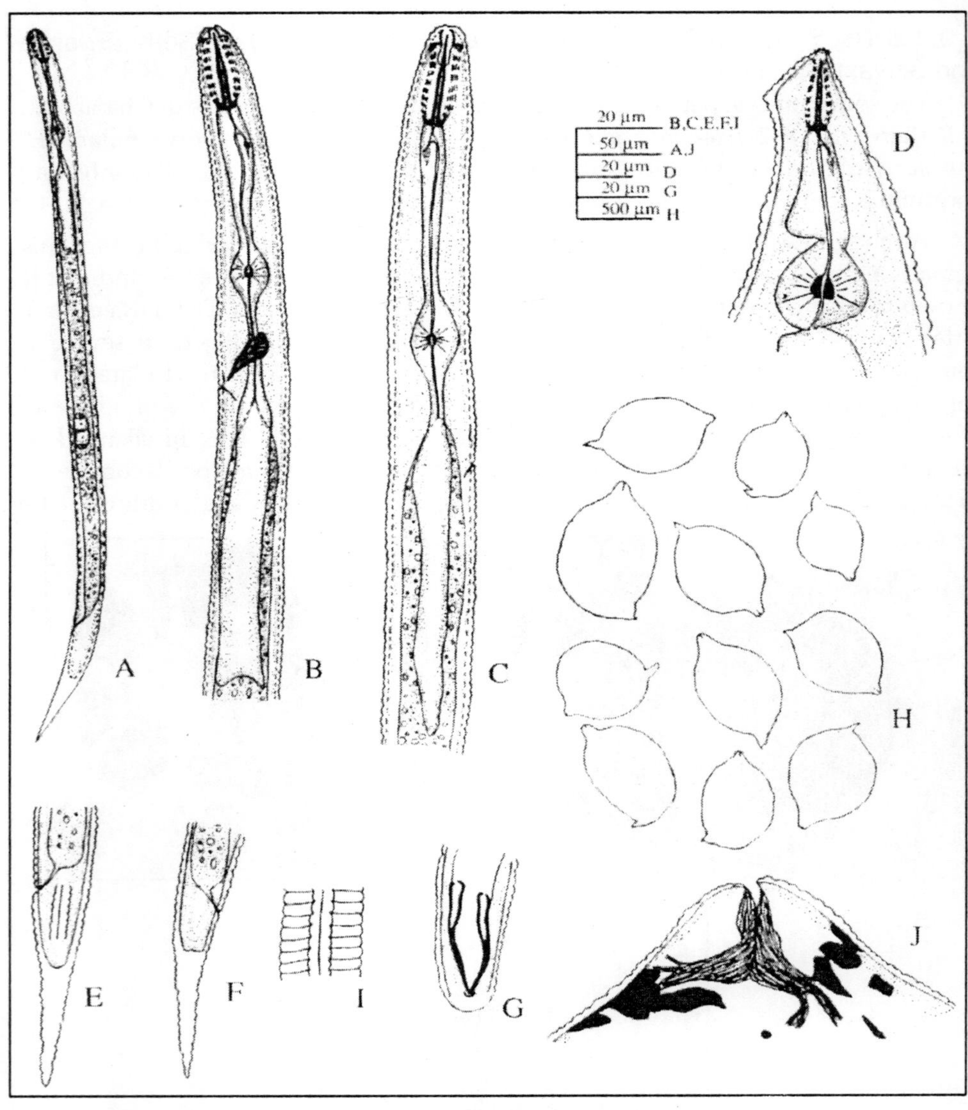

Figure 13: ***Heterodera sorghi.***
A: Entire body, of J2, B: Pharyngeal region of J2. C: Anterior region of male; D: Anterior region of female; E, F: Tails of J2, G: Tail of male; H: Cysts, I: Lateral field of male; J: Vulval cone. After Jain *et al.*, 1982.

development and reproduction of the nematode best takes place at temperature range of 25-30° C. A small egg-sac may form in the mature female which may contain few eggs. Hatching of juveniles from the cysts or egg-masses is immediate. A cyst usually contains about 250-300 eggs and juveniles in it.

Diagnostic Characters

H. sorghi characterized by the presence of dark brown cysts with a pointed, narrow neck, having ambifenestrate fenestration, large and prominent underbridge with furcating ends and peripheral but small number of roundish or thumb shaped bullae. The posterior protuberance is quite prominent, sharp and pointed. The second stage juveniles are about 450μ long. The lateral field consists of three lines only. The females may have round or posteriorly sloping stylet knobs. Males are present.

Measurements

Eggs

Length: 87-115 (109) μ; Width: 32-48 (40) μ; L/W: 2.3-3.2 (2.8).

Second Stage Larvae

Length: 420-525 (461) μ; Max. body width: 16-20 (18) μ; Stylet length: 20-22 (21) μ; AE-excretory pore: 78-108 (88) μ; Anal body width: 11-15 (12) μ; Tail length; 42-60 (52) μ; Hyaline tail length: 28-35 (31) μ; a-value: 20-26 (23); b-value: 8-12 (9); b'-value: 2.6-2.9 (3.1); c-value: 7.5-9.8 (8.8); c'-value: 3.2-5.9 (4.3).

Females

Length: 500-980 (730) μ; Width: 300-580 (483) μ; Cone top height: 40-100 (75) μ; Neck length: 90-190 (118) μ; Stylet length: 24-26 (25) μ; Female L/W: 1.6-1.9 (1.7).

Cysts

Length: 550-910 (739) μ; Width: 400-600 (492) μ; Vulval slit length: 28-39 (32) μ; Fenestral length: 40-56 (47) μ; Fenestral width: 32-56 (38) μ; Underbridge length: 70-115 (79) μ; Underbridge width: 28-40 (33) μ; Depth of underbridge: 58-70 (61) μ; Cyst L/w: 1.2-1.8 (1.4).

7.9.1.6 The Maize Cyst Nematode, *Heterodera zeae* Koshy, Swarup and Sethi, 1971

The maize cyst nematode *Heterodera zeae* of the 'schachtii' group was first detected in Chapli village of Udaipur district (Rajasthan) infesting maize and described by Koshy, Swarup and Sethi (1970) as a new species. Subsequently, Golden and Mulvey (1983) redescribed females, cysts and juvenile specimens from Maryland, USA and India, using scanning electron microscopy. It is now known to be present in almost all the maize growing states of India barring few eastern and southern states. The distribution of this nematode has been recorded in the states of J. and K., H.P., Punjab, Haryana, Delhi, Rajasthan, U.P., Bihar, Uttaranchal, Jharkhand, M.P., Chhatisgarh, Gujarat, Maharashtra and Karnataka causing good amount of damage to the maize crop. Outside India, it has been reported from Nile valley in Egypt (Aboul and Ghorab, 1982), Pakistan (Maqbool, 1981) and U.S.A (Sardanelli *et al.*, 1981; Ringer *et al.*, 1987; Eisenback *et al.*, 1993), Thailand (Chinnasri *et al.*, 1995), Nepal (Sharma *et al.*, 2001) and Portugal (Correia and Abrantes, 2005). Besides maize, other plants like barley, wheat, oat, rice, teosinte, sugarcane and millets are good hosts of this nematode (Srivastava and Swarup, 1977; Ringer *et al.*, 1987). Almond (Qasim and Ghaffar, 1986), Tomato (Shahzad and Ghaffar, 1986) *Capsicum annum*, *Corchorus capsularis* and *Raphanus sativus* from Pakistan are reported to be the hosts (around the rhizoshere and not from the roots of these plants) (Maqbool and Hashmi, 1984) and vetiver (Lal and Mathur, 1982).

The nematode has a short life cycle (15-17days) under warm conditions of 28-30°C and may complete 6-7 generations during the crop's growing season. It reproduces well in coarse-textured soil than in finer texture. The presence of this nematode in Portugal (Europe) suggests its adaptation to temperate climate though in India it has occasionally been found infecting wheat and winter maize.

Six populations of *H. zeae* collected from Himachal Pradesh (2), Delhi, Bihar, Punjab and Haryana when analyzed by PCR-RFLP of Internal transcribed spacers (ITS 1 and 2) of rDNA for detecting intraspecies variation showed that Samastipur population which is known to be a more virulent strain differed from all the other populations in having a distinct pattern with five of the eight enzymes studied. The UPGMA dendrogram also revealed that the Samstipur and Delhi populations were more variable and distinct compared to the other four populations (Uma *et al.*, 2008).

Three races of *H. zeae* have been reported from Haryana using maize and vetiver as differentials (Bajaj and Gupta, 1994). Hisar population was found to multiply both on maize and vetiver, whereas Ambala and Sonipat population multiplied only on maize and vetiver, respectively. Three *H. zeae* biotypes have also been reported from Egypt (Khair *et al.*, 1989). Ringer *et al.* (1987) reported that population of *H. zeae* from USA, India and Egypt differed in their ability to reproduce on certain hosts. Srivastava and Sethi (1984) compared populations from Pusa Bihar, Delhi and Udaipur in Rajastan for their virulence and ability to reproduce and multiply on different cultivars of maize and found that they varied significantly in their ability to reproduce and multiply on these hosts. Correia and Abrantes (2005) studied three populations of *H. zeae* from Portugal and found that the three populations were morphologically identical with similar mean values for cyst L/W ratio, and underbridge and stylet lengths. However, the S. Facundo population showed slight differences in some cyst characters, such as the bigger fenestrae and greater vulva slit lengths, and in generally smaller J2 measurements.

Diagnostic Characters

Female

Adult females are swollen, lemon-shaped and pearly-white in colour with a distinctly protruded vulva, a thin cuticle which is thickest at the neck-body junction, with a zigzag external pattern anteriorly, but appearing wavy around the vulval cone with broken transverse lines interspersed with longitudinal lines. The head has two annules, the second larger than the first. The stylet is slender with fairly developed anteriorly flattened knobs. It has a long neck and a large rounded median bulb, with well-developed valve plates. The excretory pore is placed posterior to the head tip and is at the level of oesophageal gland lobe. The oesophageal gland is in a single lobe and is variable in size and shape. Two ovaries are present. The mature females are filled with eggs in various stages of development. The vulval cone is well developed with a fairly long vulval slit. The anus is very small, indistinct and placed close to the vulva. An egg sac is present which is generally small. Most of the eggs are retained in the body and only a few in the egg sac.

Cyst

The cyst is small almost spherical, lemon-shaped, with protruding neck and vulva. It is light-brown in colour, thin walled (eggs and juveniles can be seen from outside) and without a subcrystalline layer. The cyst wall pattern on the body proper is in the form of a zig-zag pattern. The vulval cone is prominent. The end-on view of the vulval cone shows concentric lines of cuticular ridges around the vulval slit and fenestra. Fenestrae are ambifenestrate with the two semifenestrae separated by a vulval bridge and surrounded by a wide 'basin'. The anus is indistinct. Bullae are present, the characteristic arrangement of four finger-like cross bullae in a distinct formation immediately below the underbridge. The underbridge is simple, short and thin or thick. Immediately below the finger-like bullae are a number of randomized bullae mostly finger-like. The vulval slit is fairly long. Most of the eggs are in the cyst averaged in the range of 100-250 eggs/cyst.

Male

The males of *H. zeae* are very rare and detailed descriptions were made for the first time by Hutzell (1984) from a culture originating from cysts collected in Kent, Maryland, at the site of the first known infestation in the USA.

Second-Stage Juveniles

The second-stage juveniles are vermiform and cylindrical, tapering at both ends. The head has 4-5 annules, with a moderately developed cephalic framework. The stylet is slender with slightly anteriorly directed knobs. According to Koshy *et al.* (1971) the length of stylet was about 20-25 µm (23.0 + 0.26 µm), while Golden and Mulvey (1983) reported that larval stylets from India and Maryland, USA, were continuously found to average about 20.0 µm. The dorsal oesophageal gland opens close (4-5 µm) to the base of the stylet. There is a distinct median valve with well developed valve plates. The basal part of the oesophagus extends as a glandular lobe overlapping the anterior region of the intestine, mostly laterally and dorsally. The hemizonid is distinct, located closer to the excretory pore and is 1-2 annules in length. The anus is distinct. Phasmids are small but conspicuous. The lateral field has four lines. The tail is short, hyaline and terminal withan acutly rounded terminus. Mulvey and Golden (1983) compared two isolates- their measurements are as follows:

	Body Length	*Lateral Lines*	*Stylet Length*	*Tail Length*	*Hyaline Terminal Length*	*DGO*
India	410±40 (360-440)	4	23.3±0.3 (20-25)	41.0±1.1 (32-50)	24.0±0.7 (16-30)	– (4-5)
Maryland, USA	431±14 (399-460)	4	19.9±0.4 (19-20.7)	44.2±2.4 (40-49)	21.9±1.7 (16.8-25.2)	4.3±0.4 (3.4-5)

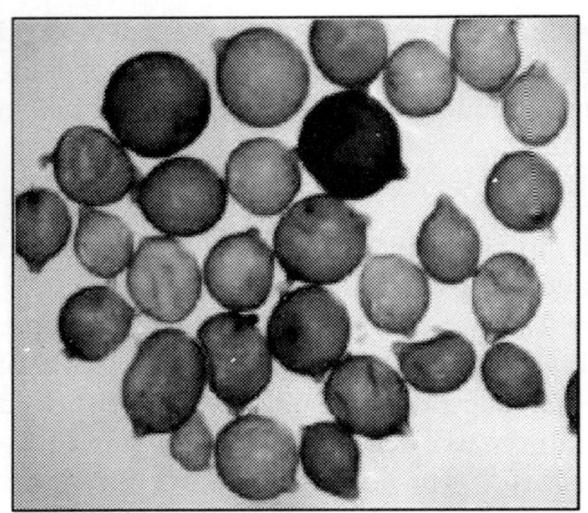
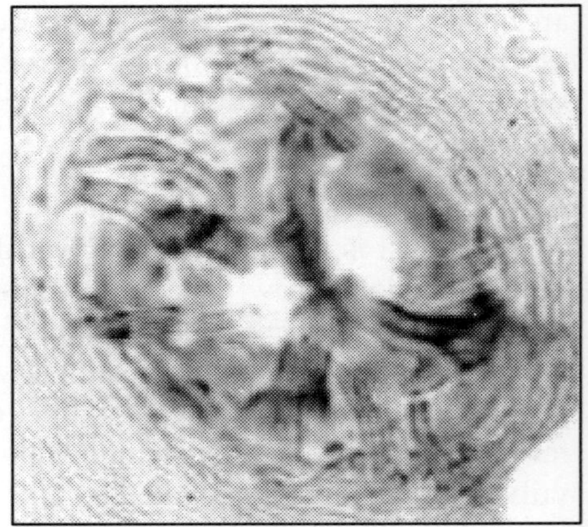

Figure 14: (a) Cysts of *H. zeae*; (b) Cone top showing cross-bullae

Measurements

Eggs

Length: 88-119 (104) µ; Width: 33-43 (38) µ; L/W: 2.2-3.3 (2.8)

Second Stage Juveniles

Length: 360-440 (410) µ; Max. body width: 15-23 (180 µ; Stylet length: 20-25 (23) µ; AE- excretory pore: 61-110 (90) µ; Tail length: 32-50 (41) µ; Hyaline tail length: 16-30 (24) µ; a: 19.0-25.4; b: 4.0-6.5; b2: 2.9-4.2; c: 8.0-13.0.

Female

Body length: 423-716 (592) µ; Body width: 325-684 (491)

Cyst

Length: 342-684 (501)µ; Width: 260-537 (396)µ; Vulval slit length: 36-58 (46)µ; Vulval bridge width: 5-9(7)µ; Underbridge length: 38-60 (47)µ; Fenestral length: 37-53 (44)µ; Fenestral width: 16-21(19)µ; Depth of underbridge: 10-23(17)µ; Cyst L/W: 1.0-1.5 (1.3).

7.9.1.7 The Rice Cyst Nematode, Heterodera skohensis Kaushal, Sharma and M. Singh, 2000

This is another species of cyst nematodes infesting rice crop and was described from a village, Skoh in Dharamsala (Kangra), H.P. (India). This nematode was also found in quite good numbers in rice fields of Palampur tehsil of Kangra district of H.P.

Diagnostic Characters

Characterized by brown to dark brown medium size lemon shaped cysts having medium vulval protuberance, ambifenestration type of fenestration with a delicate moderately developed underbridge and is without bullae. The underbridge may be provided with a 'sheaf' like structure and have bifurcating arms. The vulval bridge is thick and as long as or longer than the vulval slit. The cuticular pattern around basin loose which run parallel to each other and outer striations are zigzags lines.

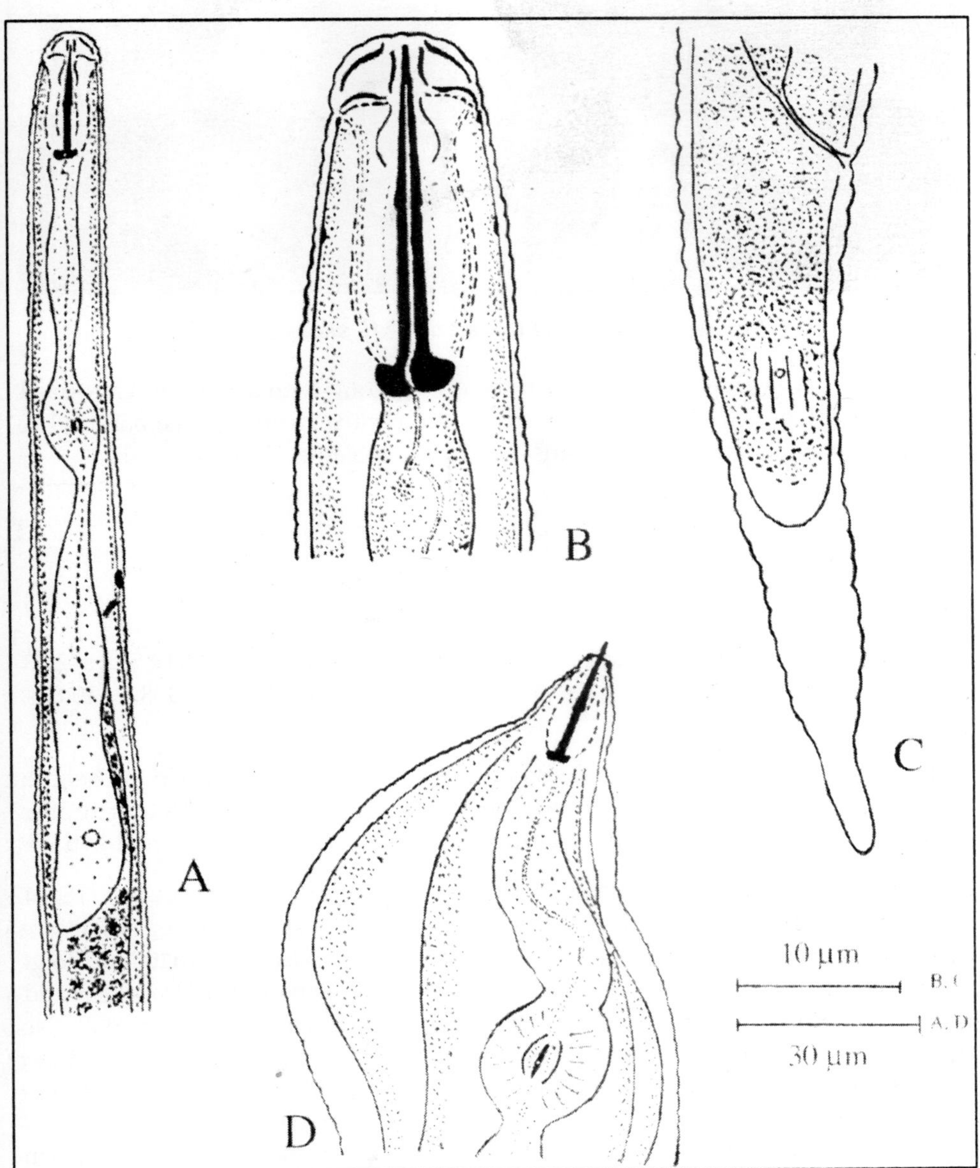

Figure 15: ***Heterodera zeae.***
A-C: J2. A: Anterior region; C: Tail, D: Anterior region of female.
After Koshy *et al.*, 1971.

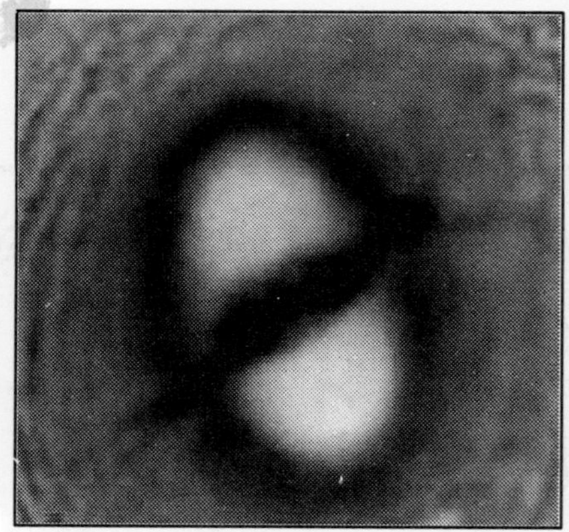

Figure 16: Cysts and Cone Top of *H. skohensis*

The second stage juveniles are 361-408 (379) µ long with a heavily sclerotized head. Lateral fields are with four incisors prominently forming three bands. The phasmids are pore-like and located just anterior to the anus. Tail is conoid.

Measurements

Eggs

Length: 92.3±5.32µ; Width: 37.9± 2.52µ; L/W: 2.12-2.71 (2.38).

Second Stage Juveniles

Length: 361-408 (379) µ; Width: 15-20 (18.8)µ; Stylet length: 18-22(19.5)µ; a: 20.3 ±1.77; b: 4.7 ±0.33; b2: 2.6± 0.19; c: 9.30 ±1.11; c2: 3.8± 0.55; AE-anus: 338.1 ±10.87.

Cysts

Length: 553± 43.7µ; Width: 372±50.8µ; Neck length: 83± 6.3µ; Cone top height: 39±7.4 µ; L/W: 1.3-1.6 (1.4); Underbridge length: 60-88 (77) µ; Fenestral length: 55µ; Fenestral width: 30.2µ; Vulval slit length: 35.2 µ; Anus to edge of fenestra: 18.4µ.

7.9.1.8 The Cereal Cyst Nematode, Heterodera avenae Wollenweber, 1924

Kuhn (1874) recorded the occurrence of the cereal cyst nematode, *Heterodera avenae* on oat in Germany and referred to it as variety of *H. schachtii* at that time. Voigt (1892) thought that these two nematodes attacking sugar beet and oat, wheat and barley are different strains having a definite host preference. Hansen (1897) also came to the same conclusion while describing the oat cyst nematode in Denmark. Later on, this nematode became so familiar and important that it came to be known as *H. schachtii* var. *avenae* (Motensen, Rostrup and Kolpin Ravn, 1908). The nomenclatorial problem was solved by Wollenweber (1924) while demonstrating the differences in the cysts of these two species and named the oat cyst nematode as *H.avenae* differentiating it from *H.schachtii*. Ignorant of this work by Wollenweber, Schmidt (1930) concluded that oat cyst nematode juveniles were larger than sugar beet nematode and should be termed as a subspecies, *H.schachtii* subsp. *major*. This subspecies, was then, raised to the specific rank by Franklin (1940) under the name

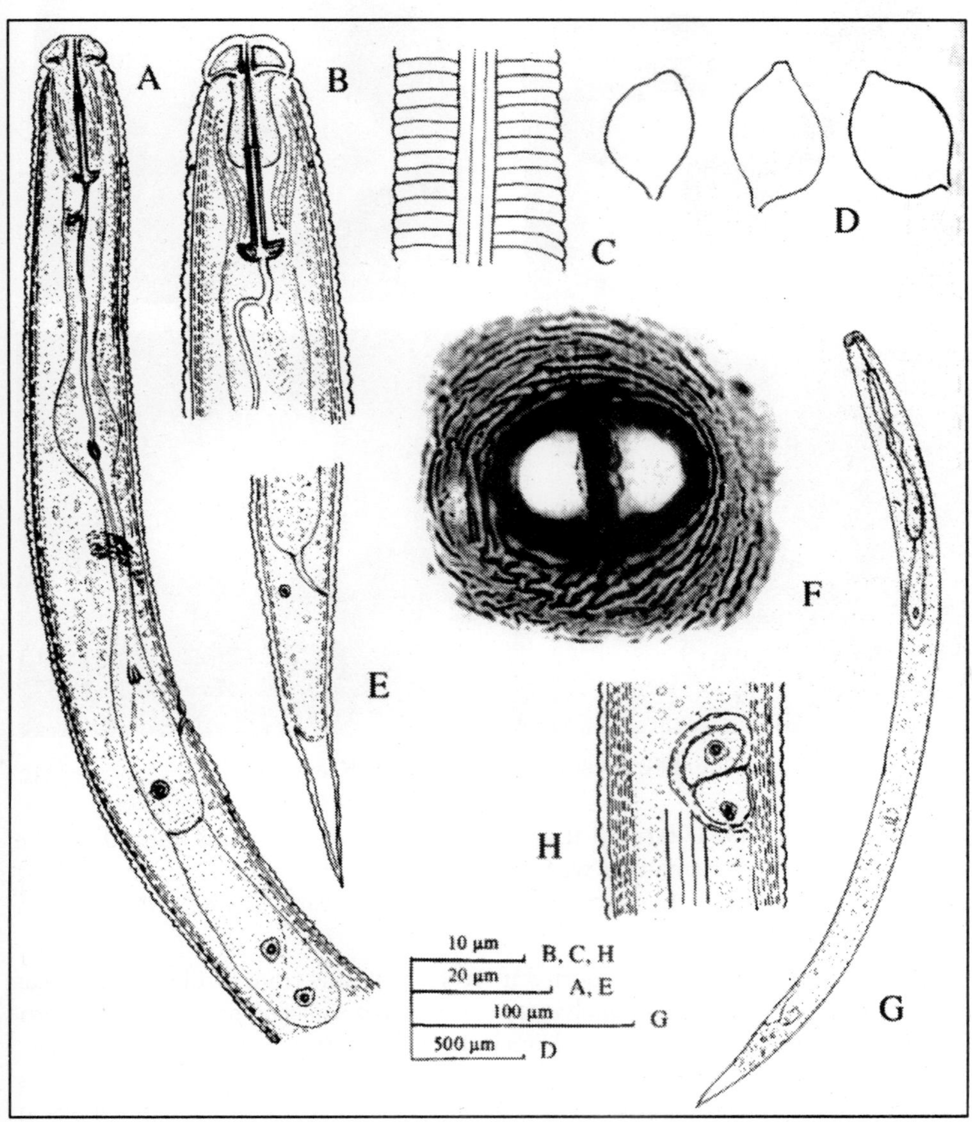

Figure 17: *Heterodera skohensis*
A: Pharyngeal region of J2, B: Anterior region of J2. C: Lateral field, D: Cysts; E: Tail J2, F: Vulval cone; G: Entire body of J2; H: Genital primordium and lateral field. After Kaushal *et al.,* 2000.

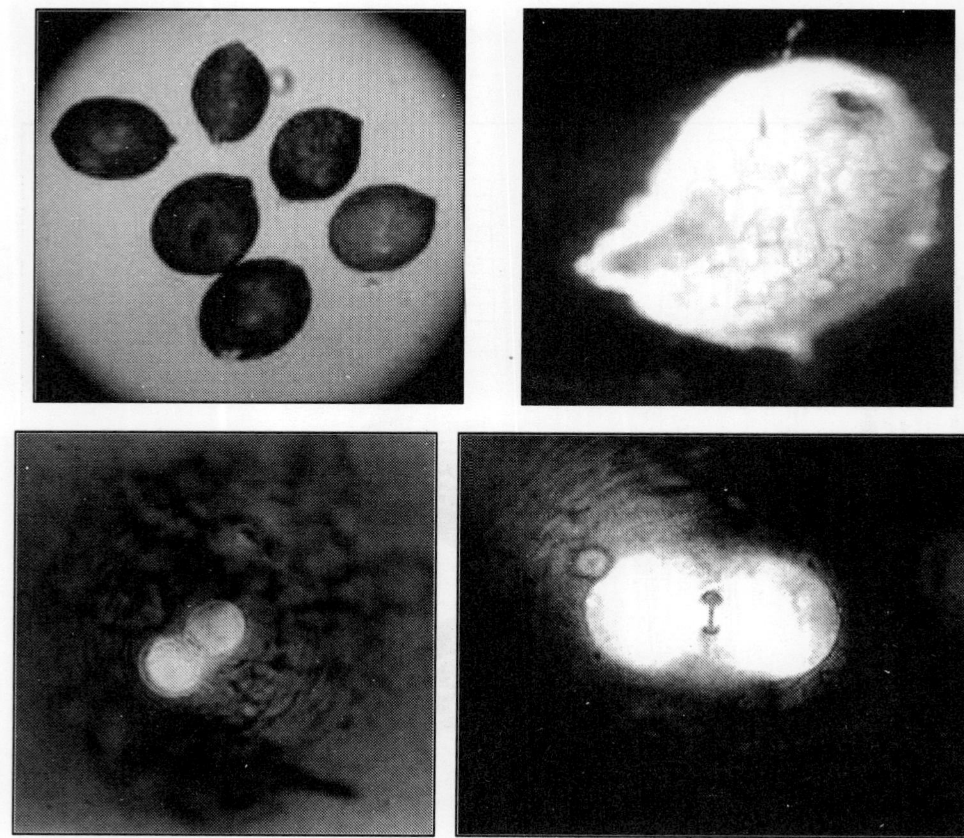

Figure 18: Cysts and Cone Top of *H. avenae*

H. major. Filipjev (1934) termed it as *H. avenae*. The controversy regarding the nomenclature of this nematode was, ultimately solved by Franklin, Thorne and Oostenbrink (1959) when they accepted and adopted the earlier name, *H.avenae* given by Wollenweber in 1924.

In India, Vasudeva (1958) reported it for the first time from a village, Neem-ka-thana in Sikar district of Rajasthan. It is now known to be present in whole of northern India where wheat is grown starting from J. and K, H.P., Punjab, Haryana, Rajasthan, Delhi, U.P., Uttaranchal, Gujarat and M.P. (Kaushal *et al.*, 2007). Only one life cycle is completed under Indian conditions in one year. After the penetration of juveniles in the wheat roots, the development of males and females takes about 23-25 days and males/females are formed in the ratio of almost 50:50. The formation of cysts can be observed after 60-65 days. A single cyst may contain about 250 eggs and juveniles on an average but as many as 900 eggs have been counted in a single cyst. A rudimentary egg sac may be seen attached with some of the cysts but is always devoid of eggs. Physiological variations have been reported in different populations of this nematode revealing two biotypes (Swarup *et al.*, 1979) but recent studies based on host differential reaction, isozyme profile and molecular characterisation of biotype II population

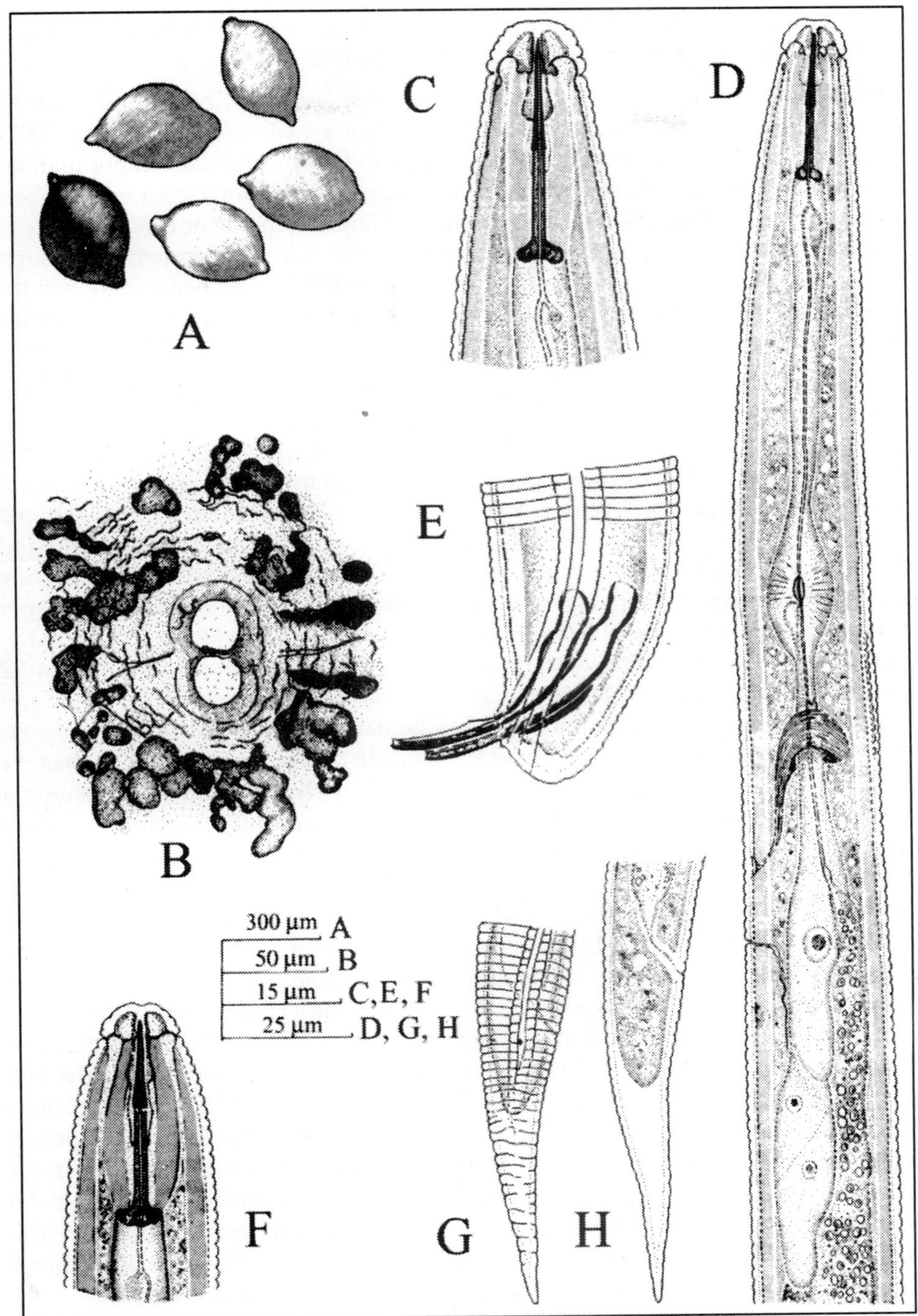

Figure 19: ***Heterodera avenae*.**
A: Cysts, B: Vulval cone; C: Anterior region of male, D: Pharyngeal region of male; E: Tail of male, F. Anterior region of J2, G, H: Tail of J2. After Williams and Siddiquie, 1972.

revealed it to be *Heterodera filipjevi,* a species in the *H. avenae* species complex (Bishnoi and Bajaj, 2004, Bishnoi *et al.*, 2004, Umarao and Sashi, 2008).

Initial studies on genetic variation among seven Indian CCN populations from major wheat growing states using RAPD–PCR Cluster analysis (Umarao *et al.*, 2007). Using 148 scorable markers generated by nine random primers, seven populations were grouped into one cluster and two out groups; populations from major wheat growing locations Sirsa, Tikamgarh, Udaipur and Jaipur formed one cluster, whereas, each of the populations from Delhi and Jhansi formed two independent out groups. Further characterisation of these populations using DNA sequence variation in the non-coding regions of ribosomal DNA indicated more intraspecific variation (Umarao *et al*:, 2004).

The cereal cyst nematodes (CCNs) are the most important group of plant parasitic nematodes attacking temperate cereals, including wheat and barley (Sikora, 1988). CCNs are a group of several closely related species which have been documented as causing economic yield loss in rainfed wheat production systems in several parts of the world including North Africa, West Asia, China, India, Australia, the United States of America and countries in Europe (Nicol and Rivoal, 2008; Baldwin and Mundo-Ocampo, 1991; Rivoal and Cook, 1993). There are several studies reporting yield losses in farmer's fields. It was observed that it could cause about 40 to 50 per cent yield loss reaching up to 60 to 65 per cent (Mathur *et al.*, 1980). In another study based on response to nematicide treatment, average avoidable losses ranging from 32 to 44 per cent in barley and 24 to 35 per cent in wheat have been reported (Handa and Yadav, 1991). *Heterodera avenae* is the principal nematode species on temperate cereals, and in Europe more than 50 per cent of the fields in major cereal growing areas are infected by this nematode (Rivoal and Cook, 1993). *Heterodera avenae* also shows a wide distribution and large genetic infraspecific variability, including many pathotypes (Andersen and Andersen 1982; Swarup *et al.*, 1979). Extensive studies have revealed the presence of several distinct species of *Heterodera* infecting cereals and grasses within studied populations primarily identified as *H. avenae*. The species of the *H. avenae* complex are differentiated from each other by small morphological and morphometrical characters. Presently, the *H. avenae* complex is considered to contain: *H. avenae*, *H. arenaria* Cooper, 1955, *H. aucklandica* Wouts and Sturhan, 1995, *H. australis* Subbotin, Sturhan, Rumpenhorst and Moens, 2002, *H. filipjevi* (Madzhidov, 1981), *H. mani* Mathews, 1971, *H. pratensis* Gäbler, Sturhan, Subbotin and Rumpenhorst, 2000, and *H. ustinovi* Kirjanova, 1969 (Wouts and Sturhan, 1995; Gäbler *et al.*, 2000; Sturhan and Krall, 2002; Subbotin *et al.*, 2002). An additional species, *Heterodera (Bidera) riparia*, belonging to this complex has been described from the Russian Far East by Kazachenko (1993). Several biochemical and molecular techniques have been used for the separation of species and populations from this complex and showed promising results like enzyme polymorphism analysis (Bergé *et al.*, 1981; Bossis and Rivoal, 1990; Ibrahim and Rowe, 1995; Romero *et al.*, 1996; Andrés *et al.*, 2001; Mokabli *et al.*, 2001), two-dimensional gel electrophoresis (2-DGE) (Ferris *et al.*, 1989, 1994; Bossis and Rivoal, 1996; Romero *et al.*, 1996); isoelectric focusing (IEF) (Rumpenhorst, 1985; Sturhan and Rumpenhorst, 1996; Subbotin *et al.*, 1996), random amplified polymorphic DNA (RAPD) (Lopez-Braña *et al.*, 1996; Sturhan and Rumpenhorst,1996; Romero *et al.*, 1996), sequences of the ITS-rDNA (Ferris *et al.*, 1994; Subbotin *et al.*, 2001, 2002), restriction fragment length polymorphism (RFLP) of ITS-rDNA (Bekal *et al.*, 1997; Subbotin *et al.*, 1999, 2000, 2002), and PCRsingle-strand conformational polymorphism(PCR-SSCP) (Clapp *et al.*, 2000).

Diagnostic Characters

The cysts are dark brown to black cysts in colour having low protuberance of vulval cone. They are marked by the absence of underbridge, small vulval slit, small and broader vulval bridge. Bullae are abundant and crowded beneath the vulval cone. Lateral field in juveniles and males have four lateral lines with outer bands usually areolated. Second stage juveniles have a well-developed stylet with large, anteriorly flattened to concave basal knobs. Lip region is rounded, offset with 2-4 annules.

Measurements

Eggs

Length: 126μ; Width: 56μ.

Second Stage Juveniles

Length: 520-610 (575) μ; Width: 200-240 (220) μ; Tail length: 45-70 (56) μ; Hyaline tail 35-45 (40) μ; Stylet length: 24-28(27) μ.

Males

Length: 1070-1590 (1380) μ; a: 32-55 (45); b: 7-12 (9); b2: 5.5-6.8 (5.8); c: 40-50 (45); Stylet length: 27-31 (28.5) μ; Spicules length: 32-55 (45) μ.

Cysts

Length: 550-770μ; Width: 360-500μ.

Physiological Variation in Cereal Cyst Nematodes Defined by an International Test Assortment of Cereal Cultivars

Biotype	*H. avenae* group pathotypes				Ha1	Ha2	Ha3				
	Ha1	Ha2	Ha3	Ha4	Ha5	Ha6	Ha7	Ha1	Ha1	Ha2	Ha3
	1	1	1	1	1	1	1	2	3	3	
					Differentials						
Barley											
Emir	S	S		S	–	R	S	S	S	S	S
Ortolan	R	R	R	R	R	R	R	S	S	S	S
Siri	R	R	R	S	S	S	R	R	S	S	S
Morocco	R	R	R	R	R	R	R	R	R	R	R
Varde	S	–	–	S	–	S	S	S	S	S	S
KVL191	R	R	R	–	S	S	S	R	–	–	–
Bajo Aragon	R	–	–	R	–	R	R	R	S	S	R

Herta	S	S	R	–	R	–	R	S	S	–	–
Martin 403–2	R	–	–	R	–	R	R	R	R	S	S
Dalmastische	(R)	–	–	S	–	R	(S)	S	S	(R)	–
LaEstanzuela	–	–	–	–	–	–	S	–	–	(R)	–
Harlan 43	R	–	–	–	–	–	R	R	–	R	S
Oats											
Sun II	S	R	R	R	R	S	R	S	S	S	S
Nidar	S	–	–	S	–	S	R	S	S	S	S
Pusa Hybrid	R	R	–	R	R	R	R	R	S	R	S
Silva	(R)	–	–	R	–	(R)	R	(R)	(R)	(R)	S
Avena sterilis	R	R	–	R	R	R	R	R	R	R	R
IGV.H 76–646	R	–	–	R	–	R	R	R	S	S	S
Wheat											
Capa	S	S	–	S	–	S	S	S	S	S	S
Loros	R	R	–	R	–	(R)	R	R	(R)	S	S
Iskamish K–2–light	S	–	–	R	–	(R)		S	S	S	S
AUS 10894	R	–	–	R	–	R	S	R	(R)	S	S
Psathias	–	–	–	S	–			S	S	S	R

S: Susceptible; R: Resistant; (S) or (R): Intermediate; –: No observation.

Source: From Rivoal and Cook, 1993; and modified from Andersen and Andersen, 1982.

7.9.1.9 The Ragi Cyst Nematode, *Heterodera delvii* Jairajpuri, Khan, Setty and Govindu, 1979

Jairajpuri *et al.* (1979) described this cyst nematode species from the roots of ragi (*Eleucine coracana*) plants grown in the experimental plots at the main research station of the University of Agricultural Sciences, Hebbal, Bangalore (India). It has not been reported from any other place in the country. Other hosts reported by Krishna Prasad *et al.* (1980) include various *Echinochloa* species, *Sorghum vulgare*, maize, pearl millet, finger millet, *Setaria italica*, wheat and *Panicum miliaceum*.

H. delvii is close to *H. cyperi*, *H. mothi* and *H. graminis*, but differs mainly in the structure of the underbridge which shows a subcircular mass attached to its centre; this is absent in the other species.

Diagnostic Characters

The cysts are light to dark brown in colour. Fenestration is ambifenestrate but older cysts may look circumfenestrate due to disintegration of vulval bridge and vulval slit. Underbridge is well developed and is provided with sheaf-like organ in the center which semicircular in shape. Bullae are absent. Lateral field in the second stage juveniles is provided with three lateral lines. Head is moderately sclerotised and is also slightly offset. Stylet knobs are anteriorly directed. Phasmids are in the middle of the hyaline tail.

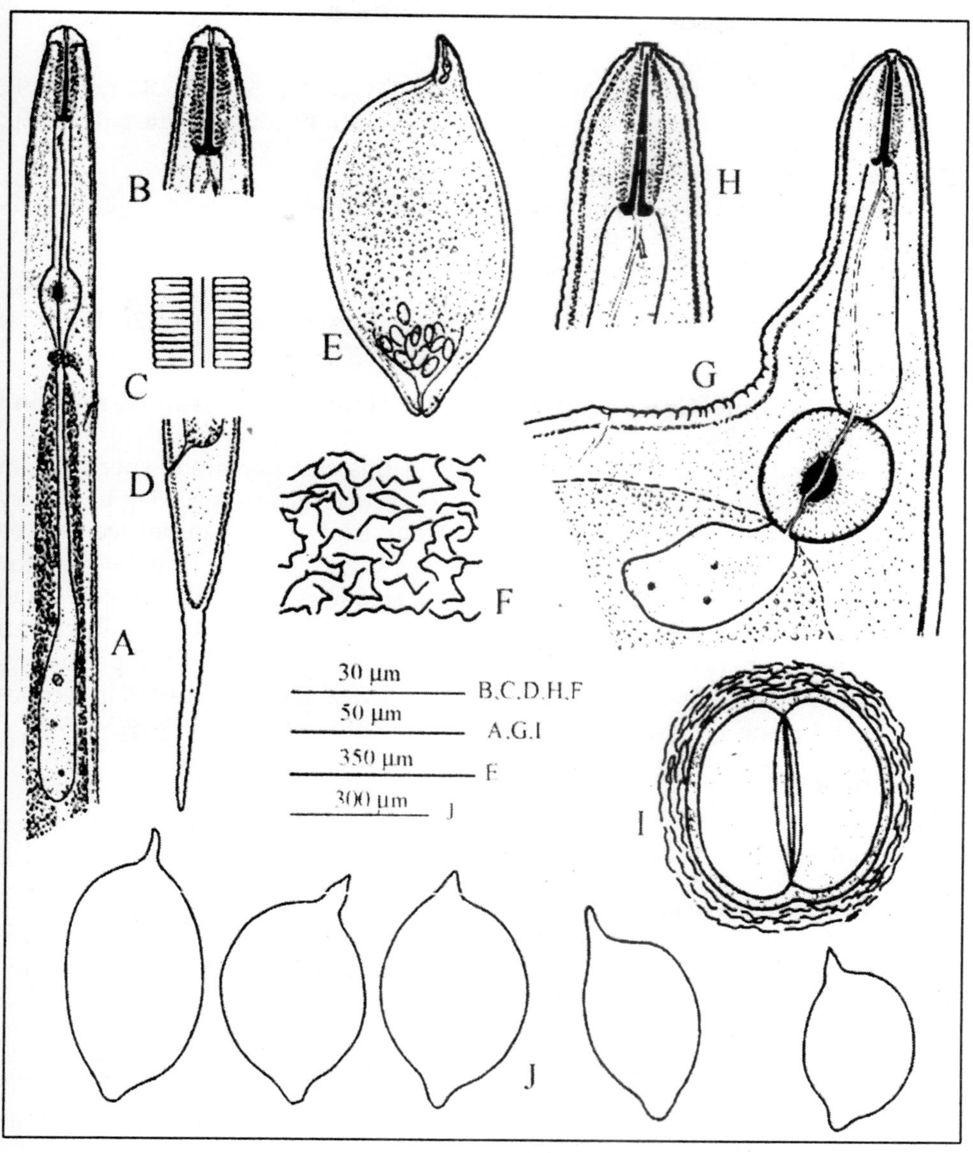

Figure 20: ***Heterodera delvii***
A-D: J2. A: Pharyngeal region, B: Anterior region, C: Lateral field, D: Tail, E-H: Female. F: Entire body, F: Cuticular markings at mid body, G: Pharyngeal regon, H: Anterior region, I: Cone top view, J: Cysts. After Jairajpuri ***et al.,*** **1979.**

Measurements

Eggs

Length: 103-138µ; Width; 36-48µ; L/W: 2.3-2.9

Second Stage Juveniles

Length: 466-520 (480) µ; a: 18-31(23), b: 4.3-5.3 (4.7) b2: 3.0-3.6(3.4); c; 8-10 (9); Stylet length: 18-20µ. Ant.- ex. pore: 84-96µ; Tail length: 49-60µ; Hyaline tail: 29-36µ.

Females

Length = 645 pm (500-800 µm); width =384 µm (250-460 pm) ; spear = 25 µm (22-28 µm). Holotype: Length=650 µm ; width=320 µm ;spear = 26 µm.

Cysts

Length: 617µ; width: 355µ; Vulval slit; 36-45µ; Underbridge: 96-112µ; Fenestral length: 40-56µ.

7.9.1.10 The Bulb Grass Cyst Nematode, *Heterodera raskii* Basnet and Jayaprakash, 1984

Basnet and Jayaprakash (1984) described this cyst nematode species from Hyderabad infesting the roots of bulb grass (*Cyperus bulbosus*) growing in the dry 'musi' river opposite Andhra Pradesh High Court (India). This species seems to be close to *H. mothi.* It has not been reported from any other place. It belongs to the 'goettingiana' group and differs from closely related *H. cyperi* by the elongate ovoid shaped cysts and females, greater fenestral length, width, vulval slit, and absence of egg sac. The stylet knob shape was round in second-stage juveniles and posteriorly sloping in females and males of *H. raskii*, while it was anteriorly directed in second-stage juveniles and spherical in females and males of *H. cyperi*.

Measurements

Eggs

Length: 100-200(105)µ ; Width: 39-45(42)µ ; L/W; 2.4-2.7(2.50.

Second Stage Juveniles

Length: 470-520(500)µ; Stylet length: 17.5-20(18.5)µ ; DGO : 5-7(5.6)µ ; a: 24.5-28.0(26); b: -78(7.5) ;b2: 3-3.5(3.2); c: 6.1-6.4(6.3); Tail length: 74-85(80)µ ; Hyaline tail: 30-43(34)µ .

Males

Length: 950-1050 (1015) µ; Stylet length: 20.5-22(21) µ; DGO: 5-7(5.5)µ; a: 30-35(33.8); b: 3.5-4.5(4; b2: 7-9.5 (8.4); c: 3117-350 (338); Spicule length: 33.5-36 (35)µ; Gubernaculum: 9-11.5(10.5)µ.

7.9.1.11 The Sugarcane Cyst Nematode, *Heterodera sacchari* Luc and Merny, 1963

H. sacchari, of the schachtii group, was originally described from sugarcane in the Niari Valley, Congo (Luc and Merny, 1963). It has subseqúently been reported from the Ivory Coast where it has been frequently found in flooded rice fields in the northern and central regions (Merny, 1970), in Nigeria on sugarcane (Jerath, 1968

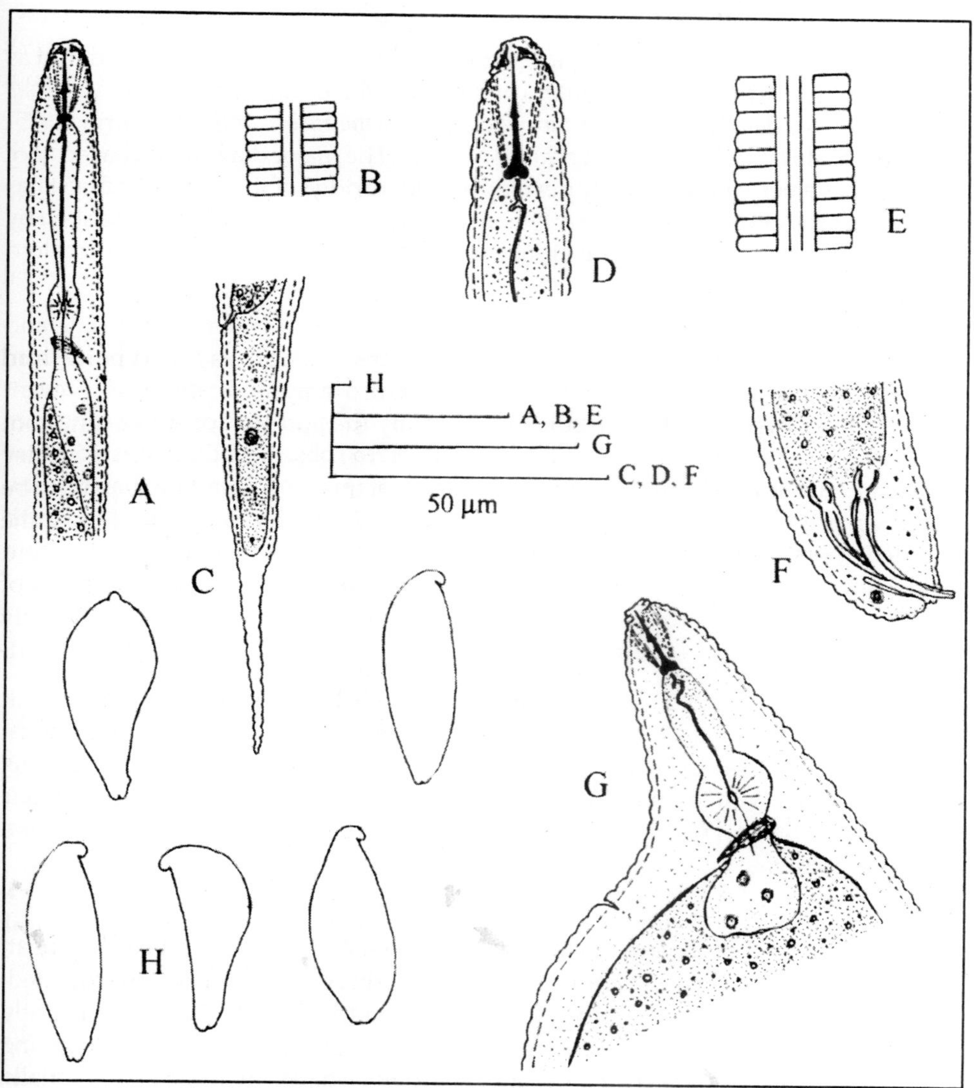

Figure 21: ***Heterodera raskii***

A-C: J2. A: Pharyngeal region, B: Lateral field, C: Tail, D-F: Male. D: Anterior region, E: Lateral field, F: Tail, G: Pharyngeal region of female, H: Cysts. After Bashet and Jayaprakash 1984.

and wild grasses (Odihirin, 1975), in flooded rice fields of Senegal and in Gambia (Fortuner and Merny, 1973) and on sugarcane in Burkina Faso (Cadet and Merny, 1978). The species is also reported to have been found on *Saccharum spontaneum* in India (Swarup *et al.*, 1964). It has also been recorded in Pakistan, Senegal, Trinidad and Tobago and Thailand.

In addition to sugarcane and rice, various wild Graminaceae are considered as hosts by Odihirin (1975): *Paspalum conjugatum, Axonopus compressus, Mariscus umbellatus, Cynodon dactylon, Eleusine indica* and *Brachiaria brizantha*. Odihirin (1975) considers that *H. sacchari* is indigenous to Nigeria. The major diagnostic character of the species is a very dark, strongly developed underbridge of the female and cyst has finger-like projections. The apparently non- functional males have been described by Netscher *et al.* (1969). Netscher (1969) has shown that *H. sacchari* is a triploid parthenogenetic species.

The hatching of juveniles from cysts is similar in sugarcane root diffusate and tap water (Garabedian and Hague, 1984) but Riversat (1981) has found potassium permangate to be useful as a hatching agent. Aldicarb, oxamyl and carbofuran prevent hatching from cysts but this effect is decreased by the addition of sugar cane root diffusate (Garabedian and Hague, 1984). Jerath (1968) observed that infested sugar cane plants are stunted and thin, and secondary roots are less abundant than healthy plants. On rice, experiments showed that infested plants are chlorotic and their growth retarded, roots are necrotic and blackened, tiller numbers are reduced and the grain yield is lower (Babatola, 1983). Cysts can be seen 30-35 days after J2 penetration in the roots.

Females

White, lemon-shaped. Lip region truncated, with 2 offset retrorse annules; weak cephalic sclerotization. Weak stylet, 23-25 µm long, with small, rounded knobs. Opening of the dorsal esophageal gland situated 4-5 µm behind the spear, procorpus cylindrical. Heavy median bulb with strong valve; short isthmus. Basal glandular part of the esophagus nearly spherical. Excretory poresituated 165-175 µm from the anterior extremity.

Males

Rarely found. Body from nearly straight to C-shaped when relaxed by heat, twisted at the posterior end, cylindrical, slightly tapering anteriorly, terminus rounded. Cuticle distinctly annulated; annules 2.5 µm in the middle of body; lateral field faintly marked by 3 longitudinal lines (=incisures) irregularly crossed by annules, occupying 1/5 of the corresponding body width. Lip region dome-shaped, with 4-5 (exceptionally 6) annules, often anastomosed, without longitudinal striations; cephalic framework heavily sclerotized, with outer margin conspicuously marked. Spear stong, anterior and posterior parts of the same length; knobs rounded posteriorly and with flat, sloping anterior surface. Dorsal esophageal gland opening situated 4 µm behind the spear. Median esophageal bulb ovoid. Nerve ring strongly marked, encircling the esophagus just behind the median bulb. Excretory pore situated 137-150 µm from the anterior end. Hemizonid flat, 7-15 µm anterior to the excretory pore and extending over one annule. Testis single. Spicules curved, notched at the tip; specimens with abnormal atrophied spicules frequent (up to 20 per cent of specimens); exceptionally, three spicules were observed. Spicular sheath present. Lamellate gubernaculum. Tail 1/4 to 1/3 of the cloacal body diameter. Phasmids not observed.

Cysts

Mature cysts brown to dark brown, lemon-shaped with prominent vulval cone and neck of medium size.Cuticle with a lace-like pattern; numerous perforations evenly but irregularly distributed on the inner layer. Young cysts with a thick subcrystalline layer. Cyst cone ambifenestrate: fenestrae rather obscurely demarcated, 45-55 µm long and 35-45 µm wide.Vulval slit 50-52 µm long, as long as vulval bridge; underbridge strongly developed with finger-like bullae: length = 100-150 µm; width 50-70 µm; depth = 22-28 µm (Mulvey, 1972); few peripheral bullae are also present.

Second Stage Juveniles

Body straight or slightly ventrally curved when heat-relaxed, slightly tapering at the anterior end, more attenuated at the posterior end. Cuticle annulated; annules 1.7 µm wide in the middle of body; lateral field composed of 3 longitudinal lines, not crossed by annules xcept in the fore-part and at the level of the phasmids. Lip region dome-shaped, with 3 annules; cephalic framework heavily sclerotized. Stylet stong, anterior part slightly shorter than the posterior; knobs well developed, rounded posteriorly, concave anteriorly. Dorsal esophageal gland opening situated 5-8 µm behind the spear Median esophageal bulb ovoid, with strong valve. Nerve ring well defined, situated immediately behind the median bulb. Excretory pore104-130 µm from the anterior end. Hemizonid immediately in from of the excretory pore, lenticular, extending over 2 annules. Genital primordium with 2 nuclei, located at the mid-body. Tail elongated, conical, with pointed terminus.Phasmids pore-like, situated 45-47 µm from the posterior end.

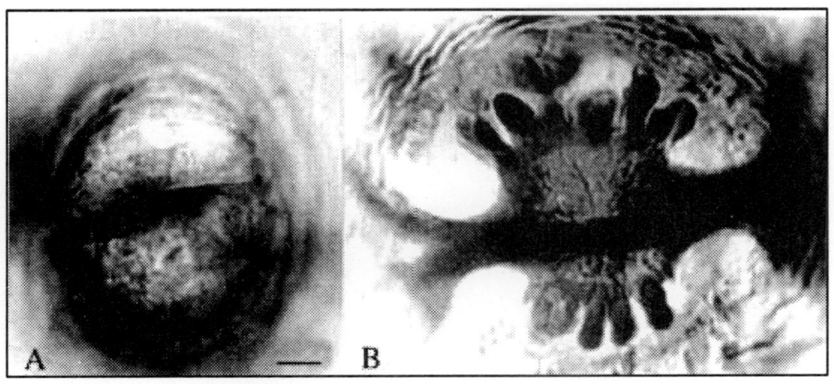

Figure 22: *Heterodera sacchari.* Vulval cone.
A: Anterior view, B: Underbridge level view (Scale bar = 10 µm).
After Mulvey, 1972.

7.9.1.12 Soybean Cyst Nematode, Heterodera glycines, Ichinohe, 1952

Soybean cyst nematode, *Hetorodera glycines* Ichinohe, is a major yield-limiting pathogen of soybean (*Glycine max* (L.) Merr.) in the world. A conservative estimate

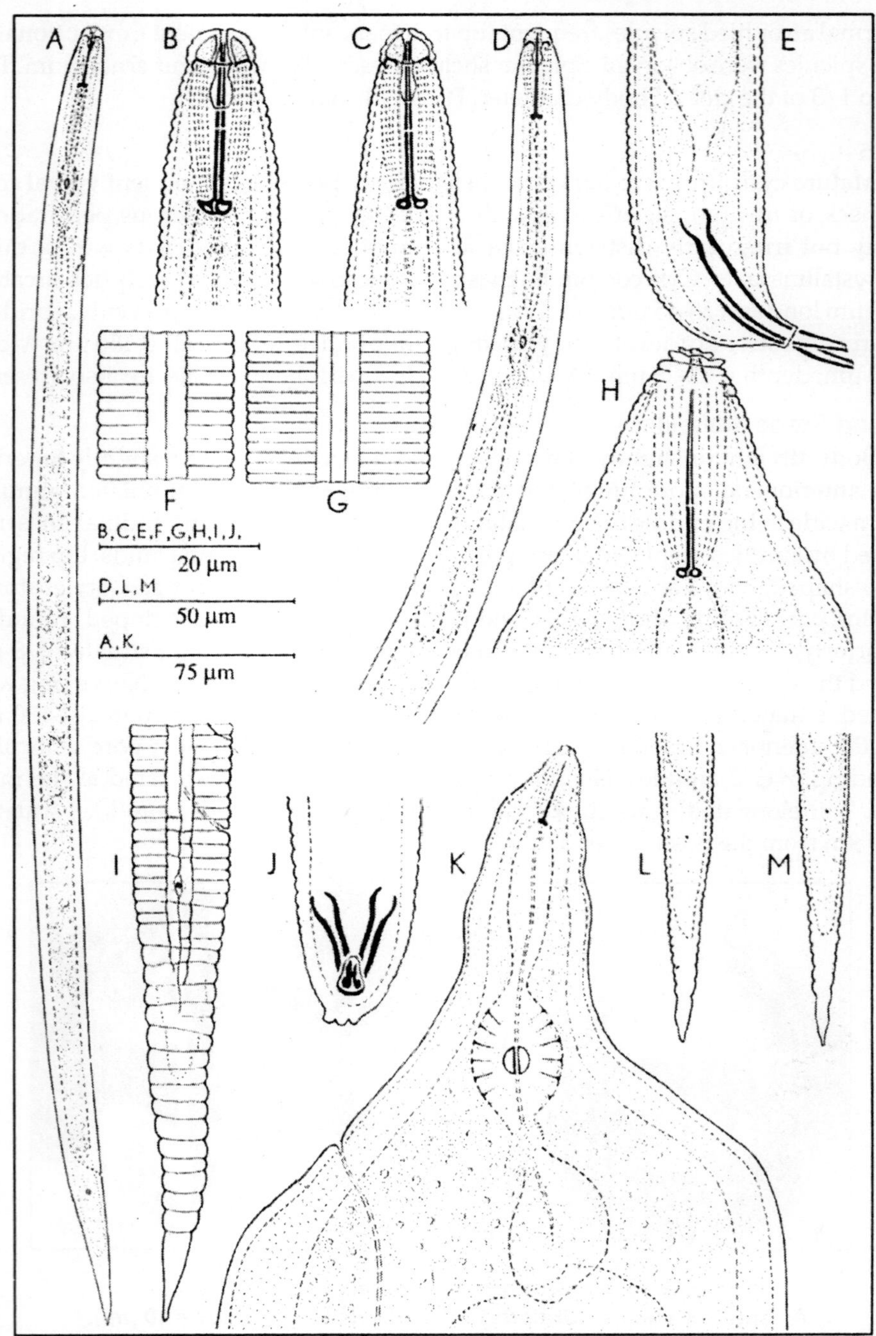

Figure 23: ***Heterodera sacchari***
A, B, F, I, L, M: J2. A: Entire body, B: Anterior region, F: Lateral field, I, L, M: Tails, C, D, E, G, J: Male. C: Anterior region, D: Pharyngeal region, E, J: Tails, G: Lateral field, H: Female anterior region, K: Female pharyngeal region. After Lic and Merny, 1963.

based on the assumption of a mean of 3 per cent direct loss from the nematode is over US$1.1 billion annually worldwide (Schmitt, 2004). It was first reported in Japan in 1915. In 1938, the nematode was reported from Manchuria (then an independent state, now in China). *H. glycines* is the most wide spread cyst nematode in the United States. It was first reported in the United States in North Carolina in 1954, but was reported in Japan in the 1880s as a race of *H. schachtii*. Soybean cyst nematode is believed to have come to the U.S. from Japan with soil imported during the late 1800's to obtain nitrogen-fixing bacteria. In India, it was recorded from the state of Madhya Prasdesh in 2004 (Kaushal *et al.*, 2004).

The problems associated with race determination, which were summarized by Niblack (1992), ultimately led to the description of all possible 16 races (Riggs and Schmitt, 1988) according to the system developed by Golden *et al.* (1970). Twelve races are currently known from the USA and races 1, 3 and 5 are present in Japan (Ichinohe, 1988). In Japan, yield loss was estimated to be 10-75 per cent (Inagaki, 1977; Ichinohe, 1988).

Life cycle of this nematode is completed within 25 days at 23 C. In the field, there are three to five generations per year. Optimum development occurs at 23-28 C; development stops below 14 C and above 34 C (Riggs, 1982; Burrows and Stone, 1985). Emergence of second-stage juveniles from eggs is enhanced by root leachates. Males are present. On maturity of the cysts some eggs (50 to 100) are deposited in the

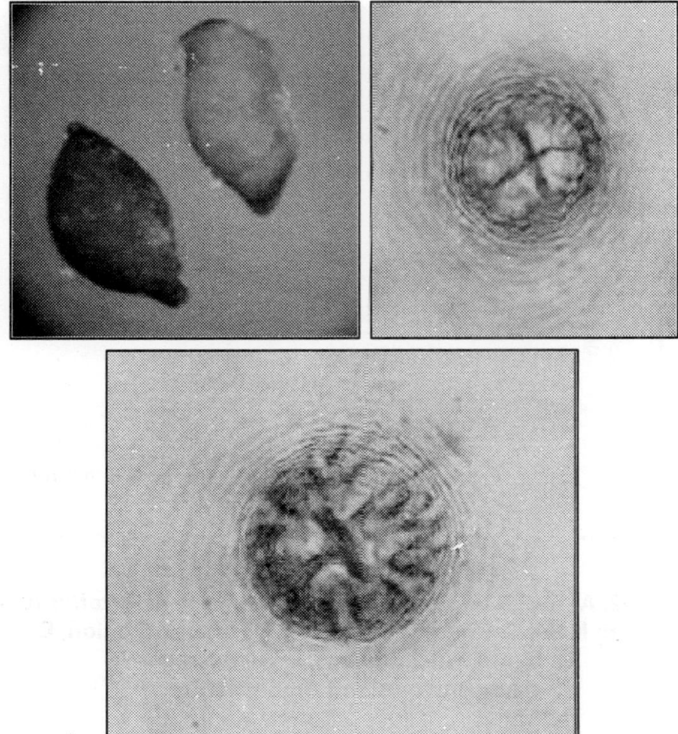

Figure 24: Cysts and Cone Tops of *H. glycines*

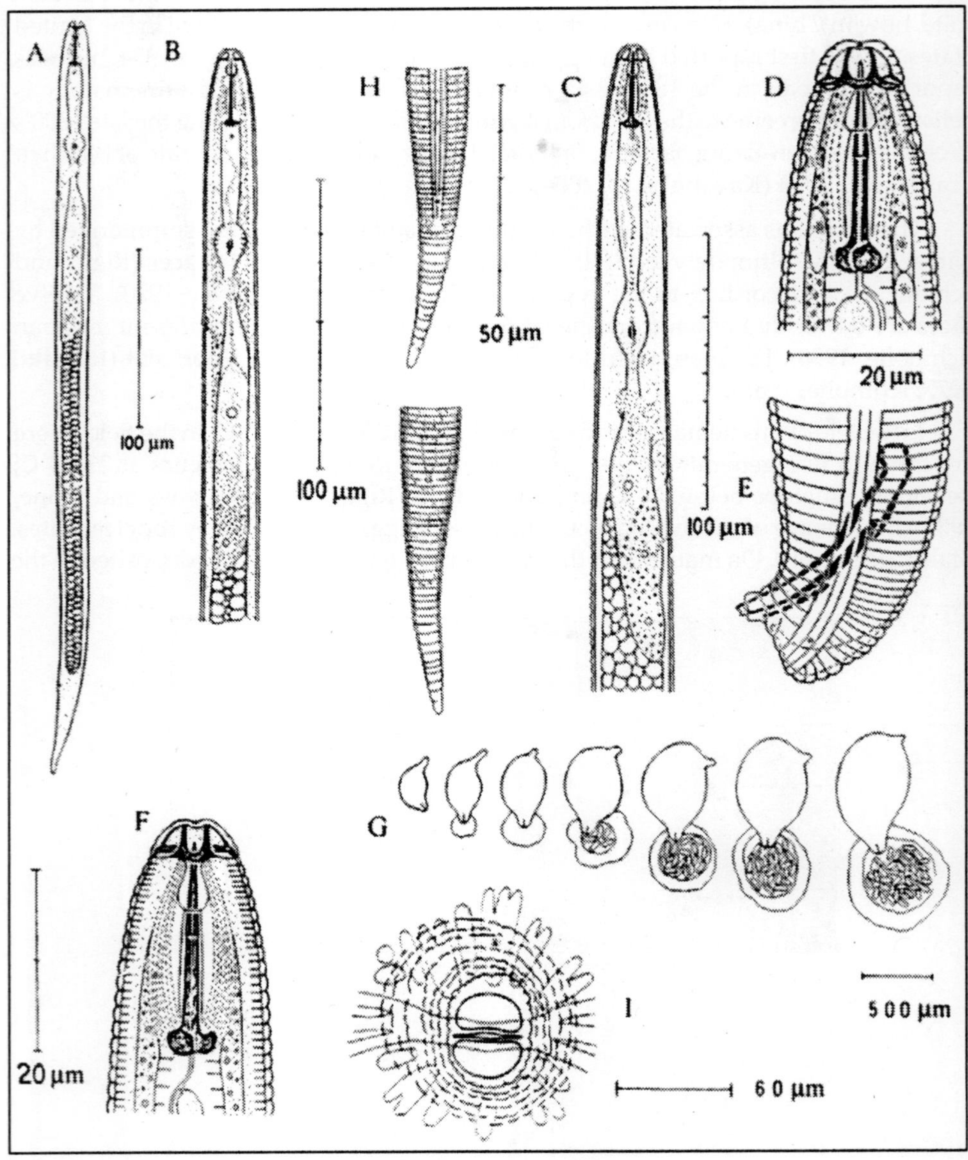

Figure 25: *Heterodera glycines*
A, B, F, H: J2. A: Entire body, B: Pharyngeal region, F: Anterior region, H: Tails, C-E: Male. C: Pharyngeal region, D: Anterior region, E: Tail, G: Cysts with egg sacs, I: Vulval cone.
After Burrows and Stone, 1985.

egg mass, but the majority of the eggs produced (150 to 300) remain within the body of the female. Infection due to this nematode causes 'yellow dwarf' disease in soybean which considerably reduces the yield. Symptoms produced are stunted, yellow plants, resulting in poor canopy closure. They are more apparent on light sandy soils where moisture stress is common or on heavier soils in years when rainfall and soil moisture are low.

Females/Cysts

They are lemon-shaped, bearing the subcrystalline layer. They pass through a pale yellow stage before developing into a dark brown cyst. Rugose pattern, a series of short, zigzag lines without orderis present. Punctations of cuticle fine but irregularly arranged. As many as 200 eggs may be deposited in the egg-sac. Vulva averaging 49.7 μ (43 to 56) long. Bullae are present.

Males

1.3 mm (1.1 to 1.4) in length, closely resembling those of other *Heterodera*. Spear knobs massive, rounded. Cephalids located dorsally and ventrally near the sixth and eighth annules. Esophageal gland nuclei about one body width apart. Spicula bidentate. Hemizonid about six annules anterior to excretory pore.

Second-Stage Juveniles

Their avarage length is 439.6 (373 to 490) μm. Cephalids are near middle of the head. Spear knobs slightly rounded. Hemizonid adjacent to excretory pore. Dorsal esophageal gland opening is 4.0 (3.3 to 5.2) μ behind spear base. About half of tail length is occupied by the hyaline portion.

7.9.1.13.1 Races of the Soybean Cyst Nematode, Heterodera glycines

Race	*Pickett*	*Peking*	*PI88788*	*PI90763*
1	–	–	+	–
2	+	+	+	–
3	–	–	–	–
4	+	+	+	+
5	+	–	+	–
6	+	–	–	–
7	–	–	+	+
8	–	–	–	+
9	+	+	–	–
10	+	–	–	+
11	–	+	+	–
12	–	+	–	+
13	–	+	–	–
14	+	+	–	+
15	+	–	+	+
16	–	+	+	+

According to the race determination schemes of Golden *et al.* (1970) and Riggs and Schmitt (1988).

Race determination is made on the basis of the pattern of "+" and """ ratings for each race. "+" rating is given if the number of females produced by an *H. glycines* population on each soybean differential is equal to or greater than 10 per cent of the number produced on the standard susceptible cultivar Lee. If the number of females is less than 10 per cent, """ rating is given.

7.9.1.13 Clover Cyst Nematode, Heterodera trifolii Goffart, 1932

It is a cosmopolitan species, occurring widely throughout northern Europe and Spain, Italy, southern France, Soviet Union, Canada, U.S., Israel, India, Australia and New Zealand. In India, its occurrence was reported by Swarup *et al.* (1964) but the cyst later on was identified as *H.cajani* Koshy (1967). Recently, Kaushal *et al.* (2008) has reported it from Palampur region of Kangra district, Himachal Pradesh infesting black gram. It severly attacks legumes (86 species in 9 families). Several forms of clover are reported as hosts of *H. trifolii* in New Zealand. A species thought to be *H. trifolii* is a serious pest of carnation in Italy.

Hatching of the eggs occurs progressively over a period of years in the soil in response to moisture and root leachates. Root leachates stimulate hatching and attract juveniles to root-tips. Emergence from juveniles in cysts may be seasonal and influenced by diapause and temperature. Juveniles emerge over a relatively wide range of temperatures (4.2 to 31.4°C), with 17.2°C being optimum.

White Females

Length = 520-650 µm; width = 320-340 µm. Lemon-shaped with prominent neck and vulval cone (posterior protuberance). Median esophageal bulb rounded, with distinct valve. Ovaries paired, greatly extended and nearly filling the body cavity of the adult female. Vulval cone covered with a gelatinous matrix containing as many as 200 eggs. Surface of female and newly-formed cysts typically encrusted with a material termed "subcrystalline layer." Transitional stage between white female and brown cyst distinctly yellow. The yellow color varies with the strain of the host plant, being constant in females developing on any one host strain.

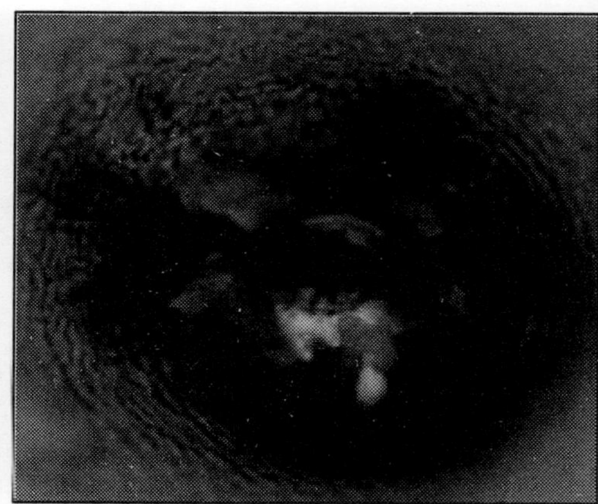

Figure 26: (a) Cysts of *Heterodera trifolii* (b) Cone Top of *Heterodera trifolii*

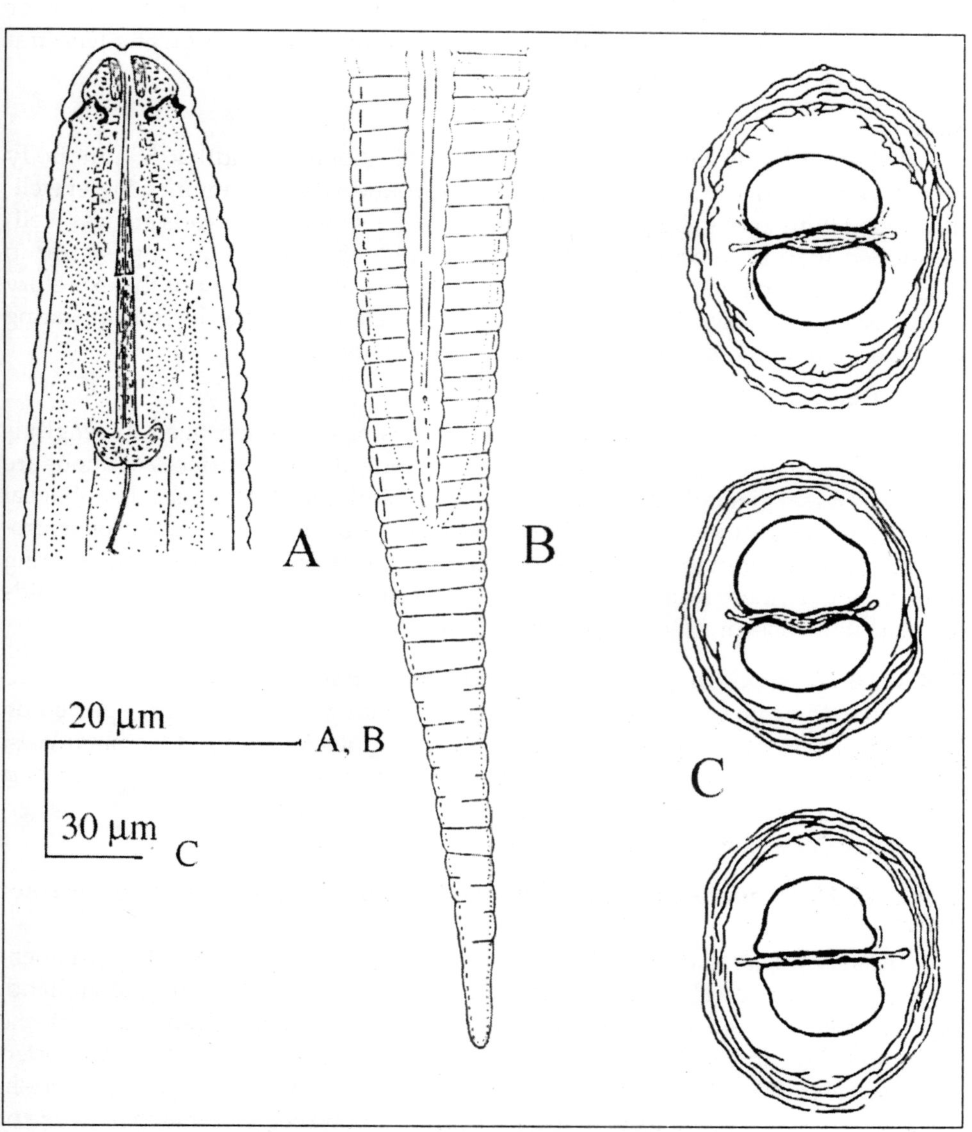

Figure 27: ***Heterodera trifolii***
A: Anterior region of J2, B: Tail of J2. C: Vulval cones.
After Wouts and Sturhan, 1978 and Subbotin, 1984.

Cyst

Brown to dark brown cysts with irregular bullae containing a few to several hundred eggs. The underbridge is 85-105µ with long fenestrae 40-69µ and a vulval-anus distance of 24-38µ. *H. trifolii* can be separated from *H. schachtii* by its heavier and more strongly pigmented underbridge, with bifurcate ends, its greater fenestral length, and shorter vulva-anus distance.

Male

Although rarely reported with length = 0.97-1.29 mm; width = 30 µm. Body robust, vermiform. Lip region set off from neck, rounded, with 6 annules. Stylet well-developed with knobs slightly backward sloping. Esophageal glands are less well-developed than in infective second-stage juvenile. Excretory pore is posterior to esophago-intestinal valve. The hemizonid is 6 to 14 annules from of the excretory pore. Testis single; spicule curved, blades with truncated or oblique lips, not appearing bidentate. Gubernaculum is simple, proximity arcuate.

Second-Stage Juvenile

Head offset, with 4 annules, head skeleton is heavily sclerotized. J2 length is usually more than 500µ. Stylet robust, 25.5-29µ long, anterior surfaces of knobs are concave. Dorsal gland duct opening is 5-9 µm posterior to stylet knobs. Esophageal glands, particularly the subventrals, well-developed, are extending ventrally or ventro-laterally well posterior to esophago-intestinal valve. Tail conoid, 54-64µ long, tapering uniformly to a finely rounded terminus. Posterior half of tail is hyaline. Phasmids small, usually obscure, located near middle of tail.

7.9.1.14 *Heterodera gambiensis* Merny and Netscher, 1976

Heterodera gambiensis, belonging to the schachtii group, has been reported on *Sorghum vulgare* and *Pennisetum typhoideum* in various locations in Gambia, West Africa (Merny and Netscher, 1976). It has, however, been recorded on *Eleusine coracana* in Karnataka, India (Narayaswamy *et al.*, 1982). There is no information on the biology, host range or pathogenicity of this species.

7.9.1.15 *Heterodera spinicauda* Wouts, Schoemaker, Sturhan and Burrows 1995

Heterodera spinicauda, was described from high-lying mud flats with permanent vegetation along the Haringvliet dyke at Zuidland, Voorne Putten, Zuid Holland, Netherlands. The reed, *Phragmites australis* (Cav.) Trin. ex Steud. (Gramineae) was the dominant plant and is considered to be the most likely host. Kaushal (1996) reported its presence from cooler regions of Palampur and Dharamsala in Himachal Pradesh, India. The cysts obtained from muddy soil with lot of weeds growing in it. In one of the samples drawn from the same area where watercress -*Nasturtium officinale*)-a broad leaves weed (used as salad) was growing, a white female of this species was recorded which showed it to be the host. The cysts were dark brown almost without any visible vulval protuberance. This new *Heterodera* species, with its bifenestrate vulval region and short vulval slit, belongs to the *H. avenae* group. Within this group *H. spinicauda* is unique because of: (1) an inconspicuous vulval cone, giving the initial impression that the cysts are spherical; (2) inside the vulval cone, immediately

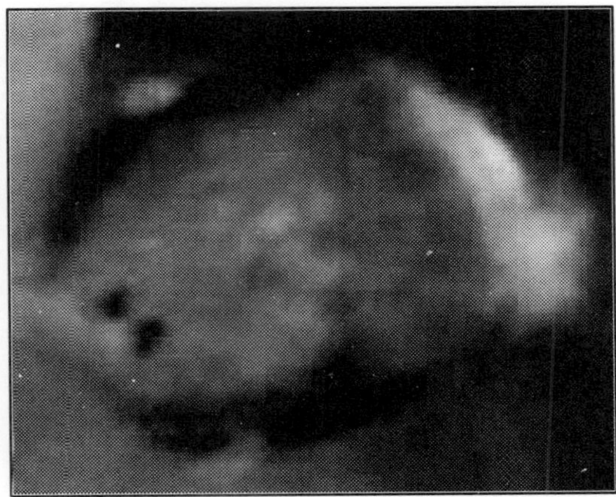

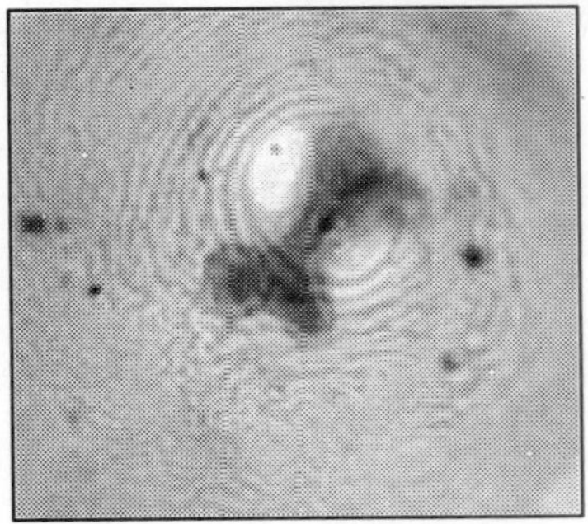

Figure 28: (a) Cyst of *H. spinicauda* (b) Bow-tie type under bridge in cyst (c) Cone-top structure

below the vulva, the presence of an obvious underbridge which in end-view appears as an irregularly multilobed 'bow-tie' type and (3) the sharply pointed tail of the second-stage juveniles. Due to the presence of these characters, it seems to be a new genus and further detailed studies may highlight this point.

Cyst (after Wouts et al., 1995)

Length (without neck):475 (340–590)μ; Width: 422 (270–540)μ; L/W ratio: 1.1 (1.1–1.2); Fenestral length: 21 (17–29) μ; Fenestral width: 23(18–29)μ; Vulva slit length: 7 (5–9) μ; vulval bridge width: 8(5–11) μ; Underbridge length: 76(46–103) μ; Underbridge width: 17(7–32)μ.

Second Stage Juveniles

Bodylength: 445(420–485)μ; Body width: 19 (17–23)μ; Stylet length: 23(22–24)μ; Lateral lines: 4; Tail length: 64(52–69μ ; Hyaline terminal Length: 34(27–40) μ; DGO: 4 (4–5) μ; a: 23(21–25).

7.9.1.16 Heterodera swarupi Sharma, Siddqui, Rahman, Ali and Ansari, 1998

H. swarupi was reported for the first time on chickpea in Dhani village, Tehsil Kishangarh in Ajmer district of Rajasthan (Sharma *et al.*,1998). The incidence on this

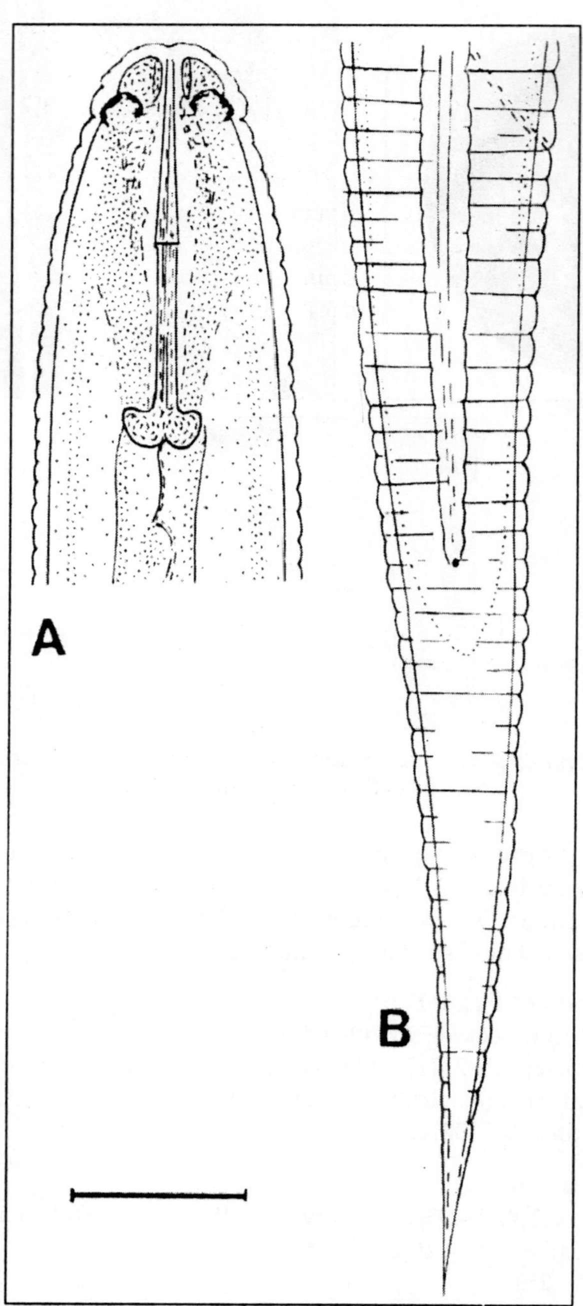

Figure 29: *Heterodera spinicauda*
J2. A: Anterior region, B: Tail (Scale bare = 10µm).
After Wouts *et al.*, 1995.

nematode on chickpea was also reported from Ajmer, Alwar, Bikaner, Jaipur, Jhunjhunu, Nagaur, Swai Madhopur, Sikar and Tonk districts of Rajasthan and in high number at a few places that indicated chickpea to be a good host. It is also found to infect pigeonpea and cowpea.This species belongs to Schachtii group and is morphologically and morphometrically close to *H. cajani* and may be a synoriym species as the morphological differences observed might be due to the dry climatic conditions of Rajasthan.

Female

Lemon shaped, white gradually turning yellow to brown. Gelatin matrix has eggs. Vulva terminal situated on a well developed cone.

Cyst

Lemon shape with distinct neck and vulval cone, generally yellowish to brown, sometime dark brown. Cuticular surface with irregular zigzag pattern of stiae or ridges. Subcrystalline layer is present. Vulval cone ambifenestrate, bullate with prominent underbidge.

Male

Body vermiform, slightly arcuate ventrally when relaxed. Labial region hemispherical, smooth marked by three faint annuli and prominent labial disc.Labial famework strongly scleritised. Lateral field with four incisors.Stylet knobs round, sloping posteriorly. Hemizonid two to three annuli long, 1-5 annuli anterior to excretory pore.Medium bulb oval. Spicules cephalate, ventrally arcuate with rounded tip.

Second Stage Juvenile

Body tapering more so posteriorly with arcuate or straight posture when relaxed.Annuli distinct. Labial region rounded, off set with two to three annuli and framework strongly developed. Lateral fields with four incisors. Stylet well developed basal knobs of equal sizeand similar shape with flat to concave anterior surface.Hemizonid distinct. Hemizonion 5-7 annuli posterior to hemizonid. Phasmids punctiform.

Measurements

Eggs

Length= 87.2 (80-110)μ, Width= 40.1 (36-45)μ, L/W= 2.2 (1.7-2.8).

Juveniles

L = 420 (400-440)μ; a= 24.1 (22.7-26.9); b=4.0(3.4-4.7); b′= 2.5 (2.3-2.9); c =8.8 (7.7-9.4); c′ = 4.2 (3.4-4.7); Stylet= 22 (21-23)μ; DGO= 5 (4.8-6.0)μ; excretory pore= 98 (87-107)μ; anterior end to esophageal gland= 166 (153-178)μ; Hyaline tail=24(36-45)μ; Tail=48(39-54)μ.

Males

L=1000 (900-1200)μ; W= 25 (22-27)μ; a = 41 (36-44); b = 7.8 (7.5-9.5); T = 63= (50-68); stylet = 26 (25-27)μ; DGO = 5 (4.5-6.0)μ; excretory pore = 134 (124-140)μ; base of esophageal glands = 188 (175-210)μ; spicules = 27 (26-29)μ.

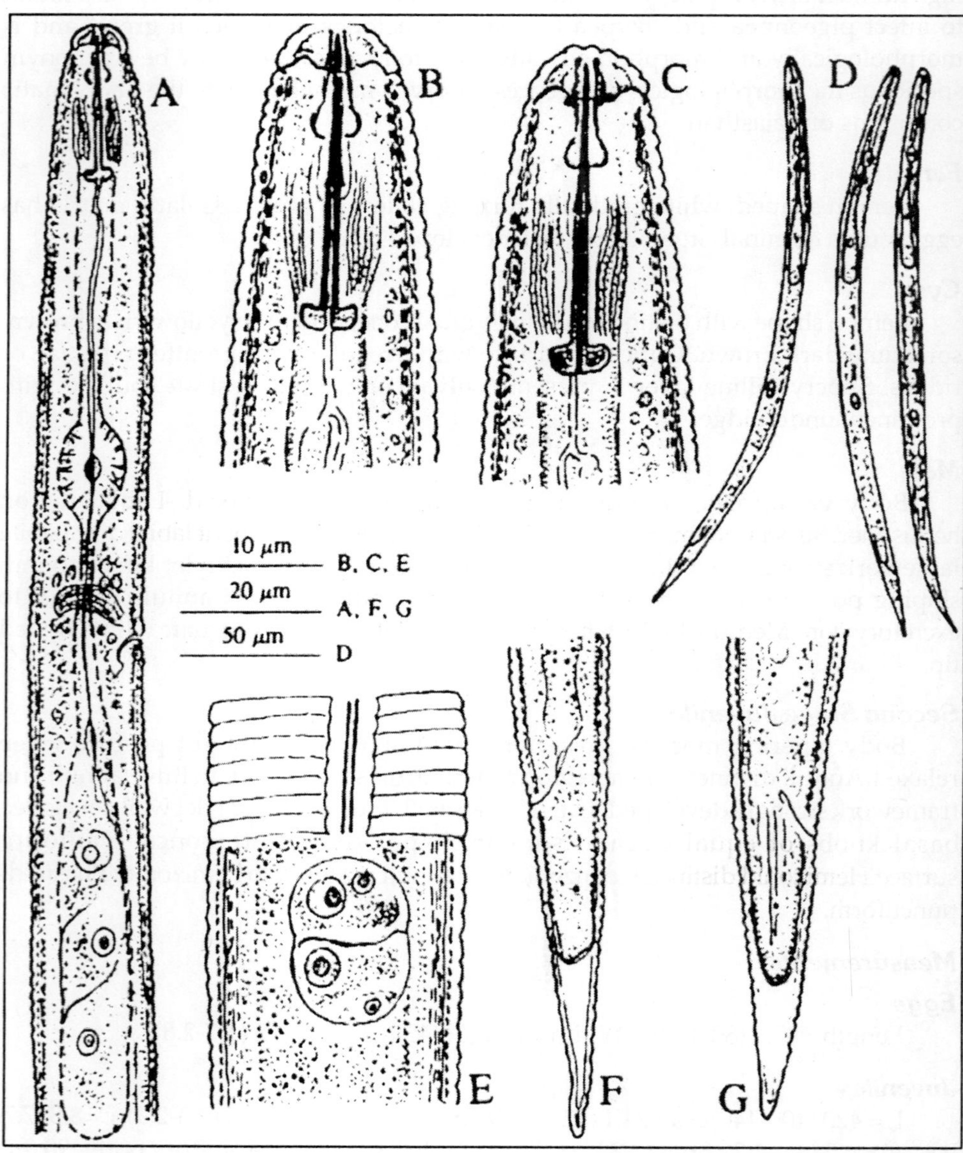

Figure 30: ***Heterodera swarupi***
A-G: J2. A: Pharyngeal region; B, C: Anterior regions, D: Three juveniles showing habitus on death, E: Genital primordium and lateral field, F, G: Tail ends. After Sharma *et al.,* 1998.

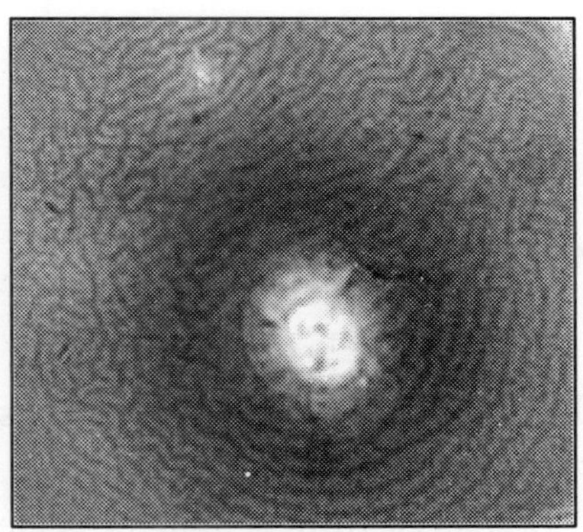

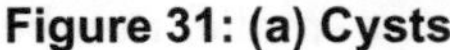

Figure 31: (a) Cysts **(b) Terminal area**

Cysts

L (including neck) = 589 (520-700) μ; W = 366 (320.475)μ; neck length = 69.1 (25-100)μ; L/W = 1.7 (1.6-2.1); cone height = 64.3 (50.80)μ; vulval bridge width = 4-6μ.

7.10 The Potato Cyst Nematodes, Globodera spp.

7.10.1 The Golden Nematode of Potato, Globodera rostochiensis (Woll., 1923) Skarvilovich, 1959

To accommodate the potato cyst nematodes and related species having round cysts, Skarbilovich (1959) erected the subgenus *Globodera* which was later elevated to generic status by Behrens (1975). The potato cyst nematodes have been reported in tropical countries but only on rare occasions. *Globodera rostochiensis* passes through a yellow stage before rupturing root cortex. *Globodera pallida* remains creamy white until dying and becoming a brown cyst. In India, potato cyst nematode *G. rostochiensis* was first reported by Jones (1961) who detected cysts from the roots of potato plants which exhibited the symptoms of yellowing of leaves in Nilgiris hills of the Tamil Nadu state. Here *G. rostochiensis* and/or *G. pallida* are considered to be serious pests of potatoes and does not have a tropical climate (Seshadri and Sivakumar, 1962; Prasad and Chawla, 1965; Logisvaran and Menon, 1969). It is also reported from Idduki district in Kerala. Recently both these species have also been reported from Shimla (H.P.) (Ganguly *et al.*, 2010).

Second-stage juveniles are vermiform and about 470 μm in length, with a strong stylet in the mouth for puncturing cell walls, and a pointed tail. Males are similar in general appearance, about 1200 μm in length, with copulatory spicules close to the tail which is short and blunt. Females are virtually spherical with a projecting neck containing the oesophagus and associated glands; diameter approximately 450 μm. The cysts are similar in shape, but with a tanned skin and degeneration of the internal organs (Golden and Ellington, 1972; Stone, 1973a; 1973b).

Measurements

Cysts

Length (excluding neck): 487-658 (556) µ; Width: 392-645 (497)µ ; Fenestral length: 8-20 (15)µ ; Fenestral width: 18µ ;Vulval slit length: 6-11 (9)µ.

Second Stage Juveniles

Length: 425-505 (468) µ; Stylet length: 21-23 (22) µ; Tail length: 42-50 (44) µ; Hyaline tail length: 24µ; DGO from Stylet base: 2.6µ.

7.10.2 The Cream Color Cyst Nematode of Potato, *Globodera pallida* Stone, 1973

This species differs from the golden nematode (*Globodera rostochiensis*) because females lack the golden phase. They are pale and have a smaller number of cuticular striae (8-20 vs. 16-31) in the perineum.This species is reported to be present in Himachal Pradesh, Kerala and Tamil nadu in India.

Cyst

Body is globose, spheroidal, with a short neck and no terminal cone. Cuticle thick, with superficial, lace-like pattern; D-layer present. Vulva is of medium length and terminal. Vulval area circumfenestrate; superficial tubercles near vulva. No anal fenestration, but anus and vulva lying both in a "vulval basin." Underbridge and bullae rarely present. All eggs retained in cyst body.

Males

Vermiform, twisted into a C or S–shape. Lateral fields with four incisures. Spicules greater than 30µm in length, distally pointed. No cloacal tubus. Tail short, hemispherical.

Second Stage Juveniles

Stylet less than 30 µm long. Lateral field with four lines. Esophageal glands fill body cavity. Tail conical pointed with terminal half hyaline. Phasmids are puntiform.

Measurements

Cyst

Length: 496-673 (587)m; Width: Fenestral length: 21µ ; Fenestral width: 19µ ; Vulval slit length: 9-15 (11)µ.

Second Stage Juveniles

Length: 440-525(484) µ; Stylet length: 21-26 (24)µ ; Tail length:42-62 (52)µ ; Hyaline length : 24µ ; DGO from Stylet base: 2.7µ.

Potatoes are by far the most important host crop. Tomatoes and aubergines (brinjal) are also attackedby these nematodes in addition to other *Solanum* spp. and their hybrids which can also act as hosts. Both species of *Globodera* have several different pathotypes (Kort, 1974). The pathotypes are characterized by their ability to multiply on certain tuberous *Solanum* clones and hybrids used in breeding. Five

Figure 32: Cysts of *G. pallida*

pathotypes are recognized within *G. rostochiensis* (Ro1- Ro5 international notation) and three in *G. pallida* (Pa1-Pa3) (Kort *et al.,* 1977). Some of these pathotypes are recognized by their almost total inability to multiply on specific cultivars of potato (single-gene resistance); for example, the most commonly grown resistant potato cultivars (based on gene H1 derived from clones of *S. tuberosum* subsp. *andigena*) are resistant to pathotype Ro1 of *G. rostochiensis* only. Other pathotypes show different levels of ability to multiply on different cultivars.

Egg hatch is stimulated by host root diffusate (60-80 per cent) - only about 5 per cent hatch in water. Some eggs do not hatch until subsequent years. J2 moves into root, establishes feeding site, and undergoes 3 more molts. Adult males do not feed. Sex is determined by food supply - males develop in adverse conditions and heavy infestations. Nematodes reproduce sexually; males are attracted to females by a pheromone sex attractant. Nematodes may mate several times. Annual population decline in the absence of a host varies from 18 per cent in cold soils (Scotland) to 50 per cent in warm soils, with an average decline rate about 30 per cent - so population decline follows this pattern: 100-70-50-35-23-etc.

Globodera pallida appears to be more responsive to potato root diffusate than *G. rostochiensis*. Consequently, where both species are present, *G. pallida* reproduces to a greater extent and predominates population levels (Devine and Jones, 2003). The problem is exacerbated when potatoes resistant only to *G. rostochiensis* are grown. Available cultivars have only partial resistance to *G. pallida* and do not prevent its increase. *Globodera pallida* is progressively replacing *G. rostochiensis* in Britain (Trudgill *et al.,* 2003).

7.10.2.1 Differences in *Globodera rostochiensis* and *G. pallida*

Sl.No.		*G. rostochiensis*	*G. pallida*
	Cysts		
1.	Cysts	Golden cysts	White or creamy cysts
2.	Number of cuticular ridges (between vulva and anus)	16-31(>14)	8-20 (<14)
3.	Granek's ratio	1.3-9.5 (>3)	1.2-3.5 (<3)
	Juveniles		
4.	Stylet length	19-23 (21.8)µ	22-24 (23.6) µ
5.	Stylet	Knobs rounded	Anteriorly pointed
6.	Length	468µ	484µ
7.	Tail length	43.9 µ	51.9 µ
8.	Contours	Rectangular contours of oral disc	Ovate contour
	Males		
9.	DEGO	5.3 µ	3.5 µ
	Females		
10.	Stylet length	22.9 µ	26.7 µ
11.	Anal-vulval distance	60 µ	43.9 µ

7.10.2.2 Pathotypes of *Globodera rostochiensis* and *G. pallid*

Differential Hosts Used for Separating Pathotypes of Potato Cyst Nematodes

Differential Host	*G. rostochiensis*					*G. pallida*		
European Scheme	*Ro1*	*Ro2*	*Ro3*	*Ro4*	*Ro5*	*Pa1*	*Pa2*	*Pa3*
S.tuberosum ssp.tuberosum	+	+	+	+	+	+	+	+
S.tuberosum ssp.andigena CPC1673	–	–	+	+	+	+	+	+
S.tuberosum ssp.andigena 289990.10							–	–
S.kurtzianum 60.21.19	–	+	–	+	+	+	+	+
S.vernei 58.1642/4	–	+	–	–	+	+	+	+
S.vernei 62.33.3	–	–	–	–	+	–	–	+
S.vernei 65.346/19	–	–	–	–	–	+	+	+
S.multidissectum P55/7	+	+	+	+	+	+	+	+

Kort *et al.* (1977).

7.10.2.3 Pathotype Status in Potato Cyst Nematodes in India

The identification of prevalent cyst nematode species from different localities in Nilgiris has indicated that both *G. pallida* and *G.rostochiensis* are prevalent at most of the localities surveyed in mixed populations. At Kodaikanal hills also both the species were encountered. Although, the cysts found at Karnataka in potato soil could not be characterised, *G. pallida* was found associated with potato at Kerala.

Initially, it was thought that PCN populations at the Nilgiris comprised *G. rostochiensis* pathotype A (Hari Kishore *et al*, 1969). Breeding and screening potato for resistance to cyst nematodes brought to light occurrence of Ro1 and Pa 2 in both species (Howard, 1977). Cultivation of nematode-resistant potato 'Kufri Swarna' (derived from *Solanum vernei*) at different localities in the Nilgiris indicated the presence of other pathotypes (Krishna Prasad, 1996). Differential host-reaction studies have shown that three pathotypes occur in each species at the Nilgiris (Krishna Prasad, 2004). Pathotype Ro1 of *G. rostochiensis* and Pa2 of *G. pallida* are the most prevalent and constitute 75 per cent of total population. Other pathtoype Pa1 accounted for 15 per cent, followed by Ro2, at 7 per cent. The least prevalent pathotypes Pa3 and Ro5 accounted for only 3 per cent but were able to develop distinctly on some of the differential hosts, indicating their virulence (Krishna Prasad, 2006). PCN populations from Kodaikanal constituted pathotypes Ro1 of *G. rostochiensis* and Pa 2 of *G. pallida*.

7.10.3 *Globodera chaubattia* (Gupta and Edward, 1973) Wouts, 1984

This species was found in the soil of an apple orchard at Chaubattia, Uttar Pradesh, India and decribed as *Heterodera chaubattia* (Gupta and Edward, 1973). On the basis of description of this species, it is now recognized as *Globodera chaubattia* (Wouts, 1984)

7.11 The Bamboo Cyst Nematode, *Afenestrata sacchari* Kaushal and Swarup, 1988

This afenestrate cyst nematode was described Kaushal and Swarup (1988) infesting 'saccharum' grass (*Saccharum munjo*) roots in village Oel of Una district of Himachal Pradesh. Subsequently, it has also been found infesting bamboo plantation in Jorhat district of Assam.

Diagnostic characters

Cyst and Female

This cyst nematode is characterized by the absence of fenestration and a sunken vulva in the cone top structure. The vulval flaps are very apparent with different patterns of cuticular configuration, which may the extension of cuticular markings on the cyst surface. The cysts are lemon shaped and light brown in colour, variable in size and are provided with a long neck. Stylet is weakly sclerotised with slightly posterior directed knobs in females. Remnants of vaginal muscles are usually present. Underbridge is present in some cysts. Bullae are absent. In second stage juveniles, the lateral field is provided with four incisors. Tail is conical with two-third area hyaline. Males are present.

Male

Body slightly curved when relaxed. Labial region with three or four annuli, labial framework heavily sclerotised. Testis single. Spicules straight with opening incloacal tubus, simple rod shaped. Tail absent.

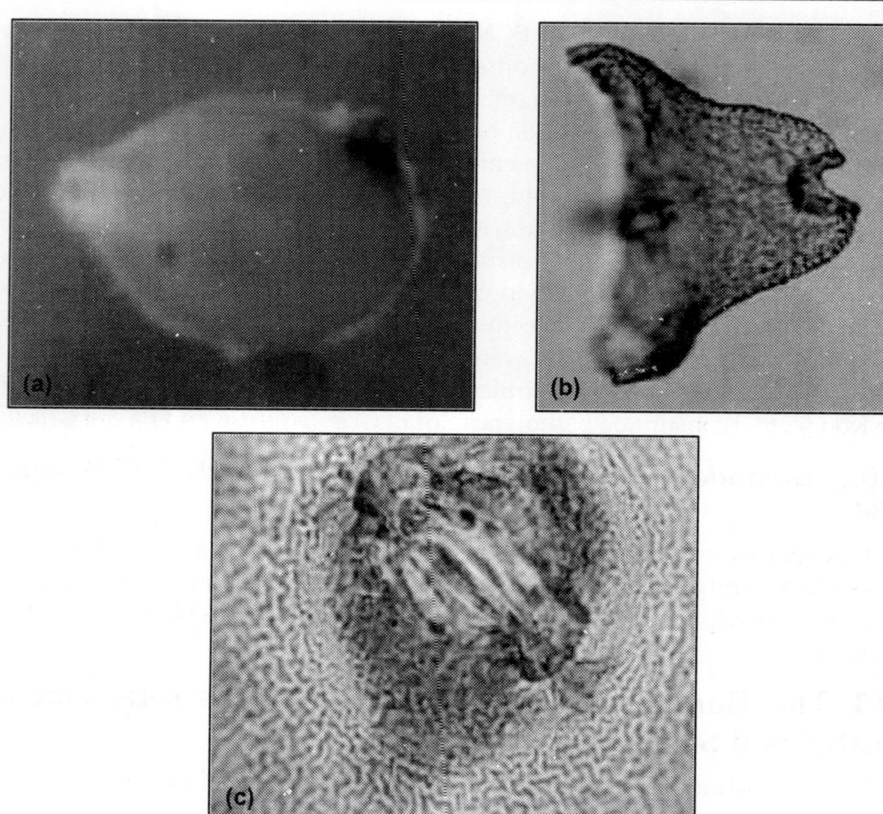

Figure 33: (a) Cyst; (b) Sunken Vulva; (c) Underbridge

Second Stage Juvenile

Body slightly curved when heat relaxed. Cuticle annulated, labial framework strongly sclerotised, labial region prominently elevated with three or four annuli. Lateral field with four incisors. Stylet strong with knobs slightly directed anteriorly. Medium bulb ovoid with well developed crescent shaped valve. Considerable overlap of intestine by pharyngeal glands that fill body cavity. Excretory located just posterior to hemizonid that is in turn near level of pharygio-intestinal junction. Tail conical,tapering posteriorly with hyaline region extending two third of tail length.

Measurements

Eggs

Length: 114-122(118) µ; Width: 34-39 (38)µ; L/W: 3.4-3.8 (3.6).

Second Stage Juveniles

Length: 392-547 (449)µ; Width: 17-19.5 (18.5) µ; Stylet length: 22-29 (24.8)µ; Ant. end –ex. Pore: 79-118 (91)µ; Tail length: 29-59 (51): Hyaline tail length: 19-49(24)µ; a: 20-25.3 (22); b: 4.2-6.6(5.3); b2: 2.1-4.5(2.5); c: 7.2-9.4(8.9).

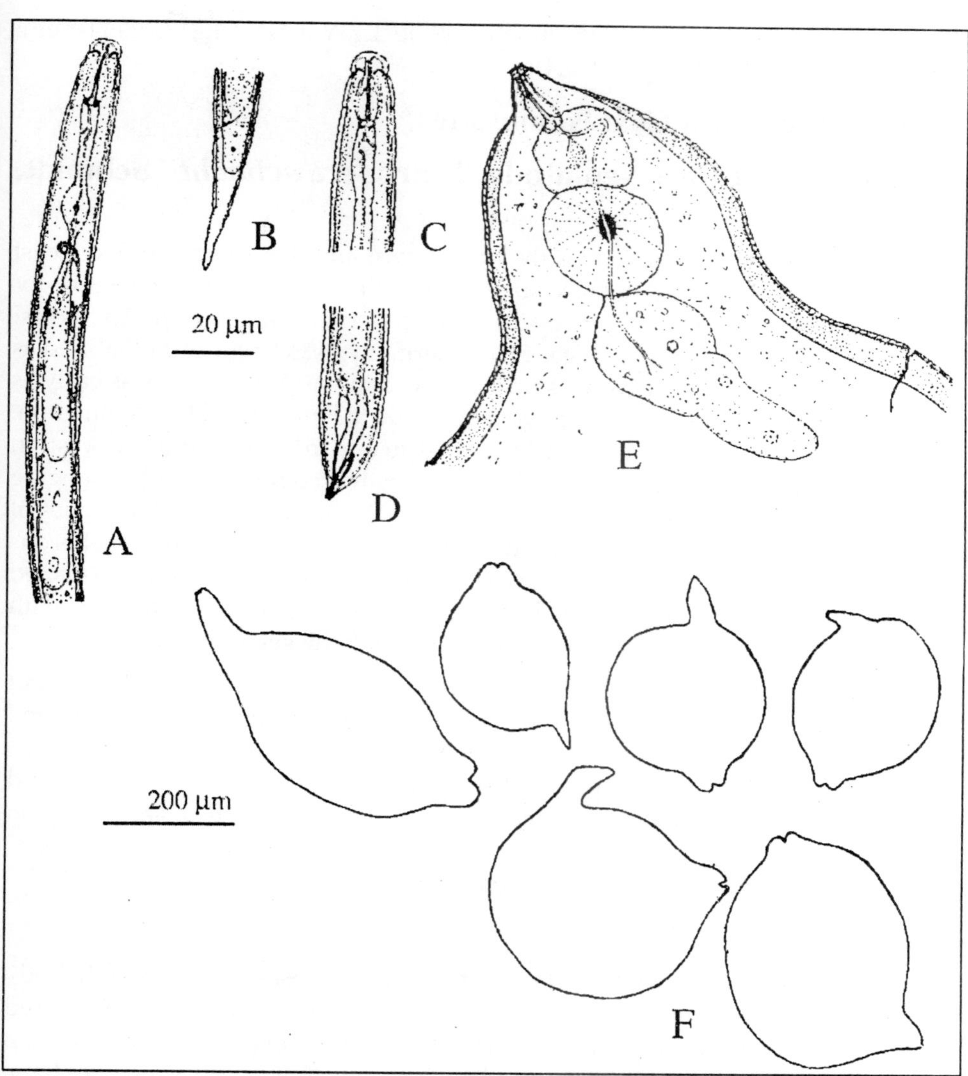

Figure 34: ***Afenestrata sacchari***
A: Pharyngeal region of J2, B: Tail of J2, C: Anterior region of male, D: Tail of male. E: Pharyngeal region of female; F: Cysts. After Kaushal and Swarup, 1988

Males

Length: 833-1120 (964) μ; Width: 22-27 (24)μ; Stylet length: 23-28 (26)μ; Spicule length: 30-44 (36)μ; Ant.end-ex.pore: 110-140 (136)μ; Gubernaculum: 10-13 (11)μ.

Cysts

Length: 378-896 (577)μ; Width: 287-812 (498)μ; L/W: 1.1-1.3(1.2); Stylet length: 23.30 (26)μ; Neck length: 112-190(134)μ.

Another Important Cyst Nematode

7.12 Sugarbeet Cyst Nematode, *Hetrodera schachtii* Schmidt, 1871

The sugarbee cyst nematode is the first known and described species of cyst nematodes which was found to cause huge losses to the sugarbeet and responsibe for closing down of many factories processing sugar in Germany and Europe. It was first identified in 1859 on sugar beets near Halle, Germany by Schacht and identified as *H. schachtii* by Schmidt (1871).It was first observed in the United States as early as 1895. It is distributed throughout Europe from Spain to Finland and Eire to Bulgaria. It is also recorded in Soviet Union, Turkey, Israel; in U.S. in both eastern and western states; in Canada, Australia, and South Africa. It mostly feeds on the plants belonging to the families Chenopodiaceae, especially *Beta vulgaris*, Cruciferae, all varieties of *Brassica oleracea*, including cabbage, cauliflower, brussels sprouts, and cruciferous weeds. It can attack over 200 plant species in 23 different plant families. Affected plants may be chlorotic and have small storage roots that are severely branched with excess fibrous roots often referred to as "bearded" or 'whiskered' roots.

The second infective stage may remain dormant in the eggs for several years, but some hatch every year and emerge from the cysts into the soil. Hatching is stimulated by exudates from roots of host and some non-host plants, but also takes place to a lesser extent in the absence of plants. The optimum temperature for growth and reproduction is 21-27 C and the life cycle requires four to six weeks, depending on soil temperature. One to 2 generations can be produced per year in temperate regions but up to 5 generations per year have been obtained.

Females

White, flask-shaped, with short neck embedded in host root and swollen body on root surface. Terminal vulva on conical protruberance ('vulval cone') covered with gelatinous matrix containing eggs. Head small, neck expanding rapidly, then cylindrical;excretory pore at 'shoulder' where body swells to nearly spherical shape terminating in vulval cone, with anus dorsally sub-terminal. Head skeleton is weak. Stylet slender with small knobs. Median esophageal bulb prominent, spherical. Esophageal glands overlap intestine latero-ventrally. Paired ovaries are long and much coiled. A few eggs are deposited in a gelatinous matrix, but most retained in body.

Males

Body is usually straight with posterior quarter spirally twisted through 90-180 degrees when heat relaxed. In length the males usually range between 1.3 and 1.6

mm. Annules of the lip region, three or four, including the labial disc. Stylet well developed with knobs concave anteriorly about 25 to 28 µ long, with strong basal knobs. Esophagus is cylindrical, expanding midway to a fusiform bulb with valve; isthmus encircled by nerve ring. Lobes of esophageal glands extending back ventrally along the intestine. Excretory pore is 2 to 3 body-widths behind median bulb;hemizonid 6 to 10 annules in from of pore.Testis single, outstretched; spicules bidentate, resting on a slightly arcuate, troughlike gubernaculum. Terminus is bluntly rounded with tail less than half body-width long. Phasmids are adanal.

Cyst

The cyst contains about 500-600 eggs. The terminal vulval slit is about as long as the 'vulval bridge' which is flanked on either side by a kidney-shaped thin area of the cyst wall that breaks down in older cysts leaving 2 apertures or 'semi-fenestrae' separated by the vulval bridge. Within the cone are the remains of the vagina attached to the side walls by the 'under-bridge' and a number of irregularly arranged, dark brown bullae situated a short distance below the vulval bridge. Vagina looks sheaflike from a dorsal view, ending in a deeply cleft vulva. Fenestrae average 45 *u* long. The surface of mature females and newly-formed cysts is encrusted with a white waxy material, the 'sub-crystalline layer' which soon falls off the cyst when it is loose in the soil.

Second Stage Juveniles

Head offset, hemispherical, with 4 annules; robust, hexaradiate skeleton; smallamphid apertures in lateral sectors close to mouth opening.Average length of juveniles is 460µ. Body annules 1.4 µm wide in spear region and 1.7 µm at midbody. Lateral field with 4 incisures. Stylet is moderately heavy with prominent, forwardly-directed knobs. Styletr averaging about 25 µ length, which usually is about the length of the hyaline portion of the tail.Esophagus as in male, but median bulb more prominent; dorsal gland duct opens 3-4 µm behind spear base. Anus is obscure, about 4 anal body-widths from terminus. Tail is acutely conical with rounded tip; a distinct hyaline terminal section 1-1.25 times the stylet length. Genital primordium is provided with 2 nuclei, located slightly behind mid-body. Phasmids are obscure, just post-anal.

Cactodera cacti (Filipjev and Schuurmann Stekhoven, 1941) Krall and Krall, 1979

This species was reported on *Echinopsis* spp. in Mysore, India (Kumar, 1964). This has also recently been observed in the Ooty region of Tamil Nadu along with potato cyst nematodes (Sharad-personal communication, 2011). The nematode could have been introduced from outside with the host plant, which is an ornamental plant.

7.13 *Cactodera johanseni* Sharma, Kaushal, Singh, Pande, Pokharel and Upreti, 2001

Sharma *et al.* (2001) described this species of cyst nematode from the radish fields in a village Nagarkot, district Bakhtapur in Nepal. Subsequently, this species

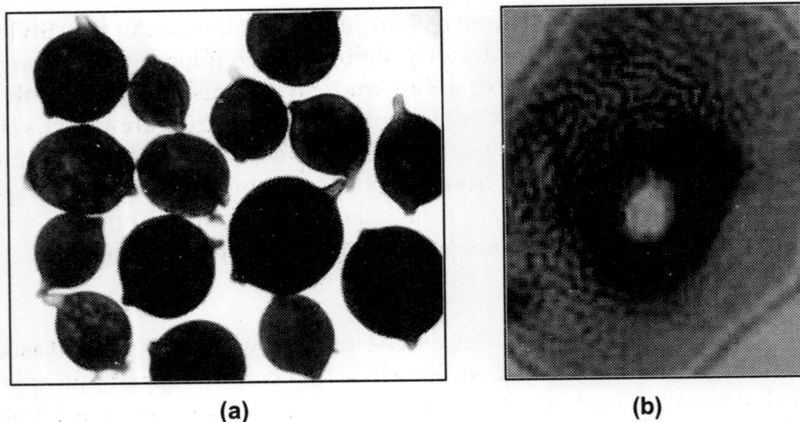

(a) (b)

Figure 35: (a) Cysts of *Cactodera johanseni* (b) Cone top *C. johanseni*

was also recorded from Shimla (Himachal Pradesh) and Almora and Champawat districts (Uttarakhand) in fields having maize-wheat cropping system (Kaushal *et al.*, 2007). Recently this cyst nematode species has been wrongly transferred to the genus *Heterodera* (*H. johanseni*) by Sturhan (2002) due to some misinterpretation of the figures given in the original description. It is indeed a Cactoderid nematode.

Diagnostics

This species is characterized by the dark brown, lemon shaped cysts with short but prominent posterior protuberance. The posterior protuberance was more in smaller than in the large sized cysts. Lateral field in the second stage juveniles has four incisors forming three bands, outer two bands crenate and often areolated.

Measurements

Eggs

Length: 85-115 (103) µ; Width: 35-40 (39) µ.

Second Stage Juveniles

Length: 420-488(452)µ; Width: 14-19 (16.3)µ; a:16-23 (19); c: 7.5-10 (8.7); c: 3.6-4.4(4); Stylet length:20-23 (21.5)µ ; DGO from stylet base: 5-6 µ; Ant. end –Ex. pore: 18-22(20)µ; Tail length: 47-61 (53)µ; Hyaline tail length: 24-31 (39)µ; anal body width :12-15 (13)µ.

Cysts

Length: 400-630 (534) µ; Width: 380-580 (508) µ; Neck length: 20-80 (42) µ; Post. protuberance height: 20-30 (22) µ; L/W:1.0-1.4 (1.4).

7.14 Some Cystoid Nematodes from India

7.14.1 The Bamboo Cystoid Nematode, Brevicephalodera bamboosi Kaushal and Swarup, 1988

Kaushal and Swarup (1988) described this as a new genus and a species of cyst nematode from the village Kumarganj in Jorhat district of Assam infecting bamboo

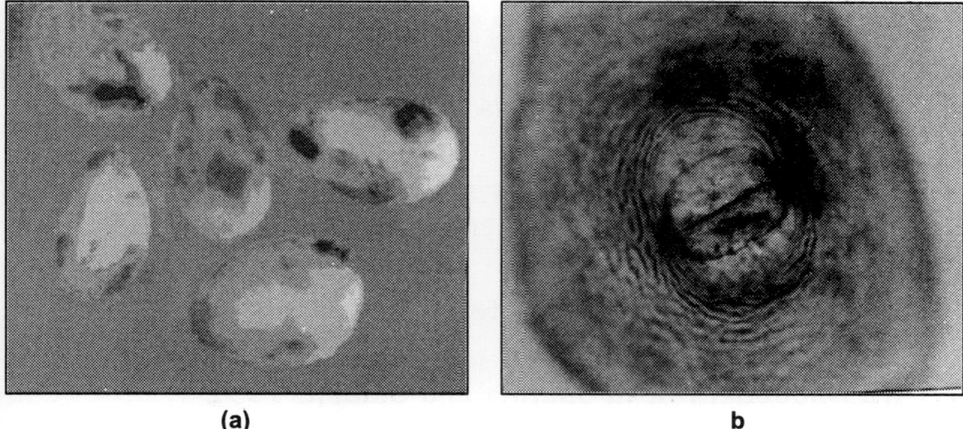

Figure 36: (a) Cystoids of *B. bamboosi;* (b) Posterior End

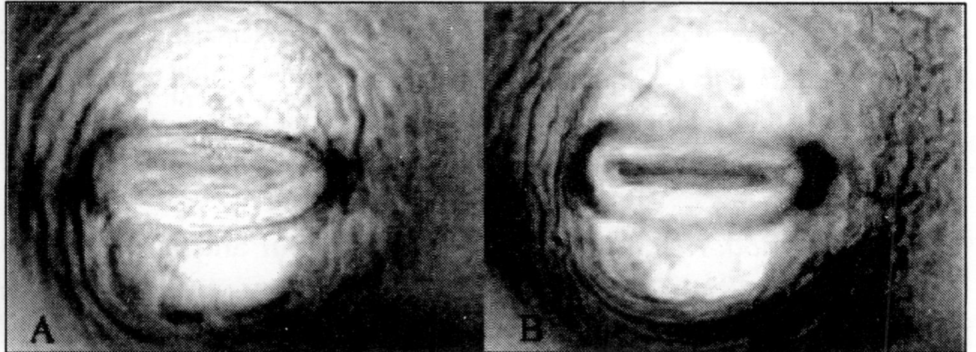

Figure 37: *Brevicephalodera bamboos*: Vulval cone.
A: Anterior view, B: View at deeper level (Scale bar = 10 µm)
(Original Z. Handoo, Nematology Collection, USDA, USA)

roots. But later, it was confirmed to be a cystoid nematode.The mature cystoid body of female does not tan properly and also does not contain eggs and larvae. This species was transferred to *Afenestrata* as *A. bamboosi* by Siddiqi (2000) and considered as *Heterodera bamboosi* by Wouts and Baldwin (1998) but here it is retained as a separate genus because the species does not conform to the definition of a cyst and *per se* the above transformations stand invalid.

Diagnostic Features

The cystoid body is provided with very thick cuticle and almost reniform shape. It is characterized by having a very short neck and a depressed vulval area with elevated vulval lips. The vulval area is trophied without any apparent fenestration. Underbridge and bullae are absent.

Measurements

Eggs

Length: 99-120 (109); Width: 32-45 (37).

Second Stage Juveniles

Length: 444-498 (472)µ ; Width : 16.5-19.0 (18)µ ; Stylet length : 19-23 (20)µ ; Ant. end –ex. Pore: 93-126 (120)µ; Tail length: 50-70 (63) µ; Hyaline tail: 37-47 (41)µ; a: 25.8-29.7 (27.4); b: 3.9-5.0 (4.7); c: 6.9-9.3 (8.2).

Males

Length: 892-1227 (1098)µ; Width: 21-25 (23)µ; Stylet length; 24-28 (25)µ; Spicules length: 31-39 (35)µ; Gubernaculum: 10-12 (11)µ; Ant.end-ex.pore: 105.138 (126)µ.

Cystoid Body

Length: 455-840 (609)µ; Width: 240-585 (370)µ; L/W: 1.4-1.9 (1.6); Stylet length: 21-26 (23.8)µ; Neck length: 40-70 (57)µ.

7.14.2 *Bilobodera flexa* Sharma and Siddiqi, 1992

Bilobodera flexa was described from soil and roots of *Allmania nodifolora* in Andhra Pradesh, India. Bilobed females characterize it with a sharp dorsal depression opposite the median vulva, a subterminal anus, a spheroidal postanal region, and eggs that are generally retained in body. There is no immature slender female stage. The male body is small, under 0.6 mm long, C-shaped when fixed; but unlike other Heteroderidae, never twisted in the posterior region; bursa absent. Second-stage juveniles are small, under 0.4 mm long and have lateral fields with four incisures and punctiform phasmids located slightly behind the anus but anterior to the middle of the tail.

7.14.3 *Cryphodera kalesari*, Bajaj, Walia, Dabur and Bhatti, 1989

Cryphodera kalesari was described by Bajaj *et al.* (1989) infesting *Colebrookea oppositifolia* Sm. from the Kalesar forests (Haryana), India.

Females

Swollen, saccate, with two to three lip annules, with vulval lips protruding.

Second Stage Juveniles

Length 384±19µm, spear=26.5±1µm, tail 46.5±6µm, hyaline portion 21.5±2.5µm, *en face* dorsal and ventral arcs of head skeleton bifurcate.

Males were not found.

Chapter 8
Root Knot Nematodes

History

After the discovery of wheat gall nematode, *Anguina tritici* in the middle of 18th century by Needham (1743), the second plant parasitic nematode known to the science was the root knot nematode found causing galls on cucumber roots in greenhouse in England (Berkeley,1855). He reported that, *'on closer examination the root was found to be covered with excrescences varying from the size of a small pin's head to that of little Bean or Nutmeg'*. Berkeley noted the enormous development of the vascular tissues within the galls and recorded the presence of 'Vibrio': *"It appeared that these cysts were regular membranous sacs, exactly resembling the sporangia of Truffles, and filled with a multitude of minute elliptic or slightly cymbiform eggs, averaging not more than 1/250th of an inch in length with a breadth of 1/600th. In many of these the nucleus already showed the form of a Vibrio, folded up once or twice, and several of the animals were free, though still of small size, having escaped from the eggs by a little circular aperture at one extremity"*.

Later on, the galled roots of many plant hosts were observed by many workers (Greeff, 1864; Licopoli, 1875; Jobert, 1878; Cornu, 1879; Muller, 1884; and Treub, 1885; Atkinson, 1889. Neal, 1889; Cobb, 1890) but all of them recognized them as the species of either *Heterodera* or *Anguillula*. To the root knot nematode found on the roots of *Dodartia orientalis* Muller (1884) found that they were related to the genus *Heterodera* but mistook them to be *Anguillulina radicicola* which had been described from the host by Greef (1872). He called them as *Heterodera radicicola*. He also studied in detail the process of gall induction and development followed by morphological observations on stylet, dorsal gland orifice, metacarpus, gonads and spicules. He was the first person to study the perineal pattern and briefly described the differences between cyst and root knot nematodes. A year later, Frank (1885) published several new hosts of *H. radicicola* and introduced the idea of using hosts for reducing the damage by root knot nematodes.

Melchior Treub (1885) named the root knot nematode infesting sugarcane in Java as *Heterodera* (*Meloidogyne*) *javanica* differentiating it from Muller's species by measurements. But Beijerinck (1887) and van Breda de Haan (1900) considered *H. javanica* identical to *H. radicicola*. In the United States, Neal (1889) worked on the gall forming nematodes of radish, peach, fig and orange trees describing them as *Anguillula arenaria* and later in the same year Atkinson (1889) gave an account of life history of root knot nematodes that he identified as *Heterodera radicicola*. Following the observations made by Jobert (1878) on the galls found on coffee roots, Emilio Augusto Goldi (1887) proposed the name *Meloidogyne exigua* for the root knot nematode he found infesting coffee roots in Rio de Janeiro, Brazil (This paper was, in fact, published in 1887 and reprinted in 1892 as Göldi also published an article entitled 'Biologische Miscellen aus Brasilien. VII. Der Kaffeenematode Brasiliens (*Meloidogyne exigua* G.)' in 1889 (Göldi, 1889). In this article he discussed '*Meloidogyne exigua* G.' and provided some morphological data of the various stages and compared it with *Heterodera*. This generic name was derived from Greek words *melois* meaning 'apple-shaped' *gyne* means 'female' and the specific name referred to the small size of the females. However, the name *H. radicicola* remained in use till Goodey (1932) reviewed the nomenclature and synonymized all the names of root knot nematodes with *Heterodera marioni*, the name given by Cornu (1979) and guided that all the root knot nematodes should be known by this name (Hirschmann,1985). He pointed that Greef's *Anguillulina radicicola* was not related to either *Heterodera* or *Meloidogyne* but was what is now known as *Subanguina radicicola*.

Shortly after Göldi's proposal of the genus *Meloidogyne*, Neal (1889), clearly unaware of the former publication, proposed the gall-forming nematode, *Anguillula arenaria*. He produced a comprehensive, highly detailed and nicely illustrated paper on the root galls of plants, including radishes, peach, fig and orange trees in Florida, USA.

He also referred to reports of this root-knot disease being recognized as far back as 'the earliest settlement of the South Atlantic and Gulf states by white people' and stated 'In 1869, I found the root-knot prevalent over Florida, and learned from old residents that as far back as 1805 it had been known'. In the same year, Atkinson (1889) found 'giant cells' in cross-sections of rootknot nematode-infected roots, although he interpreted these as dead females rather than nutritional sources. Other reports of root-knot nematodes include those of Cobb (1890) from New South Wales, Australia, and Gaston Lavergne (1901) who reported *Anguillula vialae* from the roots of vines in Chile.

Cobb (1924) recognizing that there were differences between cyst-forming and root-knot nematodes, proposed the genus *Caconema* Cobb, 1924 to contain the latter but Tom Goodey (1932) did not accept Cobb's proposal of *Caconema* as he regarded rootknot and cyst-forming nematodes as congeneric. Goodey (1932) regarded *A. marioni* Cornu, 1879, the oldest name applied to root-knot nematodes, as belonging to the genus *Heterodera*, a combination that had previously been proposed by Marcinowski (1909) and called them as *H. marioni*. Nagakura (1930) did publish an extensive study on root-knot nematodes, differentiating them in numerous ways from cyst nematodes and making observations on their morphology and life cycle, he still referred to them as *H. radicicola*. It was Chitwood (1949) who revised the genus of root knot nematodes and reallocated them to the genus *Meloidogyne* as proposed by Goeldi (1887) and recognized *M. incognita* (Kofoid and White, 1919) Chitwood, 1949, *M. javanica* (Treub,1885) Chitwood, 1949 and *M. arenaria* (Neal,1889) Chitwood,1949 as valid species. He also described two new species *M. hapla* Chitwood, 1949 and *M. incognita acrita* Chitwood, 1949. He redescribed the first three species obtaining the material from either prints or from the topotype material. For the description of *M. incognita*, he used a population from carrots from Texas, presumed to be the type locality by Kofoid and White (1919) for *Oxyuris incognito*. The original description was based on the eggs from faecal samples of 1, 40,000 soldiers at a military camp in Texas and the eggs in Chitwood's collection did not differ from those described by Kofoid and White. He did not give diagnosis for each of the above species but provided a key in which the perineal pattern, shape of the stylet knobs and stylet length in males and females and position of dorsal esophageal gland opening was given for distinguishing these species. He made some other important observations regarding susceptibility of a particular population to a particular host. For example he noticed that *M.hapla* reproduced only on strawberry and that *M. incognita* failed to infect peanut but readily reproduced on cotton. His work provided the stimulus to many workers to look more closely for the differentiating characters in their populations of root knot nematodes with the result that today around 100 species of this nematode are known.

Sledge and Golden (1964) proposed the genus *Hypsoperine* Sledge and Golden, 1964 for species of *Meloidogyne* where the mature female was characterized by a thicker cuticle and an elevated, cone-like posterior region. The history of this genus has been somewhat chequered, with many authors (*e.g.* Whitehead, 1968; Hirschmann, 1985; Jepson, 1987) regarding it as a junior synonym of *Meloidogyne*, while Siddiqi (1986) not only recognized it as a valid genus but split it into two subgenera, *Hypsoperine* (*Hypsoperine*) and *Hypsoperine* (*Spartonema*). However, in the second edition of his *magnum opus*, Siddiqi (2000) synonymized *Hypsoperine* (*Hypsoperine*) under *Meloidogyne*, but raised the former subgenus *Spartonema* to genric rank. Plantard *et al.* (2007), on the basis of 18S rDNA sequences, refuted the generic status of *Spartonema*. The monograph by Whitehead (1968), proposed four new species and recognized 23 as valid species. Franklin (1971) reviewed the genus and included some 32 species, while Esser *et al.* (1976) provided a compendium to facilitate identification of 32 species. In a later work, Franklin (1979) again reviewed the genus, recognizing 36 valid species. Five years later, this total had risen to 54 species and two subspecies (Hirschmann, 1985). Other major reference sources include the monograph by Lamberti and Taylor (1979), the compendium by Hewlett and Tarjan (1983), the two-volume treatise edited by Sasser and Carter (1985) and Barker *et al.* (1985), and the well illustrated monograph by Jepson (1987). The latest monographs are by Karssen and van Hoenselaar (1998) and Karssen (2002), both of which cover the European species of the genus, and Karssen and Moens (2006), where 89 valid species are listed

8.1 Classification

The root knot nematodes were placed in the subfamily Meloidogyninae (thereby emphasizing the differences between root-knot and cyst-forming nematodes) along with Heteroderinae Filipjev and Schuurmans Stekhoven (1941) in the family Heteroderidae Filipjev and Schuurmans Stekhoven (1941) by Skarbilovich (1959). Wouts (1972) redefined each of the subfamilies and proposed that they are all closely related except Meloidogyninae which should have a separate family, Meloidogynidae, the concept accepted by Jepson (1987), Siddiqi (1986, 2000) and Karsson (2002). Andrassy (1976) placed root knot nematodes in the superfamily Hoplolaimoidea (Filipjev, 1934) Paramonov, 1967; Siddiqi (1971) in Tylenchoidea (Orley, 1880) Chitwood and Chitwood, 1937; Golden (1971) in Heteroderoidea (Filipjev, 1934) Golden, 1971. Based on wide range of morphological data Baldwin (1992) reported that Heteroderinae are not monophyletic with Meloidogyninae. Geraert (1997) have demonstrated that Meloidogyninae has the same head end-on view as the Pratylenchidae while Heteroderidae has the same head as Hoplolaimidae. Siddiqi (2000) has classified the root knot nematodes in the superfamily Hoplolamoidea. After the classification of suprakingdom Animalia Mohn, 1984, phylum Nematoda, Potts,1932 and suborder Hoplolaimina Chizhov and Berezina, 1988, the systematic position of the genus *Meloidogyne* is:-

Kingdom	Animalia (Bilateralia)
Phylum	Nematoda
Class	Secernentea
Subclass	Diplogasteria
Order	Tylenchida
Suborder	Hoplolaimina
Superfamily	Hoplolaimoidea
Family	Meloidogynidae
Subfamily	Meloidogyninae
Genus	*Meloidogyne* Goldi, 1887

The advent of molecular diagnosis has facilitated a better understanding of phylogeny of nematodes. As a result the morphology based systematic schemes are being changed by hierarchies based on molecular phylogeny. There is a proposal for infraorders and bringing together groups which have been regarded as distinctly related (De Ley and Blaxter, 2002). So if Siddiqi (2000) and De Ley and Blaxter (2002) are to be followed simultaneously, the systematic position of the genus *Meloidogyne* will be upgraded to:

Phylum	Nematoda Potts, 1932
Class	Chromodorea Inglis, 1983
Subclass	Chromodoria Pearse, 1932
Order	Rhabditida Chitwood, 1933

Suborder	Tylenchina Thorne, 1949
Infraorder	Tylenchomorpha De Ley and Blaxter, 2002
Superfamily	Tylenchoidea Orley, 1880
Family	Meloidogynidae Skarbilovich, 1959
Subfamily	Meloidogyninae Skarbilovich, 1959
Genus	*Meloidogyne* Goldi, 1887

The root knot nematodes at genus level have always been confused with cyst nematodes for fairly long time as since their description from 1884 till 1932 they were known as *Heterodera radicicola* whereas between 1932 and 1949 the name given by Goodey (1932) was *Heterodera marioni*. In 1949, Chitwood restored the name *Meloidogyne* for root knot nematodes as was given by Goldi (1887).

8.2 Differences between Root Knot and Cyst Nematodes

Sl.No.	*Root Knot Nematodes*	*Cyst Nematodes*
	Biology	
1.	Gall forming	No galls are formed
2.	Female remain inside the root	Female comes out of the root at maturity
3.	Feeding induce giant cells	Syncytia are induced
4.	Polyphagous species	Mostly host specific
5.	Mostly parthenogenetic reproduction	Mostly amphimictic reproduction
	Second stage juveniles	
6.	Anterior body tapering and slender	Anterior body not tapering and robust
7.	Labial region weakly sclerotised and not offset	Strongly sclerotised and offset
8.	Stylet slender, <19μm long	Robust, 20-30μm long
9.	Hyaline region short near tail tip	Long and well developed
10.	3rd and 4th stages have spike tail	No spike tail
11.	Labial sectors wider than the submedian sectors	Not wider
	Males	
12.	Labial region weakly sclerotised with 2 annules	Strongly sclerotised with 4-5 annules
13.	Stylet slender	Robust
14.	Amphidial aperture is a slit with a large ampulla	Amphidial aperture is a pore with no ampulla.
	Females	
15.	Excretory pore is anterior to median bulb valvular plates	Posterior to median bulb valvular plates
16.	Eggs deposited in eggmass	Generally retained in the body
17.	Mature female remains soft on death	Mature female forms cyst on death
18.	Third and fourth stages lack stylet	Do not lack stylet

8.3 Diagnosis of Subfamily (Siddiqi, 2000)

Subfamily *Meloidogyninae* Skarbilovich, 1959

Syn. *Meloidogynini* Skarbilovich, 1959

Meloidoderellinae Hussain, 1976

Meloidoderellini Hussain, 1976

Hoplolaimidae: Root galls forming, female feeding inciting multinucleate nurse cells. Marked sexual dimorphism. Cuticle striated. Lateral fields bearing four or five incisors. Labial region low, with one to four annules. Under SCM, female labial disc dorso-ventrally elongate, dumb-bell shaped with oral opening a small round pore surrounded by six inner labial pits. Frame work moderately sclerotized, hexaradiate; lateral sectors equal to, or wider than, submedian sectors. Stylet moderately strong, male style longer and more robust than that of female. Orifice of dorsal pharyngeal gland located just posterior to style base. Median pharyngeal bulb oval or round, with larger refractive thickenings. Pharyngeal glands elongate, extending over intestine mostly ventrally but also laterally; subventral asymmetrical, extending past dorsal gland. SVN (subvental gland nucleus) always posterior to DN (dorsal gland nucleus). Excretory pore in female opposite or anterior to median bulb, in males usually posterior to median bulb. No pre adult vermiform stage (except in *Meloinema*).

Mature Female

Swollen, sedentary, round, oval to pear-shaped with a projecting neck. Cuticle moderately thick, striated generally forming typical, fingerprint-like pattern terminally. No cyst stage. Vulva subterminal or terminal. Anus located near vulval lip; tail rudimentary or absent. Stylet under 25µm long in *Meloidogyne* (but 30-35 µm long in *Meloinema*). Median bulb oval or rounded, usually offset, with large refractive thickenings. Didelphic –prodelphic ovaries coiled. Most eggs not retained in the body but laid. Large rectal glands present, gelatinous matrix present (rectal glands and egg masses absent in *M. spartinae* and *M. kikuyensis*).

Male

Vermiform, migratory, generally non-feeding, over 1mm long, posterior end twisted though 90-180°, developing by metamorphosis within a saccate juvenile. Labial region rather low and continuous; amphidial aperture large transverse slit; labial cp large, prominent; framework moderately sclerotized, lateral sectors wider than submedian sectors. Stylet strong, usually over 20 µm long, basal knobs prominent. Tail short or absent lacking a bursa (except in *Bursadera*). Spicules large (25-64µm), distally pointed. Gubernaculum linear to trough shaped, not protrusible. Cloacal lips non tuboid generally with hypoptygma.

Juveniles

Second stage migratory and infective. Third and fourth stage juveniles swollen, without stylet in type genus. In *Meloinema*, third and fourth stage juveniles are vermiform. Labial region low, anteriorly flattened or rounded. Labial sectors wider than submedian sectors, labial disc distinct in type genus. Stylet weak to moderately developed. Less than 20µm long in type genus but strongly developed in *Meloinema*.

Tail elongate, conoid, with minutely rounded tip and conspicuous hyaline portion. Phasmids dot-like, located on tail, usually anterior to middle.

Type Genus

Meloidogyne Goldi, 1887

Other genera

Meloinema Choi and Geraert, 1974

=*Nacobbodera* Golden and Jensen, 1974

Bursadera Ivanova and Krall, 1985

Genus *Meloidogyne* Goldi, 1887

=*Hypsoperine* Sledge &Golden, 1964

=*Spartonema* Siddiqi, 1986

Diagnosis of the genus, *Meloidogyne*

Mature Female

White, round to pear shaped with short projecting neck, sedentary. No cyst stage. Vulva and anus located close together, terminal, perineum with a finger-print like cuticular pattern, usually flattened, rarely elevated. Phasmid dot-like, slightly anterior to and on either side of anus. Cuticle striated. Stylet slender. Generally 12-15µm long, with small basal knobs. Excretory pore anterior to median bulb, often posterior to base of stylet. Genital tracks paired, prodelphic, convoluted. Six large rectal glands secreting gelatinous material in which eggs are deposited, eggs not retained in body.

Males

Vermiform, up to 2mm long, tail end twisted; develop by metamorphosis within a swollen juvenile. Cuticle strongly annulaed; lateral fields with four incisors. Labial region not sharply offset, with distinct labial disc and few (1-3) annules; lateral sectors wider than sub median sectors, appearing as cheeks. Stylet robust (18-25µm) with large basal knobs. Oesophageal glands lying mostly ventral o intestine. Spicules slender generally 25-33µm long, gubernaculum7-11 µm long. Testis single, but paired when sex reversal occurs. Tail rounded. Phasmid dot-like, located near cloacal aperture, which is sub terminal. Bursa absent.

Juveniles

First stage with a blunt tail tip, moulting within egg; second and third moults occur within the cuticle of second stage. Second stage vermiform, migratory, infective, straight to arcuate upon death. Labial region with coarse annules (1-4), a distinct labial disc, framework slightly sclerotized, lateral sectors wider than the sub median sectors.. stylet slender under 20µm, excretory pore posterior to hemizonid. Median bulb with large oval refractive thickenings. Tail with conspicuous hyaline region, tip narrow, irregular in outline. Third and fourth stages sedentary, swollen, sausage or flask shaped with a short blunt tail.

Type Species

Meloidogyne exigua Goldi, 1887

= *Heterodera exigua* (Goldi, 1887) Marcinowski, 1909

There are about 100 nominal species of root knot nematodes (Skantar *et al.*, 2008) infesting around 5,500 plant species in the world (Trudgill and Blok, 2001). The importance of root knot is reflected through the work carried out by various workers especially on complete sequencing of whole genome of *M. incognita*, large volume of literature with several books published on the genus and the International *Meloidogyne* project (1975-1984) (Whithead, 1968; Taylor and Sasser, 1978; Lamberti and Taylor, 1979; Sasser and Kirby, 1979; Barker *et al.*, 1985; Sasser and Carter, 1985; Karssen, 2002).

The Species

M. exigua Göldi, 1892

Syn. *Heterodera exigua* (Göldi, 1892) Marcinowski, 1909

M. acronea Coetzee, 1956

Syn. *Hypsoperine acronea* (Coetzee, 1956) Sledge and Golden, 1964

H. (H.) acronea (Coetzee, 1956) Sledge and Golden, 1964

M. actinidae Li& Yu, 1991

M. africana Whitehead, 1960

M. aquatilis Ebsary and Eveleigh, 1983

M. arabidicida Lopez and Salazar, 1989

M. ardenensis Santos, 1968

Syn. *M. deconincki* Elmiligy, 1968 (syn. by Karssen and Hoenselaar, 1998)

M. litoralis Elmiligy, 1968 (syn. by Karssen and Hoenselaar, 1998)

M. arenaria (Neal, 1889) Chitwood, 1949

Syn. *Anguillula arenaria* Neal, 1889

Tylenchus arenarius (Neal) Cobb, 1890

Heterodera arenaria (Neal) Marcinowski, 1909

M. thamesi (Chitwood *et al.*, 1952) Goodey, 1963

M. arenaria thamesi Chitwood in Chitwood, Specht and Havis, 1952

Meloidogyne arenaria arenaria (Neal) Chitwood, 1949

Meloidogyne thamesi gyulai Amin, 1993

Meloidogyne gyulai Amin, 1993

M. artiellia Franklin, 1961

M. baetica Castilo, Volvas, Subbotin and Ttroccoli, 2003

M. brasilensis Charchar and Eisenback, 2002

M. brevicauda Loos, 1953

M. californiensis Abdel-Rahman and Maggenti, 1987

M. camelliae Golden, 1979

M. caraganae Shagalina, Ivanova and Krall, 1985

M. carolinensis Eisenback, 1982

M. chitwoodi Golden, O'Bannon, Santo and Finley, 1980

M. chosenia Eroshenko and Lebedeva, 1992

M. christiei Golden and Kaplan, 1986

M. cirricauda Zhang and Weng, 1991

M. citri Zhang, Gao and Weng, 1990

M. coffeicola Lordello and Zamith, 1960

Syn. *Meloidodera coffeicola* (Lordello and Zamith) Kirjanova, 1963

M. cruciani Garcia-Martinez, Taylor and Smart, 1982

M. cynariensis Fam-Tkhan-Bin, 1990

M. decalineata Whitehead, 1968

M. dunensis Palomares Rius, Vovlas, Troccoli, Liebanas, Landa and Castillo, 2007

M. duytsi Karssen, Aelst and Van der Putten, 1998

M. enterolobii Yang and Eisenback, 1983

M. ethiopica Whitehead, 1968

M. fallax Karssen, 1996

M. fanzhiensis Chen, Peng and Zheng, 1990

M. floridensis Handoo, Nyczepir, Esmenjaud, van der Beek, Castagnone-Sereno, Carta, Skantar and Higgins, 2004

M. fujianensis Pan, 1985

Syn. *M. fujianensis* Cangsang, Jing and Shengyuan, 1988

M. graminicola Golden and Birchfield, 1965

M. graminis (Sledge and Golden, 1964) Whitehead, 1968

Syn. *Hypsoperine graminis* Sledge and Golden, 1964

H. (Hypsoperine) graminis Sledge and Golden, 1964

M. hainanensis Liao, JinLing and Feng ZhiXin, 1995

M. hapla Chitwood, 1949

Syn. *M. haplanaria* Eisenback, Bernard, Starr, Lee and Tomaszewski, 2004

M. hispanica Hirschmann, 1986

M. ichinohei Araki, 1992

M. incognita (Kofoid and White, 1919) Chitwood, 1949

Syn. *Oxyuris incognita* Kofoid and White, 1919

Heterodera incognita (Kofoid and White, 1919) Sandground, 1923

M. incognita incognita (Kofoid and White, 1919) Chitwood, 1949

M. acrita Chitwood, 1949

M. incognita acrita Chitwood, 1949

M. elegans da Ponte, 1977

M. grahami Golden and Slana, 1978

M. incognita grahami (Golden and Salana, 1978) Jepson, 1987

M. kirjanovae Terenteva, 1965 (syn. by Karssen and Hoenselaar, 1998)

M. wartellei Golden and Birchfield, 1978

M. incognita wartellei Golden and Birchfield, 1978

M. indica Whitehead, 1968

M. inornata Lordello, 1956

Syn. *M. incognita inornata* Lordello, 1956

M. izalcoensis Carneiro, Almeida, Gomes and Hernandez, 2005

M. javanica (Treub, 1885) Chitwood, 1949

Syn. *Heterodera javanica* Treub, 1885

Tylenchus (Heterodera) javanicus (Treub, 1885) Cobb, 1890

Anguillula javanica (Treub, 1885) Lavergne, 1901

M. javanica javanica (Treub, 1885) Chitwood, 1949

M. javanica bauruensis Lordello, 1956

M. bauruensis (Lordello, 1956)Esser *et al.*, 1976

M. lordelloi da Ponte, 1969

M. lucknowica Singh, 1969

M. jianyangensis Yang, Hu, Chen and Zhu, 1990

M. jinanensis Zhang and Su, 1986

M. kikuyense De Grisse, 1961

Syn. *Spartonema kikuyense* (De Grisse, 1961) Siddiqi, 2000

M. konaensis Eisenback, Bernard and Schmitt, 1995

M. kongi Yang, Weng and Feng, 1988

M. kralli Jepson, 1984

M. lini Yang, Hu and Zhu, 1988

M. lusitanica Abrantes and Santos, 1991

M. mali Itoh, Ohshima and Ichinohe, 1969

M. maritima Jepson, 1987

M. marylandi Jepson and Golden in Jepson, 1987

M. mayaguensis Rammah and Hirschmann, 1988

M. megadora Whitehead, 1968

M. megatyla Baldwin and Sasser, 1979

M. mersa Siddiqi and Booth, 1991

Syn. *M. (Hypsoperine) mersa* Siddiqi and Booth, 1991

M. microcephalus Cliff and Hirschmann, 1984 (Original spelling *microcephala*)

M. microtyla Mulvey, Townshend and Potter, 1975

M. mingnanica Zhang, 1993

M. minor Karssen, Bolk, van Aelst, van den Beld, Kox, Korthals, Molendijk, Zijlstra, van Hoof and Cook, 2004

M. morocciensis Rammah and Hirschmann, 1990

M. naasi Franklin, 1965

M. nataliae Golden, Rose and Bird, 1981

M. oryzae Maas, Sanders and Dede, 1978

M. oteifai Elmiligy, 1968 (original spelling *oteifae*)

M. ottersoni (Thorne, 1969) Franklin, 1971

Syn. *Hypsoperine ottersoni* Thorne, 1969

H. (Hypsoperine) ottersoni Thorne, 1969 (Siddiqi, 1986)

M. ovalis Riffle, 1963

M. panyuensis Liao,Yang, Feng and Karssen, 2005

M. paranaensis Carneiro, Carneiro, Abrantes, Santos and Almeida, 1996

M. partityla Kleynhans, 1986

M. petuniae Charchar, Eisenback and Hischmann, 1999

M. phaseoli Charchar, Eisenback, Charchar and Boiteau, 2008b

M. pini Eisenback, Yang and Hartman, 1985

M. piperi Sahoo, Ganguly and Eapen, 2000

M. pisi Charchar, Eisenback, Charchar and Boiteau, 2008a

M. platani Hirschmann, 1982

M. propora Spaull, 1977

Syn. *Hypsoperine propora* (Spaull, 1977) Siddiqi, 1986

H. (Hypsoperine) propora (Spaull) Siddiqi, 1986

M. querciana Golden, 1979

M. salasi Lopez, 1984

M. sasseri Handoo, Huettel and Golden, 1994

M. sewelli Mulvey and Anderson, 1980

M. silvestris Castillo, Vovlas, Troccoli, Liebanas, Palomares Rius, Landa, 2009

M. sinensis Zhang, 1983

M. spartinae (Rau and fassuliotis, 1965) Whitehead, 1968

Syn. *Hypsoperine spartinae* Rau and Fassuliotis, 1965

Spartonema spartinae (Rau and Fassuliotis, 1965) Siddiqi, 1986

M. subarctica Bernard, 1981

M. suginamiensis Toida and Yaegashi, 1984

M. tadshikistanica Kirjanova and Ivanova, 1965

M. thailandica Handoo, Skantar, Carta and Erbe, 2005

M. trifoliophila Bernard and Eisenback, 1997

M. triticoryzae Gaur, Saha and Khan, 1993

M. turkestanica Shagalina, Ivanova and Krall, 1985

M. ulmi Marinari-Palmisano and Ambrogioni, 2000

M. vandervegtei Kleynhans, 1988

Species Inquirendae

M. marioni (Cornu, 1879) Chitwood and Oteifa, 1952

Syn. *Anguillula marioni* Cornu, 1879

Heterodera marioni (Cornu, 1879) Marcinowski, 1909

Meloidogyne goeldi Lordello, 1951

M. megriensis (Poghossian, 1971) Esser, Perry and Taylor, 1976

Syn. *Hypsoperine megriensis* Poghossian, 1971

M. poghossianae Kirjanova, 1963

M. acronea apud Poghossian, 1961

M. vialae (Lavergne, 1901) Chitwood and Oteifa, 1952

Syn. *Anguillula vialae* Lavergne, 1901

Heterodera vialae (Lavergne, 1901) Marcinowski, 1909

8.4 Identification

The identification of root knot nematodes is becoming increasingly important for designing management schedules using crop rotation and plant resistance that requires precise species level identification. It is also useful in plant quarantine

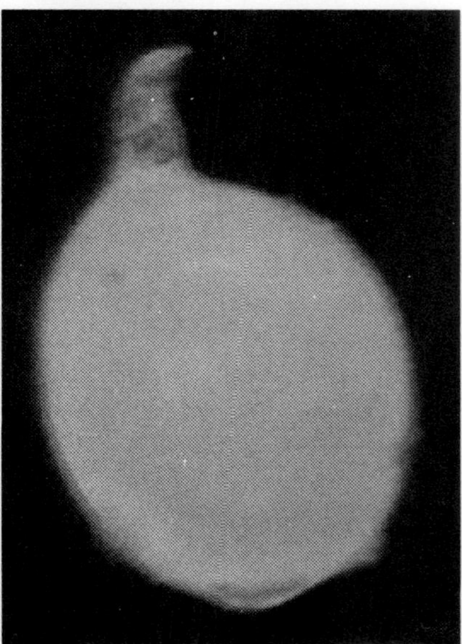

Figure 38: *Meloidogyne* Female

purposes, control measures and precise research (Hussey, 1990; Zijlistra and van Hoof, 2006). Many of the root knot species can be identified either by restricted host range or some characteristic morphological features. Although some species can easily be identified because of their characters which are typical for them but in some cases the identification becomes difficult due to different forms or aberrant individuals, overlapping characters in perineal patterns and presence of mixtures of populations.

8.4.1 Bases for Identification

8.4.1.1 Morphological Basis

The basic morphology of *Meloidogyne* species is quite similar. Nevertheless, certain distinguishing characters are useful in species differentiation. Females of all *Meloidogyne* species have a distinctive perineal pattern, that is, an external, fingerprint-like series of markings around and in the vicinity of the anus and vulva. The markings are readily visible under a light microscope. In addition to the morphology of perineal patterns, the head morphology of females, males, and second-stage juveniles, and the stylet morphology of females and males are equally important characters while describing a new specioes. Perineal patterns and head shapes of males appear to be the most helpful characters. Stylet morphology is also reliable but can be used only in specimens that are properly prepared and viewed in exact lateral position. Additional characters, such as number of lines in the lateral field, may be useful in the identification of some species. Some of the characters of male, like form of the labial region including annulations, form of the stylet and basal knobs were added later along with SEM and TEM techniques for accurate representation of these features. However, the value of these characters, in so many described species showing intra-specific variation, has been observed to be depleting and that warranted a supplementary sound characterization which isozyme electrophoresis and PCR-based molecular methodologies seem to provide.

New species are being identified and described with improved techniques but it is also a fact that in most tropical areas like India where there is so much climatic diversity with root knot disease being a great menace, lack of expertise and facilities is apparent. Most of the root knot problems are just clubbed with the threat from either *M. incognita* or *M. javanica*. This is nothing but over reliance on the perineal patterns of these species which show lot of variation but without analyzing them biochemically or at molecular level they are just labeled as *M. incognita* or *M. javanica*or any other commonly recognized species. Similar reliance on perineal pattern misdiagnosed many root knot species including *M. mayaguensis* and *M. incognita* both infesting coffee in Central America now identified as *M. enterolobii and M. paranaensis*, respectively through isozyme esterase phenotyping and RAPDs (Carneiro *et al.*, 2004; Herve *et al.*, 2005). After the first use of isozyme phenotypes by Esbenshade and Triantaphyllou (1985) to distinguish root knot species, it continues to be used widely in spite of its limitations (Brito *et al.*, 2008; Carneiro *et al.*, 2007, 2008; Handoo *et al.*, 2004; Karssen *et al.*, 2004; Palomares Rius *et al.*, 2007). The limitation being the occurrence of intraspecific variants and the difficulty in resolving the size variants between species has necessitated the use of more than one system to confirm the identity of some isolates.

8.4.1.2 Biochemical and Molecular Basis

Out of the four commonly studied enzyme patterns (non-specific estrases, malate dehydrogenase, superoxide dismutase and glutamate-oxaloacetate transaminase), esterases have been the most useful for differentiating the major *Meloidogyne* species (Esbenshade and Triantaphyllou, 1985; Cofcewicz *et al.*, 2004).

Malate dehydrogenase separates *M. hapla* from *M. incognita, M. javanica* and *M. arenaria* whereas glutamate dehydrogenase separates *M. incognita* from *M. hapla, M. javanica* and *M. arenaria* (Esbenshade and Triantaphyllou, 1985).

Unlike protein and enzyme based characterization, DNA-based diagnostic methods do not rely on the expressed products of the genome,therefore, are independent of environmental influence and independent of stage in the nematode life cycle (Hooper *et al.*, 2005). DNA analysis has been widely used in systematics and for identification of nematodes (Williamson *et al.*, 1997; Zijlstra *et al.*, 2004; Powers *et al.*, 2005). PCR assays have been used for identification of nematodes to species and are sensitive enough to identify the species of a single nematode. PCR amplification is generally carried out using purified nematode DNA or hand-picked individual nematodes (Hübschen *et al.*, 2004; Powers *et al.*, 2005). However, using PCR to detect nematodes and other organisms in soil extracts has been difficult due to the presence of inhibitors of DNA polymerase (Hyman *et al.*, 1990; Volossiouk *et al.*, 1995; Miller *et al.*, 1999; Roose-Amsaleg *et al.*, 2001).

The PCR process allows amplification of minute quantities of DNA which can be extracted from single nematode, egg or juvenile and be subjected to further analysis (Harris *et al.*, 1990). DNA sequencing is straightforward and has the added benefit of providing data useful for phylogenetic analysis. DNA markers that have aided identification of *Meloidogyne* species include the ribosomal DNA small subunit (SSU) 18S (Powers, 2004), large subunit (LSU) 28S D2-D3 expansion segments (Chen *et al.*, 2003; Palomares Ruis *et al.*, 2007), intergenic spacer (IGS) (Blok *et al.*, 1997; Wishart *et al.*, 2002), internal transcribed spacer (Powers and Harris, 1993) and mitochondrial DNA (Powers and Harris, 1993; Stanton *et al.*, 1997; Blok *et al.*, 2002; Xu *et al.*, 2004; Jeyaprakash *et al.*, 2006). Random amplified polymorphic DNA (RAPD) (Cenis, 1993; Williamson *et al.*, 1997; Dong *et al.*, 2001; Cofcewicz *et al.*, 2004; Adam *et al.*, 2007) and sequence characterized amplified regions (SCAR) markers (Zijlstra *et al.*, 2000; Randig *et al.*, 2002) have also been developed and successfully used for identifying a large number of *Meloidogyne* species. One study combined IGS PCR, SCAR markers and RAPD analysis into a diagnostic key for discrimination of seven RKN species (Adam *et al.*, 2007).The PCR mediated amplification of specific regions of the nematode genome offers a highly effective means of detecting inter and intra specific variation. PCR amplification is directed to the targeted gene using a pair of specific oligonucleotides (forward and reverse primers). Consistent variations in size or nucleotide sequence of the amplified PCR product can be used for nematode identification and characterization. Powers and Harris (1993) demonstrated this by distinguishing between *Meloidogyne* species based on amplification of mitochondrial DNA genes. The PCR fragment band size differences between the different species became apparent when PCR products were separated on agarose gels. The variation in nucleotide sequence of the different species can be further detected by sequencing or by restriction enzyme digestion of the PCR product (PCR-RFLP).

A number of species specific primers for the identification of *Meloidogyne* species have been developed based on species specific RAPD fragments (Williamson *et al.*, 1997; Zijlistra 2000; Dong *et al.*, 2001). The species specific sequence characterized amplified regions (SCAR) primers work at higher annealing temperatures as they are longer than RAPD primers (Zijlistra, 2000). The PCR reaction is also less dependent on the amount of DNA and source of template thus the PCR results are more reproducible and reliable (Williamson *et al.*, 1997). SCAR-PCR methods have the potential to be used in routine diagnostic applications using DNA extracts from single juvenile, soil samples or even infected plant materials. The SCAR primers is able to determine species identity irrespective of the developmental stage (juvenile, male, female or egg mass) and from small amounts of tissue or from mixed populations where concentration of one species may be just 1 per cent, but it is not used in the identification of new species (Zijlistra, 2000).

Though molecular techniques are not free of error, they significantly reduce the chances of bias and human error since the species identification is based on genetic information rather than morphological characteristics which require more experience and specialized skills for identification (Coomans, 2002). Molecular identification has confirmed the validity of a number of classical nematode species and necessary revisions have been made to classifications based on molecular evidence. Molecular systematics combined with related work in genetics has moved to the forefront lately and has contributed a lot to taxonomy (Barker, 2004).

8.4.1.3 Host Specific Identification

The root knot species can be identified tentatively on the basis of host response as has been done with four most important species. Differential host test gives a preliminary indication of the species involved based on the typical host response. But the differential host test cannot be relied upon entirely for identification because the population may contain more than one species or the population may comprise a species for which there is limited or no reliable data. For the identification of *M. incognita, M. javanica, M. arenaria* and *M. hapla* differential hosts has been used that gives fairly reliable information. The earliest such trial was conducted by Sasser (1955) which reported the following scheme of identifying unkown population of root knot nematode-

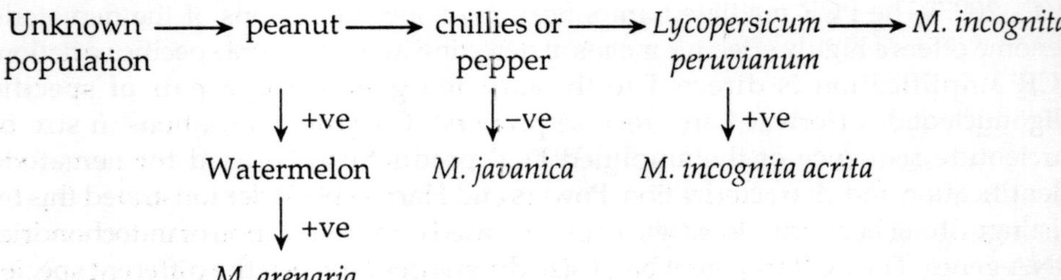

Later on, this scheme was revised and modified by the workers at North Carolina State University by including more differentials which were used against populations/species collected from different geographical areas of the world. These species were tested against six differentials: Tobacco (*Nicotiana tabacum*) NC95, Cotton (*Gossypium hirsutum*) Deltapine, Pepper (*Capsicum frutescens*) California Wonder, Watermelon (*Citrullus vulgaris*) Charleston Gray, Peanut (*Arachis hypogea*) Florunner and Tomato (*Lycopersicon esculentum*) Rutgers. The tests showed that *M. incognita* is made up of four races, *M. arenaria* had two races, *M. javanica* and *M. hapla* had one race each. Of the 222 populations of *M. incognita* collected 63 per cent (139 populations) were race 1, 22 per cent (49 populations) were race 2, 10 per cent (22 populations) were race 3 and 5 per cent (12 populations) were race 4. The host races of *M. incognita* and *M. arenaria* are distributed throughout the world.

Differential Host Tests for Four *Meloidogyne* Species

Meloidogyne sp. and Race	*Differential Host Cultivars*					
	Tobacco	*Cotton*	*Pepper*	*Watermelon*	*Peanut*	*Tomato*
M. incognita						
Race 1	–	–	+	+	–	+
Race 2	+	–	+	+	–	+
Race 3	–	+	+	+	–	+
Race 4	+	+	+	+	–	+
M. arenaria						
Race 1	+	–	+	+	+	+
Race 2	+	–	–	+	–	+
M. javanica	+	–	–	+	–	+
M. hapla	+	–	+	–	+	+

From the studies, it is also clear that there is considerable uniformity in host response among species collected from widely separated geographical regions of the world as *M. incognita* populations did not reproduce on peanut, *M.hapla* populations did not reproduce on watermelon and *M. javanica, M. hapla* and *M. arenaria* populations did not attack cotton.

8.4.1.4 Cytological Basis

Cytogenetic information can also be used to supplement morphological data in species identification and often to verify identification of certain important species. The most important cytogenetic differentiations are based on the characters like reproduction, process of maturation of oocytes, and chromosome numbers. Some species reproduce by cross-fertilization (amphimixis), others by parthenogenesis (obligatory mitotic by parthenogenesis and still others by cross-fertilization and parthenogenesis (facultative meiotic parthenogenesis). Cross-fertilizing parthenogenetic species undergo meiosis during maturation of oocytes and this involves pairing of homologous chromosomes and formation of bivalents (tetrads).

The haploid number (n) of bivalents is observed at metaphase of the first maturation division in these species. Obligatorily parthenogenetic species do not undergo meiosis and maturation of oocytes in such species consists of a single mitotic division. The diploid number of univalent chromosomes (dyads) is observed at metaphase of the single maturation division. In addition to these differences, the chromosome number may be different in various species and also may vary within species. In amphimictic species like *M. carolensis, M. megatyla, M. microtyla* (n=18-19), *and M. subarctica* the chromosome number n=18 whereas in facultative meiotic parthenogenetic species like, *M. exigua, M. grminocola, M. graminis, M. naasi* and *M. hapla* (race 1) the chromosome haploid number n varies from 14-34 but in obligatory mitotic parthenogenetic species like *M. incognita, M. javanica, M. oryzae, M. arenaria* and *M. hapla* the diploid chromosome number 2n varies from 30-56. The species *M. incognita* and *M. javanica* which exclusively reproduce by mitotic parthenogenesis have chromosome number varies from 2n=42-48.

8.5 Economic Importance

The root knot nematodes are obligate parasites of roots and one of the major limiting factors in the production and productivity of hundreds of crops in whole of the world. An average 10 per cent of loss in yield is frequently cited for vegetables (Barker and Koenning, 1998; Koenning *et al.*, 1999; Regnault-Roger *et al.*, 2002). However, much higher percentages have been recorded in local regions, depending on the genus, population level (Ornat and Sorribas, 2008). Sikora and Fernandez (2005) reported yield losses of over 30 per cent in three highly susceptible vegetable crops (egg-plant, tomato and melon). Plants affected by root-knot nematodes are characteristically stunted and yellow and appear in irregular patches within a field. Root knot is most easily identified by the knots or galls which develop on the roots of infected plants. These galls vary from pinhead size to many times thicker than the normal root. They are irregular in shape but are usually either spindle-shaped or spherical. Affected plants wilt rapidly during dry weather and may even wilt in the presence of adequate soil moisture. Plants infected with root knot often exhibit symptoms of drought or nutritional stress. Severely affected plants may die. Other soil-borne diseases such as fungal and bacterial diseases are often more severe when root knot nematodes are present.

8.6 Species Distribution

Most populations of root-knot nematodes consist of one or more of the species originally described by Chitwood (1949). These species are all highly polyphagous and have a wide geographical distribution with the exception of the type species, *M. exigua,* which apparently has a more restricted host range and distribution. This becomes obvious when the world literature on *Meloidogyne* is consulted. Nearly 90 per cent of the references dealing with *Meloidogyne* in literature between 1949 and 2010 refer to Chitwood's species. Accepting the synonymization of *M. incognita* and *M. incognita acrita* proposed by Triantaphyllou and Sasser (1960) one concludes that at present the species most widely distributed, polyphagous and amenable to selection and adaptive variation are: *M. incognita, M. javanica, M. arenaria* and *M. hapta.* One clue for this wide adaptation may be in the variable chromosomal complement of isolates of agiven species. For example, the chromosome number of *M. javanica* can range from 42 to 48 and that of *M. incognita* from 32 to 46. The variable chromosome number suggests that these species are aneuploid with different copy numbers of some chromosomes (Triantaphyllou,1985).

The International *Meloidogyne* Project (1975-1984) has collected an extensive ecological data on 662 populations of seven species from agroecosystem in 76 countries (Taylor *et al.,* 1982). The four most prevalent species were *M. incognita* (47 per cent), *M. javanica* (40 per cent), *M. arenaria* (7 per cent) and *M. hapla* (6 per cent). The first three species are widely distributed in tropical, subtropical and warm regions whereas *M. hapla* occurs primarily in cooler climates as low as 6-9°C lower than the other three species. There seem to be correlation between high pH and low rainfall cropping situation and presence of *M. incognita* and *M. javanica*. Very few populations of root knot are collected from soils which have more than 40 per cent clay or more than 50 per cent silt. Most of the other species are more specialized and generally confined to restricted geographical areas. *M. brevicauda* well illustrates this as it is known to be a parasite only of tea and is restricted to a few areas in Sri Lanka and one in South India (Sivapalan, 1972). Certain species, however, such as *M. naasi* which is found in various regions of Europe and the United States, are intermediate between very polyphagous and highly specialized species mainly found infesting cereals and sugar beet.

Species like *M. incognita, M. javanica, M. arenaria* and *M. hapla* are also wide spread in temperate climates of the world and found in great number everywhere. There are few other species like *M. artiellia,* a polyphagous species infesting brassicas, clovers and cereals, *M. mali* on apple, *M. carolinensis* on *Rhododendron, M. litoralis* on *Ligustrum, M. graminis* on cereals and grasses, *M. ardensis* on *Vinca, M. deconincky* on *Rosa* and *Solanum nigrum, M. ottersoni* on *Phalaris, M. microtyla* on *Festuca rubra* and cereals have been reported to be present in different parts of cooler regions. In the warmer regions, most of the rest of the species are present including those listed above as some of them may be present in higher and cooler spots in the topical parts.

The fact remains that they are well suited to the climate of tropical and subtropical regions of the world, thus, causing much more damage here in comparison to temperate regions.

Sasser (1977) has summarized the occurrence of root-knot nematode species in different parts of the world. His report included 11 species in Africa, 9 species in Central and South America, 18 species in the United states, 3 species in Canada,11 species in Europe and Mediterranean region, 10 species in India and Sri Lanka, 4 species in Russia, 5 species in Japan and 3 species in Southeast Asia, Australia and Fiji Islands.

8.7 Root knot Nematode Biology

Mature female root-knot nematodes release hundreds of eggs into a proteinaceous matrix on the surface of the root. The eggs which are at one cell stage when laid and cell division starts a few hours later. Juveniles hatch from eggs as vermiform second stage juveniles (J2), the first moult having occurred within the egg. Newly-hatched juveniles have a short free-living stage in the soil and in the rhizosphere of the host plant. They may reinvade the host plant of their parent or migrate through the soil to find a new host root. J2 do not feed during the free living stage, but use lipids stored in the gut (Eisenback and Triantaphyllou, 1991). An excellent model system for the study of the parasitic behaviour of plant-parasitic nematodes has been developed using *Arabidopsis thaliana* as a model host (Sijmons *et al.*, 1991). The *Arabidopsis* roots are initially small and transparent, enabling every detail to be seen. Invasion and migration in the root was studied using *M. incognita* (Wyss *et al.*, 1992). Briefly, second stage juveniles invade in the root elongation region and migrate in the root until they became sedentary. Signals from the J2 promote parenchyma cells near the head of the J2 to become multi-nucleate (Hussey and Grundler, 1998) to form feeding cells, generally known as giant cells. Each cell is initiated by the injection of salivary secretions originating from the dorsal oesophageal gland, which are thought to be of different nature than those secreted during migration (Sijmons *et al.*, 1994). These giant cells act as metabolic sinks that actively transfer nutrients from host plant to the developing nematode. Cells around the developing juvenile also become hypertrophic and more numerous, which ultimately results in the formation of the characteristic root galls, associated with *Meloidogyne* infection. Juveniles first feed from the giant cells about 24 hours after becoming sedentary. After further feeding, the J2 undergo morphological changes and become saccate. Without further feeding, they moult three times and eventually become adults. In females, which are close to spherical, feeding resumes and the reproductive system develops (Eisenbach and Triantaphyllou, 1991). The life span of an adult female may extend to three months and many hundreds of eggs can be produced. Females can continue egg laying after harvest of aerial parts of the plant and the survival stage between crops is generally within the egg. The length of the life cycle is temperature dependent (Madulu and Trudgill, 1994; Trudgill, 1995). The relationship between rate of development and temperature is linear over much of the root-knot nematode life cycle, though it is possible that component stages of the life cycle, *e.g.* egg development, host root invasion or growth, have slightly different optima. Species within the *Meloidogyne* genus also have different temperature optima. In *M. javanica*, development occurs between 13 and 34° C, with optimal development at about 29° C. Once feeding begins, the J2 loses its ability to move within the root. J2 causes little physical damage to the roots during the penetration process unlike cyst nematodes which make wounds while moving intracellularly. The penetration of J2 is most likely affected by the host plant, soil temperature and species of the root knot nematode involved. The prevailing environmental conditions act both on the nematode and plant which in totality constitutes the host-parasite relationship. The feeding J2, third and fourth juveniles staged have a spike tail. A single female may be able to lay as many as 900 eggs but on an average it lays about 250-300 eggs in the gelatin matrix it produces before laying the eggs. A gelatinous matrix is produced by six rectal glands and secreted before and during egg laying (Maggenti and Allen, 1960). The matrix initially forms a canal through the outer layers of root tissue and later surrounds the eggs, providing a barrier to water loss by maintaining a high moisture level around the eggs (Wallace, 1968). As the gelatinous matrix ages, it becomes tanned, turning from a sticky colourless jelly to an orange-brown substance which appears layered (Bird, 1958). The male remains in the root in search of a female for about 3-4 days before dying. If it finds the female, it may be able to fertile.

Reproduction of *Meloidogyne* spp. is by parthenogenesis in most of the species although other types of reproduction mechanisms also exist. There are mainly three types of reproduction systems: i) amphimictic ii) facultative meiotic parthenogenesis in which cross fertilization between male and female is there but in the absence of males meiosis within oocytes occurs but two of the nuclei with reduced chromosomal complement subsequently fuse and iii) obligate mitotic parthenogenesis in which males are not involved but one of the two nuclei produced during initial mitotic division within the oocyte deteriorates and other becomes the predecessor of the subsequent embryo. Species like *M. carolinensis, M. kikuyensis, M. megatyla, M. microtyla, M. pini, M. spartnae* and *M. subarctica* are amphimictic whereas the most commonly occurring species, *M. incognita, M. javanica, M. arenaria* and *M. hapla* (only some isolates) have obligatory mitotic parthenogenesis mode of reproduction although males are commonly found. Within the species *M. hapla* populations are exceptionally diverse cytogenetically, and reproductive modes differ dramatically between isolates. For the majority of *M. hapla* isolates, amphimixis is facultative, and reproduction can also occur by meiotic parthenogenesis (Van Der Beek *et al.*, 1998). In this species even a hermaphroditic form has been described (Triantaphyllou, 1993).

The process of sex differentiation depends on the stress and host response. The male is released outside the root without feeding, the immobile female remains in the plant tissue. Approximately three weeks after root penetration, the female becomes swollen, pear-shaped, and lays eggs. Egg production occurs without sexual cross-fertilization. Even though adult males may be present, they are not required for reproduction. The adult female deposits eggs in a gelatinous mass, and an egg mass contains from 500 to 2000 eggs. The egg masses are usually projected from the surface of young roots. The female dies soon after laying eggs. Eggs in matrices often remain attached to root fragments in the soil after the female dies. Eggs may hatch immediately and juveniles may re-infect the root, or may over-winter and hatch next spring. The number of days required for the root-knot nematode to complete its life cycle depends on soil temperature. One life cycle, from egg to reproducing adult, will occur every 28 days when the soil temperature is 25-30°C. Several generations may succeed one another under favourable conditions, infestations reaching considerable proportions. In case of heavy attacks on susceptible plants (*e.g.* tomato), galls can become very large, the root system being reduced to a swollen stump without hairs. The growth of the aerial parts is stunted with chlorosis of leaves. Infected plants are very susceptible to drought. Moreover, the presence of this nematode favours or worsens attacks by fungi and other phytoparasites. The damage to the host plant is the result of physiological and biochemical changes caused by the nematode feeding. Giant cells, which are approximately 10 times larger than normal root cells, interfere with the development of the root because transport of water and nutrients from the soil are almost cut off. In addition, some of the sugars produced by the plant photosynthesis to support normal root growth are diverted to the giant cells to sustain the developing nematode. The nematodes swell and the somatic musculature atrophies, rendering the nematodes sedentary. With changes in the nematode, striking changes also occur within the root, but the nature of these changes depends on the presence or absence of resistance in the host. In a compatible host, giant cells develop by enlargement of individual parenchyma cells in the vascular cylinder close to the nematode's head. The developing cells undergo many nuclear division without cytokinesis, and individual nuclei become highly polyploid. The cell wall is extensively remodeled, with the development of finger-like projections into the cell and a marked reduction in plasmodesmatal connections with cells other than neighboring giant cells. Giant cell development and gall formation are important parts of the complex physiological interaction that occurs between the plant and root knot nematode as the parasitic relationship progresses. It has been proposed that proteins secreted by the nematode induce and maintain the feeding site during the life of the nematode, while other unidentified proteins or degradation products may function as elicitors of plant resistance responses.

Each nematode triggers the development of five to seven giant cells. Under optimal conditions, giant cell expansion can result in a final size of 600-800 μm long and 100-200 μm in diameter (Goverse *et al.*, 2000), each containing as many as 100 enlarged nuclei (Wiggers *et al.*, 1990). These events are related to the developmental status of the nematode, and the giant cells (which serve as the obligate nutritive source for the nematode) reach maximal size. Interestingly, the transition from a parenchyma cell to a fully differentiated giant cell occurs early in the parasitic association; once the giant cells have been initiated, their characteristics do not change appreciably throughout the period of nematode feeding (apart from getting bigger, having more nuclei). This is in marked contrast to determinant nitrogen-fixing nodules in which a programme of elaborate differentiation ensue. In many hosts (but not all), cortical and pericycle cells around the giant cells enlarge and divide repeatedly (hypertrophy and hyperplasia). Swelling of roots occur due to the excessive enlargement and division of all types of cells surrounding giant cells and from enlargement of the nematode which can lead to highly disfigured and functionally adjusted roots. As the females enlarge and produce their egg sacs, they push forward and split the cortex and may become exposed on the surface of the root or lay embedded depending upon the position of the nematode in relation to the root surface. Different sizes of galls are induced depending on their host and the species of the nematode involved. On cucurbits, the nematode induces large galls, whereas in chilli small sized galls are produced. Usually the infection due to *M. hapla* produces small galls as compared to *M. incognita* and *M. javanica*. The size of galls also differs with the level of infection as in case of heavy infection large size or multiple galls or secondary galls are developed. Besides galling, forking or splittings of taproot in carrot or raddish and tubercles on potato tubers are generally observed. Plant growth regulators occur at higher levels in galled tissue. Both auxins (promoters of cell growth) and cytokinins (promoters of cell division) have been implicated - roots of susceptible tomato cultivars contain higher levels of cytokinin than resistant roots. Cytokinins are also reported from eggs, juveniles and adults; however, this may not be a source, but a result of feeding. When applied to resistant plant roots, resistance may be partially or completely reversed.

Giant cells are central to the parasitic interaction, whereas the surrounding gall is presumably a secondary response.

8.7.1 Induction of Giant Cell

Due to the large size of the cells, Treub (1887) coined the term 'giant cell' for them. The root knot feeding cell is a remarkable example of reprogramming of plant cell by any biotrophic pathogen. In spite of the fact that studies have been undertaken during the last many years but the exact nature of its built up is still baffling. A set of genes encoding secreted enzymes, in particular pectocellulolytic enzymes, has been recently identified in the salivary secretions of this nematode (Davis *et al.*, 2004). The characterization of root-knot nematode secretions as parasitism effectors and the identification of nematode-responsive plant genes constitute a major step towards understanding how this nematode dramatically alter root development to induce and maintain giant cells to its own benefit. With the aid of chemicals secreted by the three esophageal glands, the root knot nematodes initiate a fundamental change in the cells surrounding their head in the root. The secretion of effectors from the amphids or the oesophageal glands is tightly regulated during the course of infection. The analysed effectors accumulated in the root tissues along the nematode migratory path and along the cell wall of giant cells (Rosso *et al.*, 2011). These cells enlarge upto tenfold in size and become a virtual production house of cytoplasm which is a rich source of nutrients for the nematode. The dense cytoplasm has a high proportion of proteins and lipids and supported by a network of newly formed vascular tissues and by the cell wall ingrowths and have large number of nuclei (formed by repeated nuclear division without cytokinesis) underlined by additional cellular features namely small vacuoles, proliferation of endoplasmic reticulum, ribosomes, mitochondria, and plastids. The giant cells are formed from the parenchymatous cells within the stele that surround the anterior end of the nematode suggesting that esophageal gland secretions are primary molecules stimulating giant cell formation. It has been speculated that a peptide is secreted by the sub-ventral glands of the nematode at the initiation of giant cell. This peptide has a role in reprogramming (including stimulating the production of high cytoplasmic density of the cell) of the gene expression required by the giant cell formation (Davis *et al.*, 2008; Gheysen and Mitchum, 2008). The first sign of giant cell induction is cell cycle activation leading to the formation of vascular binucleate cells (Jones, 1981; Melillo *et al.*, 2006). Binucleate cells are found associated with juveniles 24 h after root infection, and become multinucleate within 48 h. Consequently, each cell has many copies of the genome which provides a large quantity of DNA to make proteins in these highly active nutrient sink cells.

More than 50 genes are upregulated to some extent in the development of giant cells (*Meloidogyne*) and syncytia (*Heterodera/Globodera*) (Gheysen and Fenoll, 2002). Both types of feeding cells have the genome amplified as a result of multiple shortened cell cycles; but the processes differ. Giant-cells go through repeated (acytokinetic) mitosis whereas the syncytia undergo repeated S-phase endoreduplication without mitosis or nuclear division.

It is common knowledge that the secretion from the subventral glands plays an important role in the migration phase of the root knot and cyst nematodes within the root. Apparently, both, subventral and dorsal esophageal glands function in concert during induction of feeding sites. Only in the later stages when feeding sites maintenance and feeding appear to be the main function, the dorsal glands secretions come in to play. The most unusual cytological feature in the cell is the formation of nematode feeding tube at the orifice of stylet inside the cytoplasm of the feeding cell elicited by dorsal esophageal gland cells secretions from the stylet, also formed in syncytia produced by cyst nematodes (Sobczak and Golinowski, 2008). The feeding tube facilitates the uptake of cell sap by the nematode and it is suggested to serve as a molecular sieve for host cell contents that are ingested by the nematode (Williamson and Hussey, 1996). It has been suggested that auxin signaling is essential in giant cell formation and implies a change in the local auxin balance upon nematode infection (Goverse *et al.*, 2000). These authors propose a range of options about the concentration of auxin at the feeding sites: i) auxin is originating from both the plant and the nematode, ii) nematode can manipulate auxin household by secreting substances, such as flavonoids (nematode secretions contain chorismate mutase, an enzyme involved in flavonoid synthesis), which inhibit auxin efflux, and iii) nematode locally increases sensitivity towards indole-3-acetic acid (IAA) and roots are very auxin-sensitive organs. The three options could be valid all-together in regulating the complex machinery of affected cells which leads to the cell changes of feeding sites.

8.7.2 Differences between Root Knot and Cyst Nematodes

The root-knot and cyst nematodes use similar strategies to enable their sedentary parasitic life styles. However, it appears that these nematodes use very different tools of fulfilling their strategies because root-knot nematodes usually have wide host ranges and cyst nematodes have a narrow one and the ontogeny of their feeding sites (giant-cells versus syncytia) is very different in certain aspects. Fully characterizing the root-knot nematode and cyst nematode parasitism genes should provide more definite answers. When assessing the currently identified series of parasitism genes

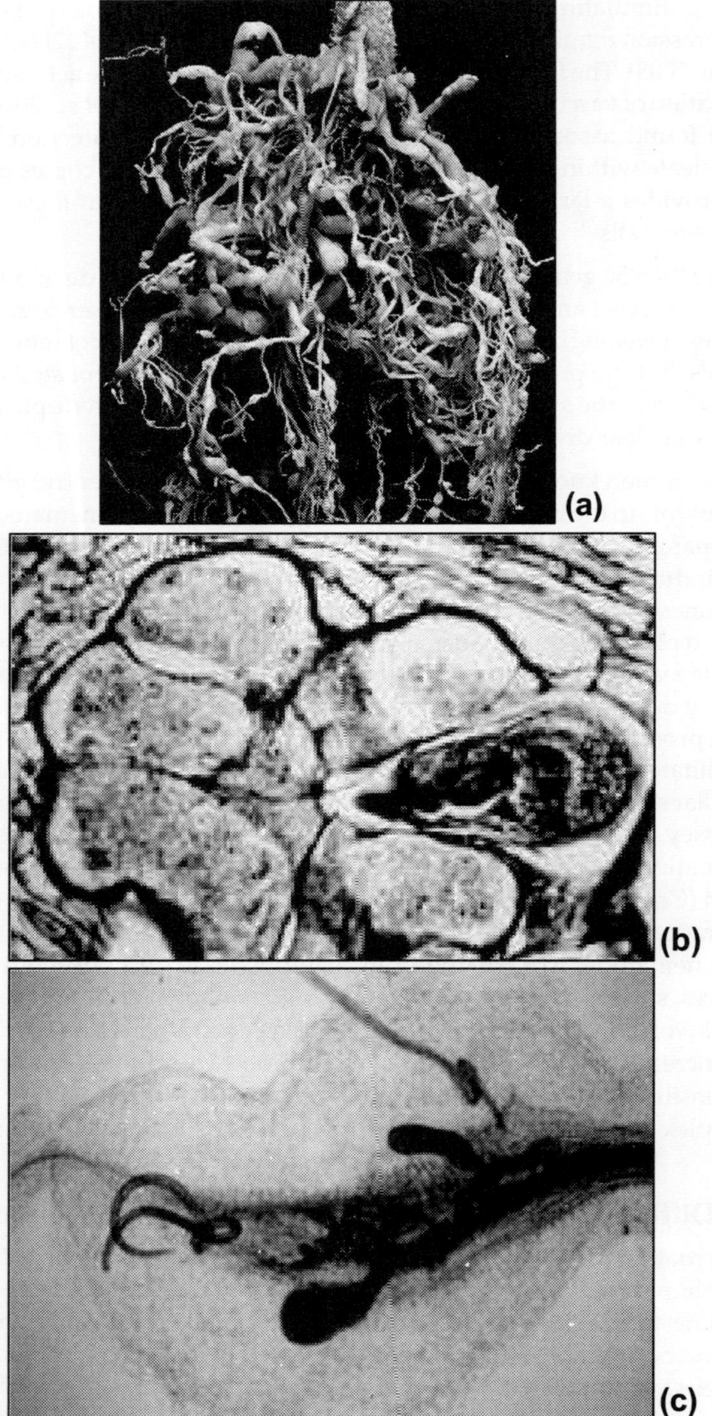

Figure 39: (a) 'Galls' on Brinjal Roots; (b) Giant Cells Formation (c) Development of Males and Females in the Infected Root

found in root-knot nematodes and cyst nematodes, one can find support for the hypothesis that root-knot nematodes and cyst nematodes use different molecular tools for their otherwise similar life habits—at least during the sedentary phase of parasitism. During the migratory phase, both nematode groups (root-knot and cyst nematodes) use cellulase and pectinase enzymes produced in their subventral glands in order to penetrate into and migrate through plant roots. Also, during the early phases of parasitism both nematode groups produce cellulose-binding proteins and venom-allergen proteins for unknown reasons and both root-knot and cyst nematodes produce chorismate mutase enzymes potentially to inactivate plant host defenses. However, as a first significant difference, a cyst nematode was shown to use an expansin parasitism protein to soften host cell walls, which is a group of proteins so far not identified in root-knot nematodes. Even more profound differences exist beyond these early stages of parasitism. While the soybean cyst nematode uses a small ligand with similarity to CLAVATA3-like proteins and appears to employ an ubiquitination pathway, no such proteins were discovered in root-knot nematodes. Instead, a large percentage of parasitism proteins without any database similarities (including cyst nematode genes) were found in root-knot nematodes. Also, while both root-knot and cyst nematodes produce a high proportion of unknown parasitism protein in their dorsal-gland, root-knot nematodes appear to produce a relatively large proportion of unknown parasitism proteins also in their subventral glands. Of course, these assessments can only rely on the current state of knowledge and can only be completely validated when all parasitism proteins of several species of both nematode groups are identified. So, the parasitism protein identities so far confirm that root-knot nematodes and cyst nematodes share certain aspects of their parasitic strategies but that key components of their arsenals of molecular tools likely are very different (Baum *et al.*, 2007).

8.7.3 Hatching

After the formation of egg sac or egg mass by the adult root knot female, the eggs are laid in it and not retained in the body unlike cyst nematodes. Immediately the physiological maturation of eggs takes place and hatching starts. Generally in root knot species, there is no cessation of development (diapause) due to unfavourable environmental conditions though there may be delay in development of various stages at low temperatures or due to other biotic and abiotic factors. There is some evidence that two kinds of egg masses (white and brown) are formed depending on the environmental stress conditions. The white egg masses are generally formed early in the in the growing host and the eggs hatch immediately to ensure more than one generation in the host growing season. As the growing season progresses and with the onset of adverse environmental conditions on the host, brown egg masses are laid which undergo diapauses and do not hatch immediately. This diapauses occurs in all the four common species of *Meloidogyne*. There are reports that incidence of unhatched J2 that enter diapause varies from less than10 per cent for the tropical *M. arenaria* to 94 per cent for the temperate species of *M. naasi* (it is the only species of the genus *Meloidogyne* that under goes diapause). The chilling period is necessary in *M. naasi* to stimulate hatching and that is why it has only one generation per year. The egg mass provides a major barrier to dessication, is a first line of defense against parasites and predators, against adverse forms of soil environment- be it low levels of oxygen, soil moisture or presence of toxins. The eggs of root knot nematodes are cylindrical containing three layers of outer vitelline, a middle chitinous and an inner lipid layer. Chitin has been estimated to be 30 per cent of the egg shell in root knot nematodes. As the activity of J2 starts for hatching there is a decrease in the inner layer of lipid also there is change in the egg shell permeability. The proteic granules present in the subventral esophageal glands are responsible for the reduction of the lipid layer. The permeability change in the eggshell is also caused by the enzyme secretion from these glands. The egg becomes flexible before the hydrated J2 starts hatching unlike *Globodera* cysts which take water after hatching and eggs remain rigid during hatching process. With egg shell becoming permeable, the unhatched J2 in the eggs become susceptible to toxins that may be present in the soil. It has been observed that the extracts of *Plantago lanceolata* and *P. rugelii* are toxic to *M. incognita* unhatched J2 and their hatching is reduced in the presence of serototin and octopamine (Meyer *et al.*, 2006; Masler, 2008). Hatching of J2 is generally not influenced by root exudates in root knot nematodes where it freely occurs in water, still there are species in which hatch is stimulated or increased by the presence of root exudates. The hatching is usually delayed or it may be deferred to the next season from the eggs which are laid late in the egg mass.

The root knot species survive, hatch and reproduce over a wide range of pH (4.0-8.0). Wallace (1966) found that hatch of *M. javanica* eggs was maximum between pH 6.4 and 7.0 and inhibited below pH 5.2. Most of the species like sandy loam coarse textured soils for all their activities from movement to development and reproduction as these conditions provide better aeration and optimum moisture availability. According to Wallace (1969) reproduction rate was highest in the coarse sand while root penetration was greatest in fine sand. Nematodes are dependent on water for their well being. In dry soil they become inactive and may die due to desiccation. In saturated soil they are unable to move due to water film thickness and show reduced hatch and activity due to lack of oxygen. Since the nematode cuticle allows passage of water rapidly back and forth, solutions of high osmotic potential cause J2 to shrink and become inactive. The J2 are able to osmoregulate and become active again when return to favourable osmotic conditions. The interactive effects of soil texture, moisture and aeration on nematode movement and survival are difficult to separate. Direct effects of moisture can be expressed in terms of suction potential, a measure of accessibility of water to the nematode. At low levels of suction potential soil pores are filled and oxygen deficiency may result which adversely affects the metabolism and movement of the juveniles and their infectivity. The relationship among various species of root knot nematode, their development and pathogenesis, rate of reproduction and even growth of the host in the soil is as complex as it is being influenced by environmental factors.

8.8 Diagnostic Characters of Four Common Species of Root Knot Nematodes

It is now abundantly evident that all kinds of taxonomic approaches have supported the validity of four most common species (*M. incognita, M. javanica. M.*

arenaria and *M. hapla*) of the root knot nematodes. Among these species there may be some morphological variations, presence of races which may be physiological or cytological. Identification of these species is primarily based on the morphological features and that too more on the perineal patterns. In the perineal patterns of these species, *M. incognita* shows the maximum variation and least variation is observed in other species although in mixed populations also it becomes quite difficult to identify the species involved. The perineal pattern of root knot nematodes is comparable with vulva anus regions of *Meloinema*, *Cryphodera* and *Nacobbus* except for the vulva-anus distance. Within *Meloidogyne* this distance is so small that only few species like *M. coffeicola*, *M. indica*, *M. decalineata*, *M. propera* and *M. nataliei*, have some continuous striae between anus and vulva. The other characters which are valuable in the identification of these species include heads and stylets of males and J2 and female stylet in addition to the various morphometrics of the body of these stages.

8.8.1 Important Characters of Male Head Shape and Stylet of Four Most Common Species

	Head Cap	*Stylet Region*	*Stylet Cone*	*Stylet Shaft*	*Stylet Knobs*	*Stylet Length*	*DEGO*
M. incognita	Flat to concave labial disc raised above the medial lips	Not set offusually marked by 2-3 incomplete annulations	Tip blunt blade-like often narrows near knobs	Usually cylindrical to transversely elongate sometimes indented anteriorly	Set off rounded	24 µm mean, 23-25 µm range	Short 3 µm mean, 2-4 µm range
M. javanica	High rounded set off from head region	Not set off smooth or marked by 2-3 incomplete annulations	Tip pointed cone straight	Usually cylindrical	Set off, low and very wide	20 µm mean, 18-22 µm range	Short, 3 µm mean 2-4 µm range
M. arenaria	Low sloping posteriorly	Not set off smooth or marked by	Tip pointed cone broad and robust 2-3 incomplete annulations	Usually cylindrical often broadens near knobs	Not set off, backward sloping, merging with shaft	22 µm mean 20-25 µm range	Long 5 µm mean, 4-7 µm
M. hapla	High and narrow	Set off smooth, larger diameter than first body annule	Tip pointed, cone narrow and delicate	Cylindrical often wider or narrower at its base	Set off small and round	20 µm mean, 18-22 µm range	Moderately long, 5 µm mean, 4-6 µm range

8.8.2 Heads and Stylets of Males (Eisenback *et al.*, 1981)

The shapes of heads and stylets of males are good taxonomic characters in the identification of root knot species (Chitwood, 1949 and Whitehead, 1968). The morphology of stylets of these stages has been found to be a stable character and is species specific. The heads and stylets of males are extremely valuable characters and particularly with SEM observations it has been possible to identify each species with some amount of certainty even in mixed populations. Jepson (1983) has tried to put some 24 species in groups (1-5) on the basis of variability in head morphology of males.

8.8.2.1 *Meloidogyne incognita*

In this species the shape of head characterized by a large, round labialdisc that is raised above medial lips. The labial disc is concave to flat, and the high head cap is nearly as wide as the head region in lateral view.The head region is marked by 2-3 incomplete annulations but may be completely smooth.The head region is not distinctly set off from the rest of the body. The stylet is distinguished by a blunt, blade-like tip which is wider than the medial portion of the cone. The opening is often marked posteriorly by a pronounced prolection. Th shaft is usually cylindrical and often narrows near the knobs.The knobs are large, rounded to transversely ovoid, sometimes anteriorly indented and set off from the shaft. Stylet length varies from 23-25 µm with a mean of 24µm.Generally, the distance from the DEGO to the stylet base is usually short, 2-4µm with a mean of 3µm.

8.8.2.2 *Meloidogyne javanica*

Male heads of this species are distinguished by high, rounded head cap that is distinctly set off from the head region. The labial and medial lips are fused and form one smooth, continuous structure which is almost as wide as the head region in lateral view. The head region may be marked by 2-3 incomplete annulations. The head region is not distinctly set off from rest of the body.

The stylet of males is characterized by very broadly elongate stylet knobs. The anterior two-third of the cone gradually increases in width, while the posterior one-third rapidly widens. The shaft is cylindrical. The knobs are low and wide, sometimes anteriorly indented and always set off from the stylet. Stylet lengths range from 18-22µm and average 20µm. The distance from DEGO to the base of the stylet is relatively short, 2-4µm with a mean of 3µm.

8.7.2.3 *Meloidogyne arenaria*

Male heads of *M. arenaria* are characterized by a low head cap that slopes posteriorly. The labial disc and medial lips form one smooth, continuous structure that is nearly as wide as the head region in a lateral view. The head region may be smooth or marked by 1-3 incomplete annulations. The head region is not distinctl set off from rest of the body. Stylet is relatively straight, broad and robust. Th posterior two-third of the cone gradually increases in width and the posterior one-third widens rapidly. The posterior portion of the cone is much broader than the anterior portion of the shaft. The shaft is cylindricalbut sometimes widens near its junction with the

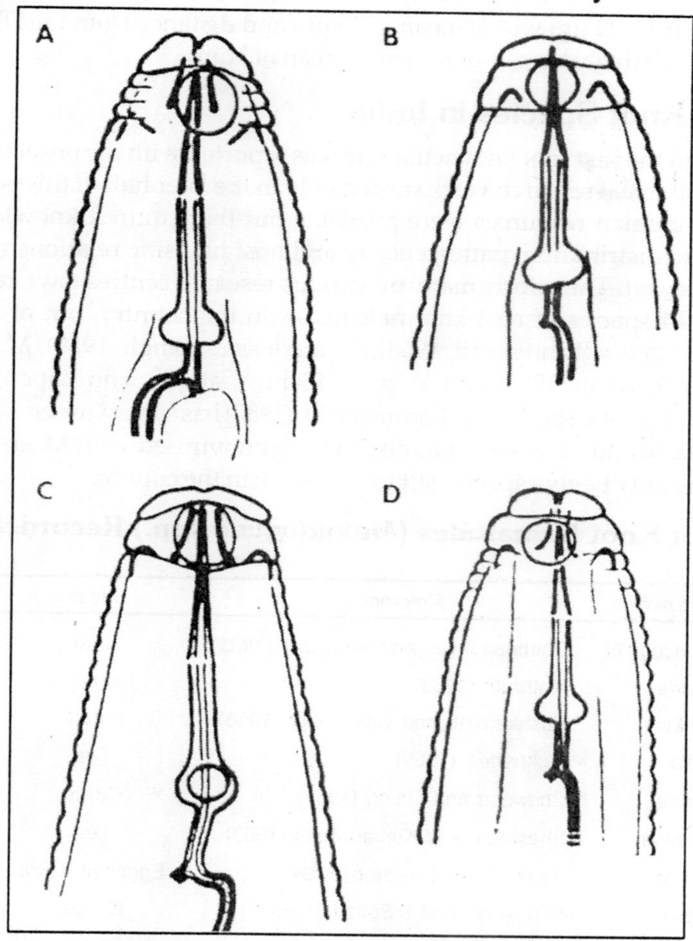

Figure 40: Heads of *Meloidogyne* Species
A. *M. incognita*, B. *M. javanica*, C. *M. arenaria*, D. *M. hapla*

knobs. The knobs are large, posteriorly rounded and gradually merge with the shaft.Stylet length varies from 20-25μm but average around 22μm. The distance from the DEGO to the stylet base of the stylet is relatively long 4-7μm and averages 5.5μm.

8.8.2.4 *Meloidogyne hapla*

In this species the heads of males are characterized by a high and narrow head cap and a set off head region. The labial disc and medial lips are fused and form one smooth, continuous structure that is much narrower than the head region. The head region is without annulations and distinctly set off from the body annulations.The body annules increase in width and diameter posteriorly. The stylet of males are very narrow and delicate in comparison to the other three species.The cone gradually increases in width and the base of the cone is only slightly wider than the anterior region of the shaft.The stylet knobs are small, rounded and set off from the shaft.

Stylet length is 18-22 µm with a mean of 20 µm and distance from the DEGO to the stylet base is relatively long, 4-6µm with a mean of 5µm.

8.9 Root Knot Species in India

Although the first root knot nematode was reported with the onset of twentieth century, but the real research work started only in the later half of this period when not only the human resources were available but the required knowledge about identification, distribution, pathogenecity and host parasitic relationship was also being worked out. The efforts made by various research centres have revealed the presence of 14 species of root knot nematodes in the country, out of which four species *i.e. M. indica* (Whitehead, 1968), *M. lucknowica* (Singh, 1969), *M. triticoryzae* (Gaur,Saha and Khan, 1993) and *M. piperi* (Sahu, Ganguly and Eapen,2000) have been described as new species. As Fortuner *et al.* (1987) has synonymised *M. lucknowica* with *M. javanica* and *M. thamesi* has also been synonymised with *M. arenaria*, so, in fact, there are only twelve species left to be known in the country.

8.9.1 Root Knot Nematodes (*Meloidogyne* spp.) Recorded in India

Name of the Species	*Reference*	*Host*	*State*
Meloidogyne incognita	Chattopadhyay and Sengupta (1955)	Jute	West Bengal
M. arenaria	Mathrani (1955)	Tobacco	A.P.
M. javanica	Pushkarnath and Chaudhury (1958)	Potato	U.P.
M. hapla	Mukherjee (1960)	Tea	Assam
M. africana	Chitwood and Toung (1960)	Citrus	Delhi
M. brevicauda	Sheshadri and Kumarswami (1963)	Tea	Tamil Nadu
M. thamesi	Sethi, Gill and Swarup (1964)	Eggplant, Okra	Delhi
M. exigua	Srinivasan and D'Souza (1965)	Coffee	South India
M. graminicola	Patnaik (1969)	Paddy	Orissa
M.graminis	Nayak, Ray and Routray (1986)	Wheat	Orissa
*M. indica**	Whitehead (1968)	Citrus	Delhi
*M. lucknowica**	Singh (1969)	Spongegourd	U.P.
*M. triticoryzae**	Gaur, Saha and Khan (1993)	Paddy, Wheat	Delhi
*M. piperi**	Sahu, Ganguly and Eapen (2000)	Pepper and Brinjal	Kerala

8.9.2 *Meloidogyne incognita* (Kofoid and White, 1919) Chitwood 1949

This most important species of root knot nematods was first reported from India by Chttopadhayay and Sengupta (1955) infecting jute in West Bengal. Now, this species is present throughout the country and the most worked upon root knot nematode.

Perineal Pattern (Eisenback, 1985)

It is characterized by the presence of high, squarish dorsal arch that often contains a distinct whorl in the tail terminal area. The striae are smooth to wavy, sometimes zigzagged. Distinct lateral lines are absent but the lateral field may be marked by breaks and forks in the striae.

Stylet of Female

It is characterized by a dorsal curve. The anterior half is cylindrical and the posterior half is conical. The shaft is slightly wider posteriorly. The knobs are set off from the shaft, anteriorly indented (sometimes deeply indented) and transversely elongate. The average stylet length is 16 μm and the range is 15-17 μm.

Head of Male

It is characterized by a large, round labial disc that is raised above the medial lips. The labial disc is concave to flat and high head cap is nearly a wide as the head region in lateral view. The head region is usually marked by 2-3 incomplete annulations but may be completely smooth. The head region is not distinctly set off from rest of the body.

Stylet of Male

The stylet of males is distinguished by a blunt blade like tip which wider than the medial portion of the cone. The opening is often marked by a pronounced projection. The shaft is usually cylindrical and often narrows near the knobs. The knobs are large rounded to transversely ovoid sometime anteriorly indented and set off from the shaft.

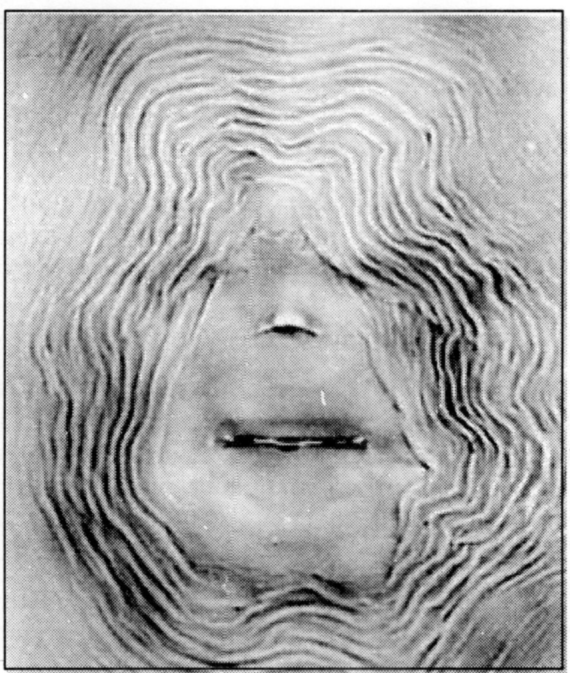

Figure 41: *Meloidogyne incognita*

Stylet length varies from 23-25 µm with a mean of 24 µm. Generally, the distance from the dorsal esophageal glands orifice to the stylet base is relatively short 2-4 µm with a mean of 3µm.

Second-Stage Juvenile

Length= 346-463 µm (405); tail length= 42-62 µm (52); head end to stylet base 14-16µm (15).

M. incognita populations reproduce exclusively by mitotic parthenogenesis. There are two chromosomal forms within this species. One form has 2n=32-36 chromosomes and is considered to be diploid; the other form has 2n=40-46 chromosomes and probably represents a triploid. The triploid form is by far the most common and widely distributed around the world. This species constitutes about 52 per cent of the *Meloidogyne* species collected through the IMP. From a total of 423 populations studied (in IMP), 72 per cent were host race 1; 15 per cent race 2; 11 per cent race 3; and 2 per cent race 4. Thus, 87 per cent of *M. incognita* populations do not reproduce on cotton, 83 per cent do not reproduce on resistant tobacco and none of the four races reproduces on peanut.

8.9.3 *Meloidogyne javanica* (Treub, 1885) Chiwood, 1949

The species was reported from U.P. infecting potato by Pushkarnath and Chaudhury (1958). It is the second most important species in India.

Perineal Pattern

In this species the perineal pattern is unique because of the presence of lateral ridges that divide the dorsal and ventral striae. Generally the ridges run the entire width of the pattern but gradually disappear near the tail terminus. The dorsal arch is low, and rounded to high and squarish and often contains a whorl in the tail terminus. The striae are smooth to slightly wavy and some striae may bend toward the vulval edges.

Stylet of Female

It is similar that of *M. incognita* except the cone is slightly curved dorsally and the shaft is more cylindrical. The knobs are also transversely elongate but they are not deeply indented anteriorly. The stylet length varies from 14-18 µm with a mean of 16 µm.

Head of Male

In this species the heads are distinguished by a high rounded head cap that is distinctly set off from the head region. The labial disc and medial lips are fused and form one smooth continuous structure which almost as wide as the head region in a lateral view. The head region may be smooth or marked by 2-3 incomplete head annulations. The head region is not distinctly set off from rest of the body.

Stylet of Male

It is characterized by broadly elongate stylet knobs. The anterior two-third of the cone gradually increases in width, while the posterior one-third rapidly widens. The

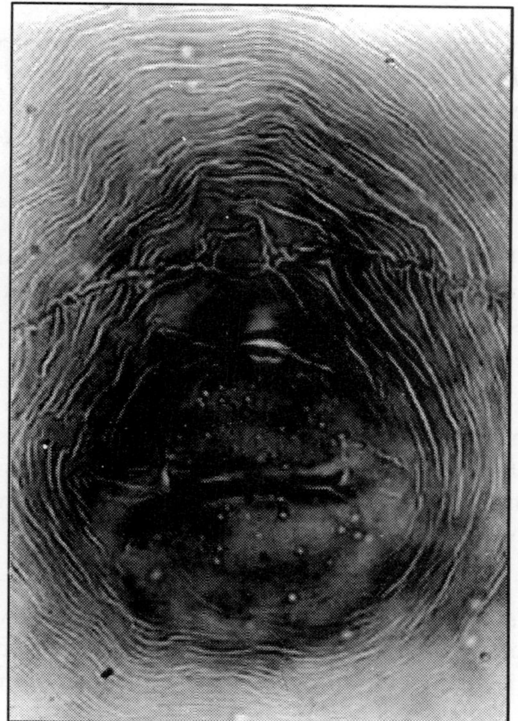

Figure 42: *Meloidogyne javanica*

shaft is cylindrical. The knobs are low and wide, sometime anteriorly indented and always set off from the shaft. The stylet length ranges from 18-22 μm and average 20 μm. The distance from the DEGO to the base of the stylet is relatively short, 2-4 μm with a mean of 3μm.

Second-Stage Juvenile

Length= 402-560μm (488); tail length= 51-63μm (56) pm; head end to stylet base 14-16 (15) pm.

Some populations of *M. javanica* produce male intersexes that are characteristic for the species. These intersexes show different degree of female secondary sex characters varying from a small ventral protuberance anterior to the spicules to a large protuberance marked by a rudimentary vulva. Populations of *M. javanica* reproduce exclusively by mitotic parthenogenesis. The chromosome number varies from 2n=43 to 48. All populations belong to the same chromosomal form, which may represent a triploid. *M. javanica* does not reproduce on cotton or peanut and rarely on pepper. In RFLP and isozyme analyses, *M. javanica* most often clusters with *M. arenaria*.

8.9.4 Meloidogyne arenaria (Neal, 1889) Chitwood, 1949

Mathrani (1955) reported this species infesting tobacco in Andhra Pradesh. It is quite prevalent in areas where groundnut and tobacco are grown in the country specially in Gujarat and Karnataka states.

Perineal Pattern

It is distinguished by a low arch that is slightly indented near the lateral field to form rounded shoulders. The dorsal and ventral striae often meet at an angle. Distinct lateral lines are absent but short, irregular and forked striae mark the lateral field. The striae are smooth to slightly wavy and some may bend toward vulva. Sometime the patterns are extended laterally and form one or two wings.

Stylet of Female

The female stylet is quite characteristic as the whole stylet is broad and robust and the wide, rounded backward sloping knobs gradually merge with the shaft. Stylet length ranges from 13-17µm with a mean of 15.5µm.

Head of Male

In this species the male head is characterized by a low head cap that slopes posteriorly the labial disc and medial lips form one smooth continuous structure that is nearly as wide as the head region in a lateral view. The head region may be smooth or marked by 1-3 incomplete annulations. The head region is not distinctly set off from rest of the body.

Stylet of Male

It is relatively straight, broad and robust. The posterior two-third of the cone gradually increase in width and posterior one-third widens rapidly. The posterior portion of the cone is much wider than the anterior portion of the shaft. The shaft is

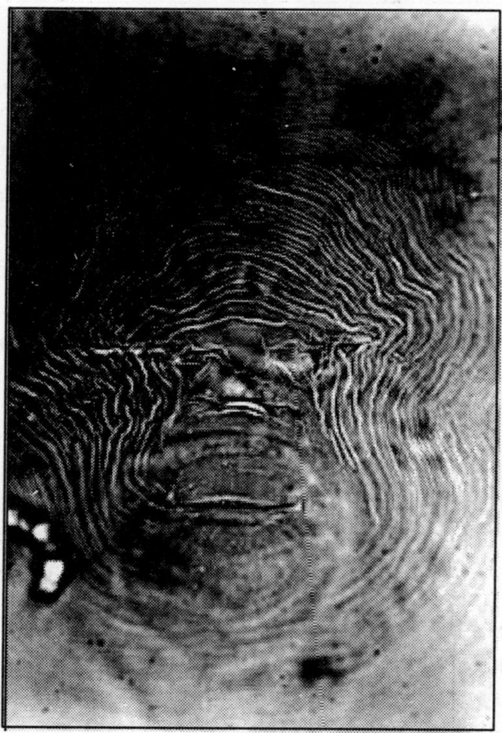

Figure 43: ***Meloidogyne arenaria***

cylindrical but sometime widens near its junction with the knobs. The knobs are large, posteriorly rounded and gradually merge with the shaft. Stylet length varies from 20-25µm, averages around 22µm. The distance from the DEGO to the base of the stylet is relatively long 4-7 µm and averages around 5.5 µm.

Second-Stage Juvenile

Length, 398-605µm (521)); tail length= 44-69 µm (58); head end to stylet base= 14-16µm (15).

M. arenaria populations often produce many small bead-like galls that do not form short lateral roots. All populations of *M. arenaria* reproduce by mitotic parthenogenesis. Two chromosomal races are recognized in this species. Race 1 is the most common and includes triploid populations with 2n=50 to 56 chromosomes. Race 2 is the diploid race with 2n=34 to 37. It constitutes about 8 per cent of the populations encountered in samples collected through the IMP. Two host races are recognized. Race 1 reproduces on peanut; race 2 does not. Race 2 will not reproduce on pepper and neither race reproduces on cotton.

8.8.5 *Meloidogyne hapla* Chitwood, 1949

This species was first recorded by Mukherjee (1960) from tea plantation in Assam. It has also been reported from Shimla in H.P. infesting potato.

Perineal Pattern

The shape of the perineal pattern is rounded hexagonal to flattened ovoidal shape, very fine striae and subcuticular punctuations in the smooth tail terminus area. The dorsal arch is usually low and rounded but may be high and squarish. Lateral ridges are absent but the lateral fields are marked by irregular striae. The dorsal and ventral striae often meet at an angle and striae are smooth and slightly wavy. Sometime pattern may form wings on one or both lateral sides.

Stylet of Female

The stylet is narrow and delicate as compared to other species. The cone is only slightly curved dorsally and the shaft is little broader posteriorly. The knobs are small, round, and set off from the shaft. Stylet length varies from 13-17 µm with a mean of 15.5 µm.

Head of Male

The head is characterized by a high and narrow head cap and set off head region. The labial disc and medial lips are fused and form one smooth, continuous structure that is much narrower than the head region. The head region without annulations and distinctly set off from the body annulations. The body annules increase in width and diameter posteriorly.

Stylet of Male

The stylet is very narrow and delicate compared to the other three species. The cone gradually increases in width and the base of the cone is only slightly wider the anterior portion of the shaft. The stylet knobs are small, rounded and set off from the shaft. Stylet is 17-23 µm long with a mean of 20 µm. The distance from the DEGO to the base of stylet is relatively long, 4-6 µm with a mean of 5 µm.

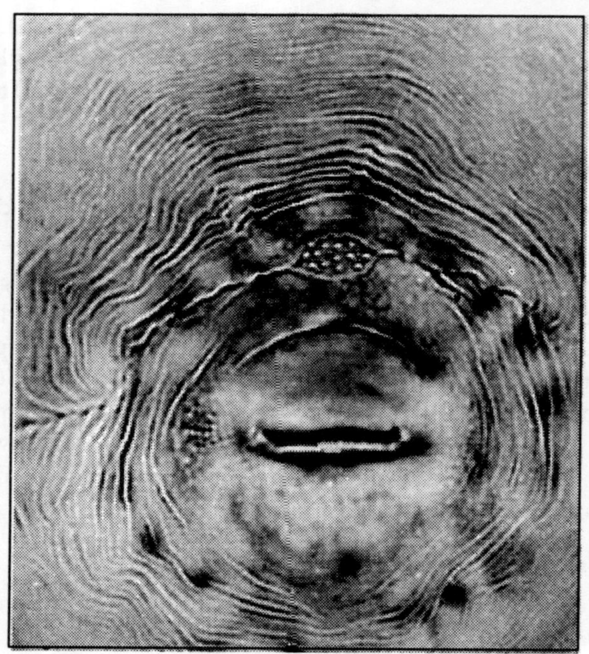

Figure 44: *Meloidogyne hapla*

Second Stage Juvenile

Length=357-467 (413) μm, tail length= 46-58μm (53), head end to stylet base= 14-16 μm (15).

Populations of *M. hapla* B populations reproduce on resistant tobacco, pepper, peanuts, and tomato. Unfavorable hosts of *M. hapla* include cotton and watermelon, according to the North Carolina Differential Host Test. *M. hapla* is made up of populations belonging to two distinct cytogenetic races (A and B). Race A is the most common and includes populations that reproduce by facultative meiotic parthenogenesis. Most of them have haploid chromosome numbers of n=17 or 16 and some have n = 15 or 14. Race B populations reproduce exclusively by mitotic parthenogenesis. Some of them are diploid with 2n=30 to 31 but most are triploid with 2n=43 to 48 chromosomes. *M. hapla* was found in 8 per cent of the populations collected in IMP. It is known to occur and cause losses in cooler climate grown crops.

8.9.6 Meloidogyne africana Whitehead, 1960

This root-knot nematode is a tropical or subtropical species reported from India, Kenya, and Zaire (Campos *et al.*, 1990; Sikora and Greco, 1990; Swarup and Sosa-Moss, 1990). The host range of this species includes: coffee (*Coffea arabica*), corn (*Zea mays*), and cowpea (*Vigna unguiculata*). Decline of coffee, cowpea, and sugarcane has been associated with this root knot species.

Chitwood and Toung (1960) reported the existence of a root knot nematode (Asiatic pyroid citrus nema) infecting citrus in Taiwan and New Delhi and they found its posterior to resemble *Meloidogyne africana* Whitehead, 1960 but they never were able to finish its morphological studies. Vovlas and Inserra (1996) suggested that pyroid citrus nema reported by Chitwood and Toung (1960) may be the same species as *M. indica* described by Whitehead (1968) from New Delhi, India as they have common host, *Citrus sinensis*.

Female

M. africana have an oval cuticular perineal pattern. It has a low dorsal arch and lack prominent lateral lines. No shoulders are present. This root-knot nematode has sedentary endoparasitic habits. Second-stage juveniles (J2) in the soil penetrate host roots where they establish a specialized feeding site (giant cells) in the stele. As J2 develop, they cause root swellings and become swollen females. Females rupture root cortex and some time protrude with the egg masses from the root surface. Galls produced are spherical to elongate. J2 emerge from the egg masses and migrate in the soil. This species is also known to infect *C. reticulata, C. grandis* and *C. aurantinum.*

8.9.7 Meloidogyne brevicauda Loos, 1953

Meloidogyne brevicauda Loos, 1953 was described nearly 40 years ago as a root-knot species parasitizing tea in Sri Lanka. It has been reported only in Sri Lanka and India (Rao, 1970; Silvapalan, 1972, 1978; Lamberti *et al.,* 1987). The labial disc of females and males is distinctly separated from the lips by a deep groove. In the second-stage juvenile, (the labial disc and lips are divided by a groove but are joined by a small band of cuticle. Head annulations are absent in all three life stages and the lips extend posteriorly to the first body annule. The tail of the male is marked by a

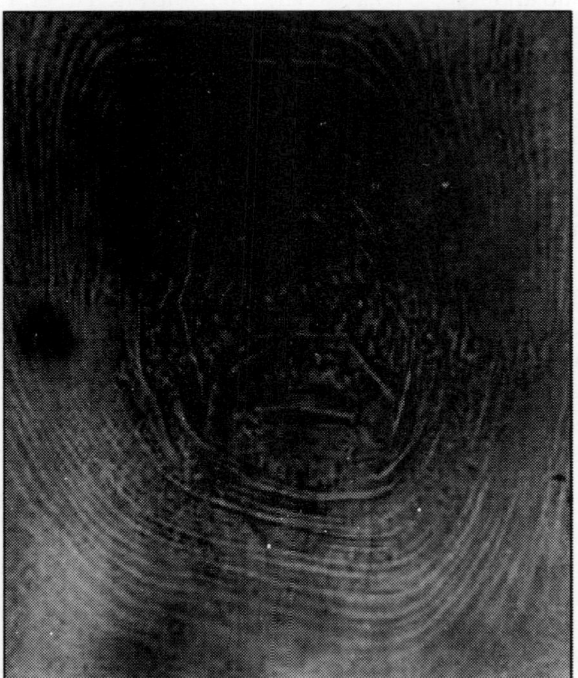

Figure 45: *M. brevicauda*

crescent-shaped fold around the posterior portion of the cuticle surrounding the cloacal opening. The tail of second-stage juveniles of *M. brevicauda* is very different from the four common species named above. The very short tail is the character from which the species name was derived. The location of the phasmidial opening on the ventral edge of the lateral field is most unusual because in other root-knot species it generally occurs in the middle of the field.

The perineal pattern of *M. brevicauda* is quite distinctive. The striae are smooth, coarse, and continuous; they completely encircle the perineum and extend anteriorly for some distance. The dorsal arch is high and squarish to rounded. The lateral fields are usually indistinct, but they may be marked by irregularities, or small, wing-like projections. The striae may completely encircle the large, rounded tail terminus.

8.9.8 *Meloidogyne graminicola* Golden and Birchfield, 1965

The rice root knot nematode, *Meloidogyne graminicola* Golden and Birchfield, is recognised as one of the constraints to rice production in Asia (Bangladesh, India, Laos, Veitnam, Myanmar, Srilanka, Phillipines and Thailand) and causes significant yield losses in upland and rainfed lowland rice production due to rice cropping intensification and increasing scarcity of water (Jairajpuri and Baqri, 1992; Prot and Matias, 1995; Soriano *et al.*, 2000). It has been recorded from rlce growing regions in Laos (Manser, 1968), India (Israel, Rao and Rao, 1963; Roy, 1973), Thailand (Buangsuwon *et al.*, 1971), U.S.A. (Golden and Birchfield, 1968; Yik and Birchfield, 1979), and Bangladesh (Hoque and Talukdar, 1971; Page *et al.*, 1979). It has been found mainly on rice growing in upland conditions and inurseries (Buangsuwon *et al.*, 1971; Israel, Rao and Rao, 1963; Manser, 1968; Rao and Israel, 1971, 1972), and is reported to be absent when the rice crop is grown in flooded fields (Buangsuwon *et al.*, 1971; Manser, 1968) and to occur in small numbers in poorly drained soils (Rao and Israel, 1971). There have been few reports of *M. graminicola* or other *Meloidogyne* species infesting deep water rice apart from the observations that a species of *Meloidogyne*, referred to as *M. exigua*, occurs in the deep wate rice region of Thailand (Hashioka, 1963 ; Kanjanasoon, 1962 ; Ou, 1972). However, in Bangladesh *M. graminicola* commonly caused galling of deep water rice roots in flooded conditions sometimes to a depth of 1.5 m during the early growth stages of the crop (Page *et al.*, 1979). It is also present in U.S.A. in 20-30 per cent of rice growing fields in Texas, Louisiana and Georgia. In India, it poses serious problems in boro and kharif nursery particularly in sandy loam and alluvial soils of West Bengal. Though it is a problem mostly in upland rice, it is also becoming problem in transplanted rice grown in waterlogged conditions and has been found occur widely in Assam, West Bengal, Gujarat, Orissa, Karnataka and Tripura.

Meloidogyne graminicola is the most damaging species in rainfed upland rice in India (Panwar and Rao, 1998) causes a loss of 17-30 per cent or more in affected fields (Prasad *et al*,. 1986, Phukan, 1995) and lowland and deep-water rice in Bangladesh (Bridge *et al.*, 1990). A complete failure of boro rice nursery in 'simurali' in the district of Nadia, West Bengal has been noticed (Anon, 2001). In upland rice, there is an estimated reduction of 2.6 per cent in grain yield for every 1000 nematodes present around young seedlings. In irrigated rice, damage is caused in nurseries before

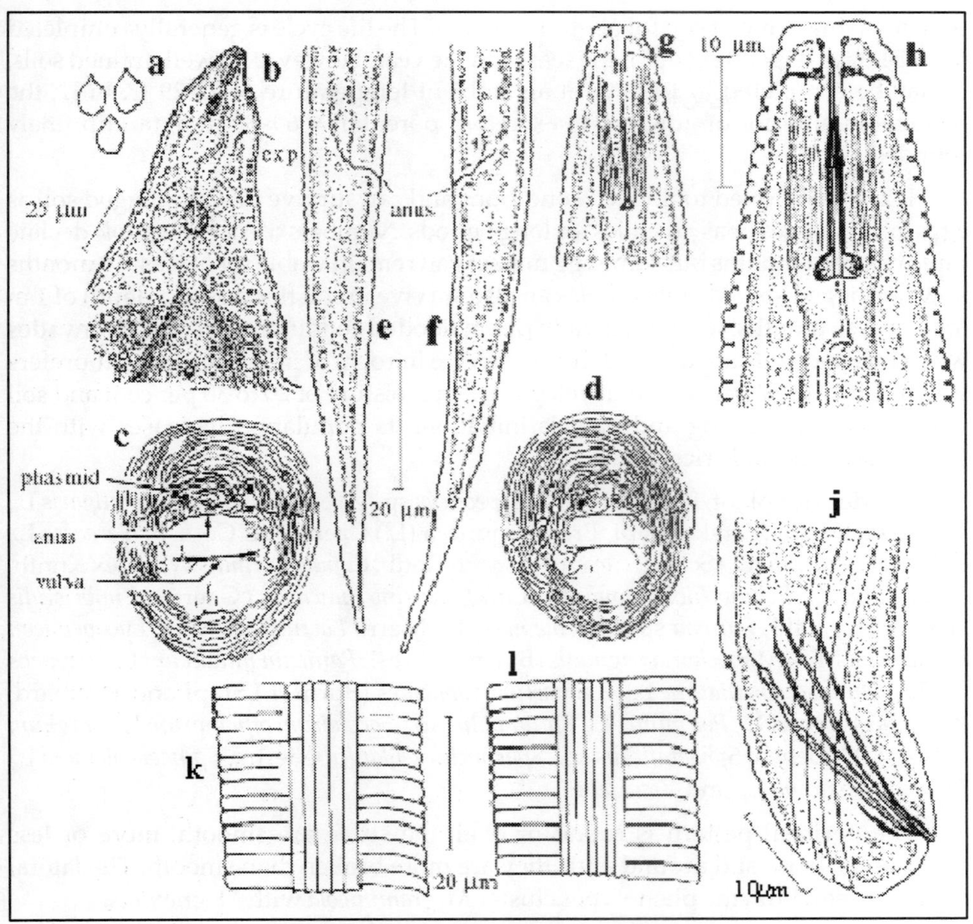

Figure 46: *Meloidogyne graminicola*
a: Females, b: Anterior end of female, c and d: Perineal patterns, e and f: Posterior region of J2, g: Anterior region of J2, h: Anterior region of male, j: Posterior region of male and k and l: Lateral lines

transplanting or before flooding in the case of direct seeding. Experiments have shown that 4000 juveniles per plant of *M. graminicola* can cause destruction of up to 72 per cent of deepwater rice plants by drowning out. Yellowing and curling of leaves towards midrib and delay in flowering are common symptoms of nematode attack. Swollen and hooked root tips are characteristic of *M. graminicola. M. graminicola* significantly reduced growth of deep water rice before flooding and to a greater extent after flooding.

On submergence, the majority of plants (72 per cent) infested with large numbers of the nematode were unable to grow above the water and were thus drowned out. The juveniles (L = 410–480 µm) after penetrating the roots within five hours of inoculation, cause stuntedness, chlorosis and reduced tillering followed by hyperplasia and hypertrophy in the affected roots. The galls formed are very small that may not be visible through naked eyes. The mature female lays eggs inside the root tissue which remain viable in moist soil for about 12 months.The life cycle is generally completed in 26-51 days depending upon the season in the year. However, in well drained soils, it may be completed in 19 days at an ambient temperature of 22-29°C. In J2, the hemizonid anterior or adjacent to excretory pore, tail = 60–80 mm, tail tip finely rounded.

It is well adapted to flooded conditions and can survive in waterlogged soil as eggs in eggmasses or as juveniles for long periods. Numbers of *M. graminicola* decline rapidly after 4 months but some egg masses can remain viable for at least 14 months in waterlogged soil. *M. graminicola* can also survive in soil flooded to a depth of 1 m for at least 5 months. It cannot invade rice in flooded conditions but quickly invades when infested soils are drained. It can survive in roots of infected plants. It prefers soil moisture of 32 per cent. It develops best in moisture of 20 to 30 per cent and soil dryness at rice tillering and panicle initiation. Its population increases with the growth of susceptible rice plants.

Besides rice plant, it also prefers *Alopecurus* sp., *Avena sativa* L., *Beta vulgaris* L., *Brachiaria mutica* (Forsk.) Stapf, Brassica juncea (L.) Czem. and Coss, *B. oleraceae* L., *Colocasia esculenta* (D.) Schott, *Cyperus procerus* Rottb. *C. pulcherrimus* Willd. ex Kunth, *C. rotundus* L., *Echinochloa colona* (L.) Link, *Eleusine indica* (L.) Gaertn., *Fimbristylis miliacea* (L.) Vahl, *Fuirena* sp., *Glycine max* (L.) Merr., *Lactuca sativa* L., *Lycopersicon esculentum* Mill, *Monochoria vaginalis* (Burm. f.) Presl, *Panicum miliaceum* L., *P. repens* L., *Paspalum scrobiculatum* L., *Pennisetum typhoides* (Burm. f.) Stapf and Hubbard, *Phaseolus vulgaris* L., *Poa annua* L., *Ranunculus* sp., *Saccharum officinarum* L., *Sorghum bicolor* (L.) Moench, *Sphaeranthus* sp., *Sphenoclea zeylanica* Gaertn., *Spinacia oleracea* L., *Triticum aestivum* L., and *Vicia faba* L.

The perineal pattern is round or oval, the striae are smooth, more or less continuous, loose and around anus they are more broken than smooth. The lateral field is absent. Enzyme phenotypes cluster *M. graminicola* with *M. graminis*.

8.9.9 *Meloidogyne exigua* (Goeldi, 1887) Chitwood 1949

Meloidogyne exigua was the first root-knot species described under this genus name by Göeldi (1887) and was found to be the causal agent of coffee decline in Brazil. It was reported by Jobert in 1878 on coffee in Brazil speculating the involvement of *Anguillula*. It is known to be distributed in coffee-producing areas of Central and South America (Bolivia, Brazil, Colombia, Costa Rica, Dominican Republic, *El* Salvador, French Guiana, French West Indies, Guatemala, Honduras, Peru, Suriname, Trinidad and Tobago, and Venezuela) and southern India (Srinivasan and D'Souza, 1965). Beside coffee, this nematode also infects *Allium cepa*, black nightshade *Solanum nigrum*, *Capsicum annum*, watermelon *Citrullus lanatus*, Banana, Citrus and *Camellia sinensis*, sugarcane and tomato. In South India, it is known to infect two weeds, Spanish needle, *Bidens pilosa* and *Pilea* sp.(rice) growing in coffee plantation but not on coffee roots.

M. exigua feeding causes yellowing of leaves, leaf fall, destruction of root hairs and rootlets, root lesions, small root galls, vascular disruption, and secondary invasion by other fungal and bacterial organisms. In some cases *M. exigua* does not produce

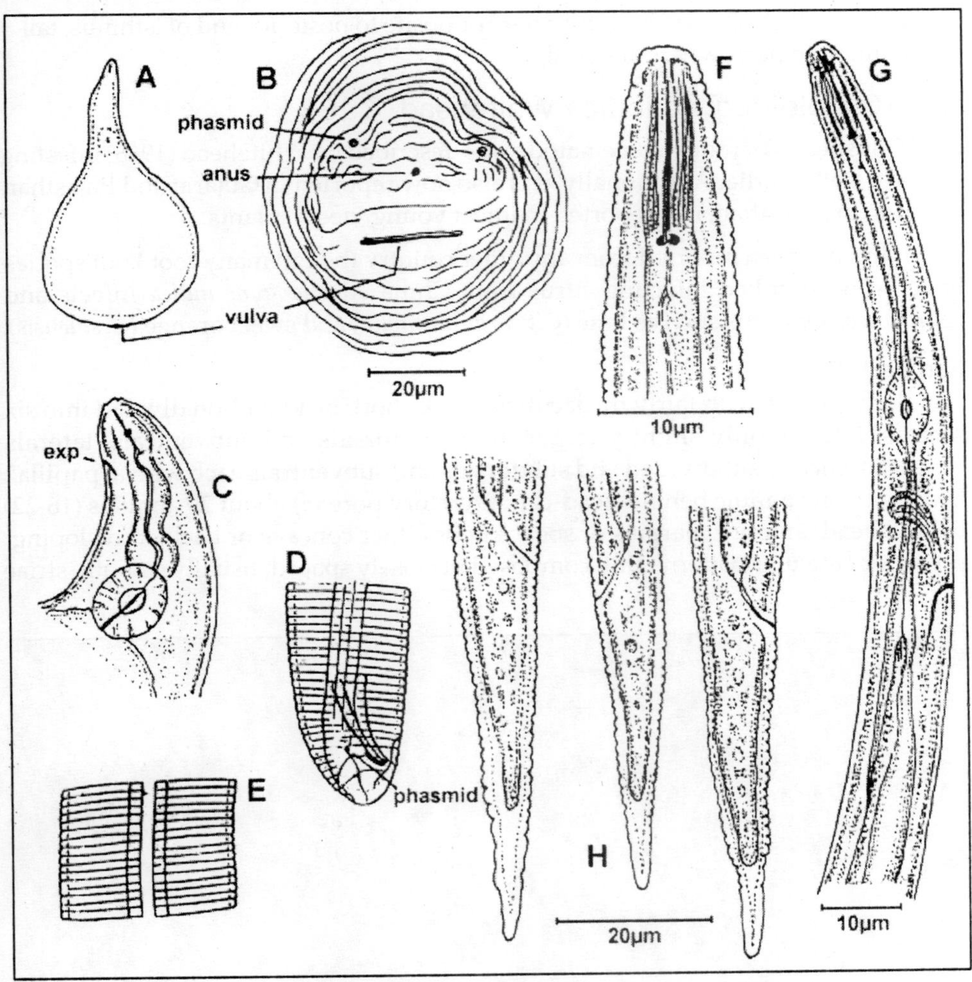

Figure 47: *M. exigua*
A: Female, B: Perineal pattern, C: Anterior region of female, D: posterior region of male, E: Lateral lines, F and G: Anterior region of juveniles, H: Posterior region of juveniles

galls, only the mature female protrude from the root surface and lay brownish or yellowish egg masses. Yields of non-infested plants may be twice as high as those infested with *M. exigua* (Lordello, 1986). Root necrosis and defoliation are greater when roots are infected by both *M. exigua* and *Rhizoctonia solani* than by either organism alone. Decline and dieback of coffee trees and yield suppression up to 50 per cent are associated with nematode infection in Brazil and South America (Campos *et al.*, 1990; Lehman and Lordello, 1982).

The perineal pattern is round or oval, dorsal arch is low; striae are smooth and widely spaced, coarse, broken and folded in the lateral region. The lateral field is absent.

In J2, L = 290–370 µm, excretory pore opposite to posterior end of isthmus, tail = 39–50 µm with narrowly rounded tip.

8.9.10 Meloidogyne indica Whitehead, 1968

This species of root knot nematode was described by Whitehead (1968) infesting citrus in Delhi, India. Additionally, it is also now reported in Gujarat and Rajasthan in India. In Rajasthan, it is reported to infect young 'neem' plants.

It has not been reported from any other country though many root knot species have been described infecting citrus from China. *Meloidogyne indica* infects and reproduces on citrus such as lime (*Citrus aurantifolia*) and sweet orange (*C. sinensis*).

Female

Body saccate with fairly thick cuticle, neck short; head end-on divided into six sectors, laterals only slightly larger than subdorsals and subventrals, laterals characteristically subdivided and subdorsals and subventrals each with a 'papilla'; head with one annule behind head-cap; excretory pore (5) about 20 annules (16-22) behind head; anterior margins of spear knobs either concave or backward sloping; posterior cuticular pattern faint, composed of closely spaced, usually smooth, striae

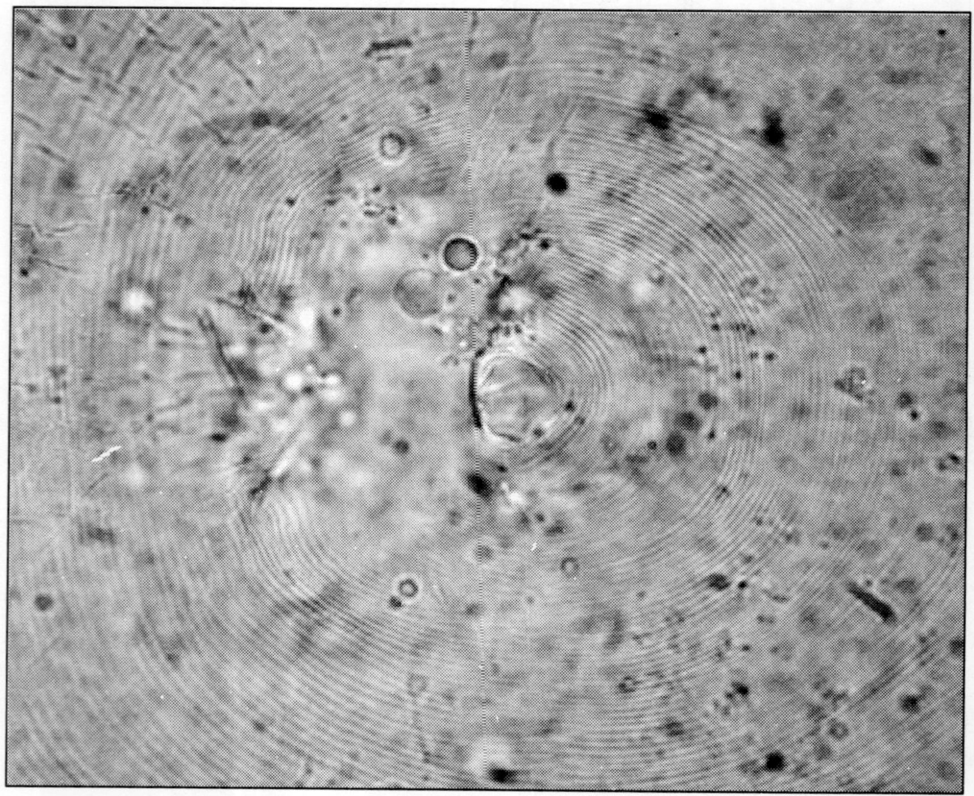

Figure 48: *Meloidogyne indica* Perineal Pattern

forming a distinct tail whorl with low, rounded dorsal arch, phasmids not as close to tail terminus as in *M. africana* and lateral field usually absent; posterior cuticular pattern resembles that of *M. decalineata, M. coffeicola* and to a lesser extent *M africana*.

Male

Head in lateral view hemispherical or truncate-cone shape, not offset, with two annules behind head-cap, anterior annule longer than posterior, which appeared to be 'tilted' in incident illumination; basal plate fairly thick; anterior part of stylet longer than shaft, stylet knobs small and rounded triangular shape, anterior margins well swept back; posterior cephalid about seven annules behind head; oesophagus overlaps intestine ventrally; excretory pore near anterior end of posterior oesophageal region and two to five annules behind hemizonid; hemizonid one annule long; one testis; lateral field with four incisures for greater length of body, outer bands fully areolated, inner band occasionally cross-striated towards posterior end of body, where lateralfield fully areolated; tail rather long and apparently concave ventrally; striae pass around terminus; phasmids opposite cloaca.

2nd Stage Juveniles

Head not offset, low truncate-cone shape, in lateral view of head head-cap outline slightly posterior to general contour of head, two, sometimes three annules behind head-cap; head annules appear marked by longitudinal striae; distal sclerotisation of head thicker than proximal; cephalids not seen; stylet knobs fairly prominent rounded, anterior margins often backsloped; rectum not inflated; tail conoid, terminus usually blunt, unstriated.

Females

Stylet= 12-16 µm (14); Width stylet base= 4-5 µm (5); DGO to stylet base= 2-4 µm (3); Length median bulb =31-43 µm (38); Width median bulb =33-46 µm (39); Length median bulb valves= 13-16 µm (14); Width median bulb valves= 7-11 µm (9).

Second Stage Juveniles

Body length (L) =381-448µm (415±4.5); stylet = 10-14µm (12±0.9); tail length =13-20µm (16.8±1.88); body length: tail length ratio = 21.2-31.0 (24.9±1.36); tail length: body width (seen laterally at level of anus) = 1.06-1.78µm (1.57±0.012).

Eggs

Length (L) = 71-88µm (77 ±4.7); width (W) = 26-35µm (30±1.9); L/W ratio = 2.6

Diagnosed by the wide stylet base, the saccate body and the posterior cuticular pattern of the female, the low rounded head with two annules behind the head-cap on the sublateral head sectors, the anterior longer than the posterior, the well swept back stylet knobs and the fairly long ventrally concave tail of the male and the short bluntly rounded conoid tail of the juvenile.

8.9.11 *Meloidogyne graminis* (Sledge and Golden, 1964) Whitehead, 1968

Meloidogyne graminis was described by Sledge and golden (1964) as *Hypsoperine graminis* but with synonymization of *Hypsoperine* with *Meloidogyne,* Whitehead (1968)

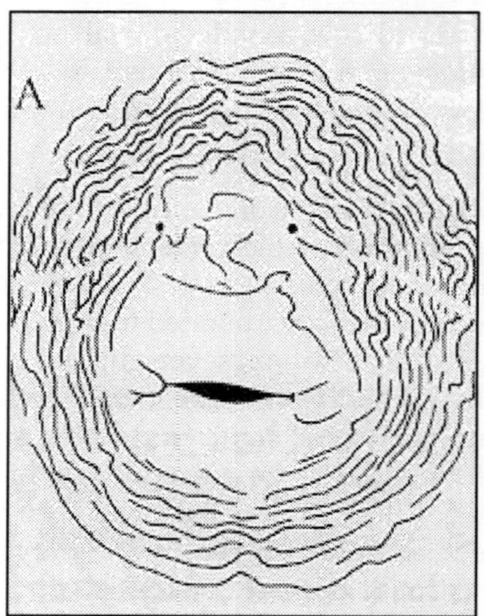

Figure 49: *Meloidogyne graminis*

named it as *M. graminis*. In India, it was reported by Nayak, Ray and Routaray (1986) infesting wheat in Orissa.

Female

Body oval and asymmetrical shaped, relatively large with short neck. Perineal pattern located on a posterior protuberance. Short stylet, cone slightly curved dorsally, knobs ovoid and slightly backwardly sloping. S-E pore less than one stylet length behind head end. Perineal pattern rounded to oval shaped, with coarse striae and angular arch, tail. Terminus area free of striae, lateral field visible.

Male

Body anteriorly tapering. Head slightly set off from body. Head cap relatively small, labial disc not elevated, lateral lips absent. Head region without transverse incisures. Short stylet with rounded knobs, slightly backward sloping. DGO to stylet knobs relatively short. Lateral field with four incisures, clear areolation not observed.

Second-Stage Juveniles

Relatively long and slender body. DGO close to stylet knobs. Hemizonid posterior to the S-E pore. Long slender tail with inflated rectum and relatively long gradually tapering hyaline tail part, ending in rounded terminus.

Measurements

Female

L=463+31.4µ(403-506), Body width =13.7 +0.7µ (12.6-15.2), Stylet length=12.4 + 0.4 (12.0 - 13.4), DGO = 2.5 + 0.3 (1.9 - 3.2), Tail length=76 + 4.9 µ (66 - 82), a=33.9 + 1.9 (31.5 - 37.3),b′=8.1 + 0.6 (6.9 - 8.9), c=6.1 + 0.3 (5.7 - 6.4), c′= 7.7 + 0.5 (6.8 - 8.3), S-E pore / L x 100 = 16.4 + 1.4 (15.2 - 19.4).

Males

L= 1248 + 96 μ (1152 - 1344), Body width = 34.8 + 3.2 μ (31.6 - 37.9), Stylet length= 18.1 + 0.4 μ (17.7 - 18.3), DGO = 2.7 + 0.4 μ (2.5 - 3.2), anterior end to metacorpus = 61 + 4.9 μ (57 - 66), Spicule = 27.4 + 3.6 μ (25.3-31.6), Gubernaculum = 8.0 + 0.4 μ (7.6 - 8.2), a = 36.0 + 0.5 (35.5 - 36.5).

Females

Stylet length = 12.8 + 0.4 μ (12.6 - 13.5), DGO = 4.1 + 0.6 μ (3.3 - 5.0), S-E to anterior end = 11.4 + 1.6 (8.2 - 12.6), Anterior end to metacarpus = 99 + 7.3 μ (92 - 112), Metacorpus length = 36.0 + 5.0 μ (30.3 - 44.2), Metacorpus diameter = 35.3 + 3.1 (31.6 - 41.1).

It is known to parasitize grasses, including different cultivars of both *Cynodon dactylon* (L.) Pers. and *Zoysia japonica* Steud. *Paspulum notatum* Fluegge, *Stenotaphrum secundatum* (Walter) Kuntze, *Oryza sativa* L., *Digitaria sanguinalis* (L.) Scop. and *Ammophila arenaria* (L.) Link. (Jepson, 1987). It has also detected on the cereals like *Sorghum vulgare* Pers. and Zea *mays* L.

8.9.12 *Meloidogyne triticoryzae* Gaur, Saha and Khan, 1993

This species of root knot nematode was described by Gaur *et al.* (1993) from the farm area of Indian Agril. Research Institute, New Delhi infesting rice and wheat. They observed numerous small, rounded, elongated galls and club shaped root tips containing different stages of the nematode and also eggmasses embedded in the root tissue. The infestation due to this nematode exhibited unthrifty, stunted growth, chlorosis, poor tillering and declining yield.

Females

Body elongate, lemon shaped to nearly round with elevated perineum; neck 110-185μ long, asymmetrically located and bent to the ventral side of the longitudinal axis. Head continuous, weakly sclerotised, stylet with posteriorly sloping rounded knobs, metacarpus rounded to slightly ovoid.The excretory pore is located near or little behind the level of DGO. In the perineal pattern, dorsal arch is hemispherical with widely spaced striae. Striae smooth and continuous except in the region of tail tip. Tail whorl is indistinct without punctuations. Lateral field marked by lateal lines with boken striae, ventral arch marked with smooth but has closely knit striae.

Males

They are variable in length. Head high with 2-3 annules, continuous with body contour, head cap narrower. Caphalic framework slightly sclerotised, amphidial opening slit like. Stylet moderately sclerotised with elongate posteriorly sloping rounded knobs. Oesophagus well developed with ovoid metacarpus. Excretory pore 90-128μ from head and well behind the level of nerve ring, hemizonid one annule preceding it. Tail rounded to asymmetrically conoid with a smooth terminus, cloaca slightly bulging, spicules ventally arcuate. Phasmid on middle of tail. Latral field with four incisors, no areolation seen but outermost incisors crenated. Males are rare.

Second Stage Juveniles

The J2 assumes slightly arcuate shape upon killing with gentle heat. Head weakly sclerotised, hemispherical continuous with body contour and smooth. Stylet weakly

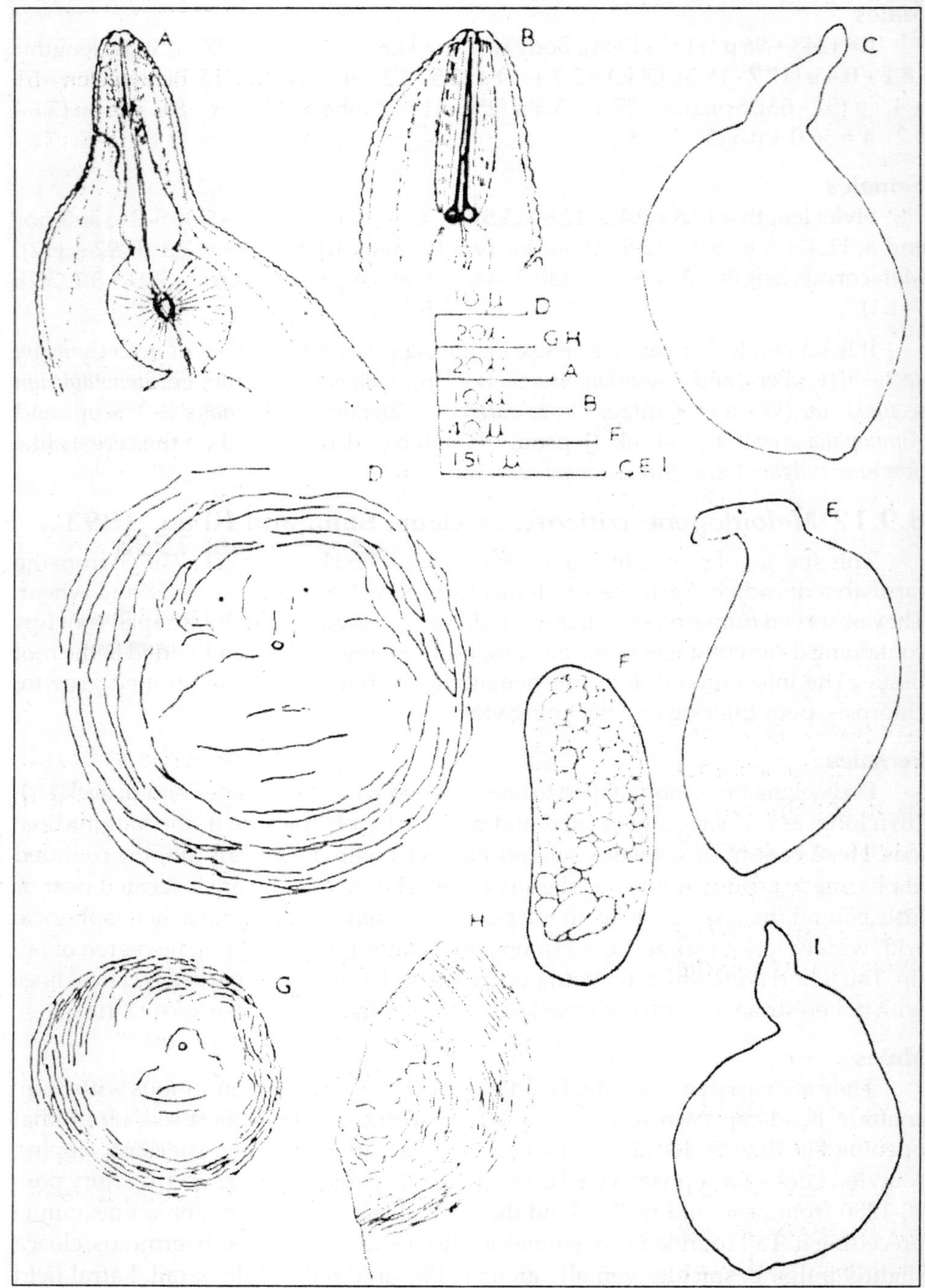

Figure 50: ***Meloidogyne triticoryzae*** **sp.n.: Female**
A, B: Anterior end, C, E, I: Body shapes, D, G: Perineal patterns;
H: Laterial view of perineal pattern, F: Egg (After Gaur ***et al.,*** **1993)**

sclerotised with posterior sloping stylet knobs. Metacorpus weakly sclerotised with weak cresentric valve. Lateral field marked by four incisors. Tail long with a fine rounded terminus.

Measurements

Female

Length (with neck) = 330-480 (425) μ; Body width = 200-320 (242)μ; a = 1.43-2.2 (1.78); Neck length = 110-185 (139)μ; Stylet length = 12-14 (13.2)μ; DGO = 2-4 (2.9) μ; Metacorpus length = 24-28 (26.1) μ; Metacorpus width = 21-23 (21.7) μ; Vulval slit length = 16-21 (18.0) μ; Interphasmidial distance = 8-11 (9.4) μ.

Male

Length = 1100-1590 (1305) μ; a = 55-74 (63.5) ; Stylet length = 17-19 (17.5) μ; Metacorpus length = 17-20 (18.3) μ; Metacorpus width 7-11 (9.0) μ; Tail length = 3.5-6.5(4.6) μ; c ratio = 216-397 (293); Spicule length = 26-32 (29.0) μ.

Eggs

Length = 76-93μ; Width = 28-35μ.23.

8.9.13 *Meloidogyne piperi* Sahoo, Ganguly and Eapen, 2000

This root knot species was described by Sahoo *et al.* (2000) from the roots of black pepper (*Piper nigrum*) in Calicut, Kerala.

Mature Female

Body opaque, pyriiform with short neck. In few specimens, the dorsal curvature of the body was more than the ventral. Posterior protuberance is absent. Lip cap present. Head with three annules, the first annule being anteriorly directed, the second one low and regressed while third annule wide and outwardly directed. Stylet slender with its dorsally curved conus. Stylet knobs short. Excretory pore located about 2 stylet lengths posterior to spear knobs almost at the level of median bulb.

Perineal Pattern

The perineal pattern oval with high dorsal arch in most of the specimens, having broken wavy striae in both dorsal and ventral arches. Lateral field and lateral lines absent. The lateral forking of a few striations of the ventral arch seen near the lateral region on one side. Broken striae in the ventral arch directed toward right lateral region in one side only and sometimes spreading up to the middle of dorsal arch. Also, broken striae spread toward the perineum making the perineum region very narrow. Tail whorl absent, phasmids closely spaced located near the tail tip.

Second Stage Juveniles (J2)

Body slender, tapering at both the ends. Head hemispherical slightly set off from body having a lip cap followed by two head annules. Basal plate thin. Dorsal gland orifice located more than one fourth of the stylet length behind the spear base, cephalids absent, hemizonid 3 annules long. Excretory pore located 8-9 annules posterior to hemizonid. Oesophageal glands overlapping the intestine ventrally, lateral field with four incisures, phasmids located at the mid region of tail, rectum inflated, tail tapers with a notch just posterior to hyaline region and ends with subacute terminus.In an

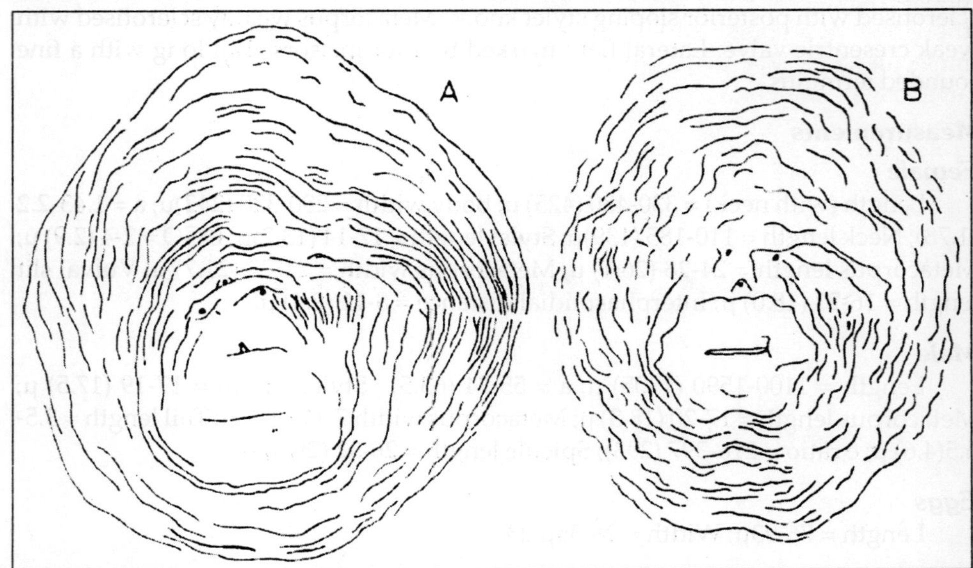

Figure 51: Perineal Patterns of *M. piperi*

aberrant form tail narrows abruptly below the hyaline region forming a distinct notch and then ending in a digitate tail terminus.

Male

Note found

Females

Length = 720μm ± 45 (555-874), W =533μm ± 43 (400-685), Stylet = 14.3μm± 0.5 (13-16), DGO= 5μm ± 0.3 (4-6), Neck length = 154μm ± 13.6 (120- 207), Width-neck = 36μm ± 2.8, Length of Median Bulb = 38μm ± 2.1 (30-45), Width of Median Bulb =36±2.8 (30-50), a =1.4±0.1 (1.1-1.6) b = 15.4 ± 2.3 (7.9-26.8), vulval slit length =19.9μm ± 1.3 (18-22), Anus to vulval slit distance = 13.3μm ± 0.8 (12-14.5), Anus to tail terminus distance= 10.8 μm ±1(10-12), Inter phasmidial distance = 19μm ± 1.9 (1.9 (15-22).

Second Stage Juveniles

Length = 356μm± 15 (310-400), stylet length= 12.2μm± 0.4 (12-14), Maximum body width = 12.9μm ± 0.3 (11-13), DGO = 3.6μm ± 0.2 (3-4), H-Exp=74μm±2.8(62-81), Tail= 47μm ± 1.9 (40-53), a=28.4±1.1 (26-33), b= 4.9 ± 0.1 (4.3-5.4), b′=2.5±0.2 (1.9-3.1) c= 7.8 ± 0.2 (7.5-8.5), c′= 5.5 ± 0.1 (5.1-6).

Eggs

Length. = 85μm ± 2.4 (77-92), width = 37μm ± 1.1 (33-42), L/W ratio = 2.3 ± 0.1 (2.1-2.7).

These species of rootknot nematodes are mainly reported from cultivated crops in India. There may be many more species present in the country but for ignorance and lack of knowledge about the correct identification of the species, they remain undescribed. There are many populations of root knot nematodes which are just thought to be among the four important species but none of them is characterized at molecular level. Then there are many areas hitherto unexplored particularly the vast areas covered under the forests which may reveal the presence of many new root knot species in India.

Chapter 9

Management of Root Knot and Cyst Nematodes

A fundamental principle of pest management is that the management or control is unnecessary unless the pest is at a level at which it is expected to damage the crop (damage threshold). In some pest management disciplines, such as entomology, the availability and use of damage thresholds and economic injury levels are common and at times leads to the assumption that this concept is equally valid for other disciplines. Although the concept is likely just as valid in nematology as in entomology, the ability to develop and utilize damage thresholds has proven to be fairly elusive. Cousens (1987) suggested that there are three distinct pest management strategies: i) eradication- this is a strategy in which extensive efforts and costs are provided in the short term to completely remove the pest and therefore provide unhindered produce development in future periods; ii) prophylaxis- this is a strategy of insurance, in which pest controls are applied systematically, periodically and generally preventively regardless of the pest population; iii) containment- the intention is to ensure the pest population stays below a specific level. The producer in this situation accepts some loss of yield (and therefore revenue) and controls the pest when it is cost-effective to do so. Usually no single pest management strategy is dominant for any given pest.

Nematode population dynamics are also density dependent and are influenced by host growth, the reproductive potential of the species and by various environmental factors. Consequently, modelling nematode population dynamics is an equally impressive science. Again, good field data are required but the complicating effects of biological control agents, host susceptibility differences and environmental factors, and errors associated with measuring initial population densities, may mean it is practically impossible to predict reliably the multiplication rates of most nematodes, especially those with several generations per season.

9.1 Economic Damage

Pest population assessment and decision making are among the most basic elements in any integrated pest management (IPM) program. In fact, these activities characterize state of the art approaches in pest technology and differentiate IPM from other strategies. Forming the basis of assessment and decision making is bioeconomics, the study of the relationships between pest numbers, host responses to injury, and resultant economic losses (Pedigo, 1996). An important outcome of bioeconomics is the formation of decision rules that are used with management options.

Southwood and Norton (1973) presented a practical mathematical expression that has been used widely to define economic damage. That expression is:

$$C(a) = Y[s(a)] \times P[s(a)] - Y(s) \times P(s)$$

where,

Y = yield, P = price per unit of yield, s = level of pest injury, a = control action [s(a)] is level of injury as modified by the control action], and C = cost of the control action. This equation simply states that cost of the control tactic equals yield times price when the tactic is applied minus yield times price without the tactic. Consequently, economic damage begins at this point, *i.e.*, when the cost of damage equals the cost of suppression.

9.2 Economic Injury Level (EIL)

EIL is another of the basic elements, was defined by Stern *et al.* (1959) as the lowest population density that will cause economic damage. The EIL is the most basic and theoretical value that, if actually attained by a pest population, will result in economic damage. Therefore, the EIL is a measure against which we evaluate the destructive status and potential of a pest population.

Although the EIL is expressed as a pest density, it is actually a level of injury that is indexed by pest numbers. Because the EIL is actually a degree of injury, it is sometimes useful to think of it in terms of injury equivalents (Pedigo *et al.*, 1986). An injury equivalent is the total injury produced by a single pest over an average lifetime. It is a potential value, *i.e.*, a pest dying prematurely will obtain only a partial equivalent. The EIL is governed by five primary variables: cost of the management tactic per production unit, (C), market value per production unit (V), injury units per pest(I), damage per injury unit (D), and the proportional reduction in pest attack (K). If the relationship of these variables is linear or roughly so, the EIL can be given as:

$$EIL = C/VIDK$$

In instances where D is strongly curvilinear, it would be necessary to replace D with a complex function relating damage to total injury from a population (Pedigo *et al.*, 1986).

9.3 Economic Threshold

Economic threshold is based on initial soil population levels that will multiply over the growing season and cause economic damage to the crop. The economic threshold is expressed as the number of nematodes in a particular amount (250gm or 1Kg) of soil and is often different for each crop and each nematode species. It may be one juvenile per gm of soil for root knot nematode infesting vegetables or four juveniles for cyst nematodes infesting wheat. The economic threshold differs from the EIL in that it is a practical or operational rule, rather than a theoretical one. Stern *et al.* (1959) defined the ET as "the population density at which control action should be determined (initiated) to prevent an increasing pest population from reaching the economic injury level." Although measured in pest density, the ET is actually a time to take action, *i.e.*, numbers are simply an index of that time. Damage due to nematodes is proportional to the intensity of attack; this is often proportionally greater in sandy soils where nematodes can move more freely, than in heavier soils where movement is restricted. Adequate soil moisture is essential for free movement so attack is often limited as soils dry out later in the season. Temperature also influences the rate of nematode movement, but plant growth is usually equally affected.

For many nematode-crop associations, damage thresholds have not been developed. Yield losses are influenced by the pathogenicity of the species of nematode involved, by the nematode population density at planting, by the susceptibility and tolerance of the host and by a range of environmental factors. Because of this, available models only estimate yield losses as proportions of the nematode-free yield. Estimating threshold levels further involves various economic calculations. Consequently, predicting yield losses and calculating economic thresholds for most nematode/ crop problems is quite difficult. What is needed is more field-based information on the relationship between nematode population densities and crop performance, and various approaches to obtaining such data. Measuring the population density, especially of cyst and rooknot species, is a major problem which needs addressing.

Damage thresholds are often thought of as absolute numbers. For nematodes, at least, the temperature at the time of planting can be a major factor in determining rather or not damage will occur. If the temperature at the time of planting is below the activity or infectivity threshold of the particular nematode of interest, less damage will be experienced than if temperatures are warmer. Root-knot nematodes are found in all agricultural regions worldwide. They can survive in temperate climates and can devastate crops grown in the tropics. Most root-knot nematodes also have extremely wide host ranges. Although it is difficult to ascertain the number of hosts for any one root-knot nematode species, it is likely that some root-knot nematodes can survive on hundreds of different plant species. This can make it extremely difficult to control a root-knot nematode problem, particularly if the nematode can survive on weeds. In addition, root-knot nematodes have repeatedly been shown to predispose their host plants to infection by other crop pathogens, increasing the potential for crop loss.

Since eliminating nematodes is not possible, the goal is to manage their population, reducing their numbers below damaging levels. The various methods in vogue for the management of different plant parasitic nematodes have been practiced and discussed by many workers are elaborated below. Nematode management is a spectrum of techniques with considerable overlap between methods. Each practicing nematologist works out their own categorization in a way that helps them best plan, explain and carry out a management programs.

9.4 Cultural Practices

Cultural practices is a way of nematode management technique under which various practices like crop rotation, use of resistant varieties, fallowing, flooding, trap crops, cover crops, date of planting and harvesting, physical removal of infected plants can be considered. However, this is also true that no two nematologists would include all these techniques under cultural practices. As some consider use of resistant varieties under biological control and still other may take it as entirely separate category of management. Also techniques like fallow, trap crops, cover crops and even resistant varieties may be considered under crop rotation by some workers.

The presence of nematodes in soil *per se* does not mean that crop yield will be adversely affected. Many factors such as climate, soil type, pattern of nematode distribution, previous cropping history, presence and distribution of pathotypes, and species and cultivars determine whether the damage and yield loss are likely to occur. Cultural practices play a vital role in nematode control strategies for subsistence farming since low value crops rarely receive any major input of fertilizer and pesticides. Various cultural practices that can be used for effective control of nematode pests are discussed.

9.4.1 Cropping Systems and Crop Rotations

A cropping system can be defined as the sequence of growing crops in a given field along with the required technologies for their production. The sequence could be two types: spatial and temporal. The spatial sequence refers to the arrangement of different crop species in a piece of land. Mixed cropping, intercropping, and strip cropping are examples of this system. Temporal sequence refers to distribution of crops over a period of time. Crop rotation, relay cropping, discontinuous planting, and fallow rotation are examples of this system. Studies on crop diversity through mixed or intercropping have been carried out and insect pest or disease damage relationships have been estimated. Similar work relating to nematodes is needed. The damage caused by the root-knot nematode (*Meloidogyne* sp) is reduced when susceptible hosts are grown as a mixed crop with nematode antagonistic crops. For example intercropping of marigold (*Tagetes* sp) adversely affects various nematode populations including *Meloidogyne* sp that would have severely damaged the susceptible hosts. Marigold is a nematode antagonistic crop and its roots release nematicidal compounds such as α-terthienyl (Gommers *et al.*, 1980). Cultivation of tomato (*Lycopersicon esculentum*) with castor results in considerable reduction in the number of galls on tomato (Hackney and Dickerson, 1975). The number of galls on the susceptible eggplant (*Solanum melongena*) is reduced when it is planted in association with marigold (Choudhury, 1981). Intercropping of onion (*Allium cepa*) and maize (*Zea mays*) reduce the galling by *M. incognita* on potato (*Solanum tuberosum*) roots. Crop rotation is the best known method of management to reduce damage by nematodes. It is the most widely practiced method of disease control in agriculture. The principle behind crop rotation is that nematode reproduction is restricted and plant damage is reduced with breaks in cultivation of nematode susceptible crops. The principal aim of nematode control by crop rotation is to grow susceptible crops at long intervals so that nematode population levels fall below damage thresholds between the crops, and to prevent buildup of nematode populations to damaging levels (Johnson *et al.*, 1975; Rao *et al.*, 1984). Nematodes such as the root lesion nematodes (*Pratylenchus* spp) and root-knot nematodes (*Meloidogyne* spp) have a wide host range, and implementation of effective rotations to control these species may be difficult. Sometimes complications arise with nematode species (*e.g.*, *Heterodera* spp) which have races or pathotypes. Some crop cultivars may be susceptible to one pathotype but resistant to or tolerant of others. Rotation of upland rice with soybean reduces populations of *Heterodera elachista* and increases rice yields. Population density of *H. oryzae* is reduced when rice is grown in rotation with *Sesbania rostrata* (Germani *et al.*, 1983). Rotation of wheat with various legume species reduces the population density of cereal cyst nematode (Brown, 1987). Inclusion of cowpea (*Vigna unguiculata*), mung bean (*Vigna radiata*), and pigeonpea greatly increases the population densities of *Heterodera cajani* and suppresses the population densities of other plant-parasitic nematodes on Vertisols (Sharma *et al.*, 1996). Sorghum, safflower (*Carthamus tinctorius*), and chickpea rotations markedly reduce the population of *H. cajani* and increase the population of *Helicotylenchus retusus*. Addition of cowpea to the system results in a significant increase in the densities of *Rotylenchulus reniformis*. Intercropping, sequential cropping, fallow, and crop rotation have a regulatory effect on nematodes in the pigeonpea-based cropping system and can be used as an effective nematode management tactic. In the Philippines, rice-mung bean-rice, maize-cabbage (*Brassica oleracea* var. *capitata*)-rice, rice-tobacco (*Nicotiana tabacum*)- rice, rice-cotton-rice, and rice-watermelon-rice rotations reduce the root-knot nematode populations dramatically (Myint, 1981).

At the face of it, crop rotation may appear to be a simple method for managing nematodes. However, when one starts to implement a crop rotation program, it soon becomes apparent that there are a number of aspects, such as nematodes present in the field, host range of specis present, availability of resistant varieties, time of planting and harvesting, nematode threshold level and importance of weeds should be considered. In many situations, it may also become apparent that our knowledge of some of the information needed to optimize a crop rotation program is limited. At the very least, one must know what nematodes are present in a field in order to choose a nonhost crop and this is true only for endoparasitic nematodes which are more or less host specific. This requires sampling the field and analysing the nematodes present. If root-knot, cyst or lesion nematodes are present, identification to species is needed. If dagger nematode is present, it should be determined whether or not the species is *Xiphinema index*.

Crop rotation with a non-host crop is often adequate by itself to prevent nematode populations from reaching economically damaging levels. However, it is necessary to positively identify the species of nematode in order to know what plants are its host(s) and non-hosts. A general rule of thumb is to rotate to crops that are not related to each other. For example, pumpkin and cucumbers are closely related and rotating between them would probably not be effective to keep nematode populations down. A pumpkin/bell pepper rotation might be more effective. Even better is a rotation from a vegetable to a grain crop. Corn, onions, garlic and small grains are good rotation crops for reducing root-knot nematode populations. Crotalaria, velvet bean, and grasses like rye are usually resistant to root-knot nematodes. (Wang, *et al.*, 2004; Yepsen, 1984; Peet, 1996). Rotations like these will not only help prevent nematode populations from reaching economic levels, they will also help control plant diseases and insect pests. Once the host and nematode pest status are known, a rotational sequence of crops can be recommended along the following lines:

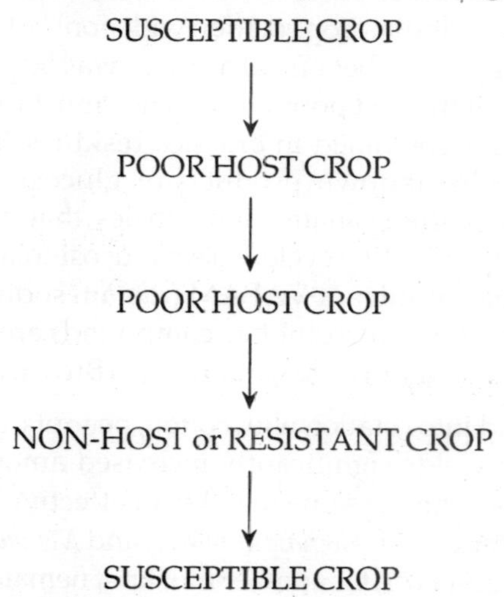

Cultural management basically involves all crop husbandry practices that deprive the nematodes from infesting potato thereby reducing their population (Evans and Stone, 1977; Evans, 1993). The most common practice is use of healthy seed and intercropping with antagonistic plants. Surveys conducted to document species complex, distribution and subsequent mapping of nematode intensity, have helped identify nematode-free zones for seed potato production in India. This has helped in obtaining healthy seed potatoes and thus potato from North- West plains of Punjab and Haryana States and the high hilly districts of Kulu, Kinnaur and Lahual and Spiti in Himachal Pradesh are more favored as a source for reducing RKN infection. On the other hand, potato from Nilgiris and Kodaikanal hills is avoided, to restrict the movement of PCN. Growing non-host crops such as maize, wheat and the trap-crop, marigold helps minimize RKN in northern India (Gill, 1974; Deshraj, 1983). A two-year crop rotation and application of nematicide (carbofuran @ 2 kg ai /ha in two equal splits, once at planting and the other at earthingup) increased potato yields by 45 per cent and brought down RKN tuber-infection by 96 per cent in Shimla hills (Krishna Prasad, 1986). Crop rotation with non-solanaceous vegetables (beetroot, cabbage, carrot, cauliflower, french bean, garlic, radish, turnip, etc.) between September-November reduced PCN cysts and its propagules in South Indian hills, ranging from 24 and 76 per cent respectively (Birhman *et al*, 1998; Gaur *et al*, 1999).

Continuous breeding and selection of resistant lines of potato brought to light existence of variations within the species of RKN and PCN which were designated as biotypes and pathotypes. Pathotype-specific, major genes control resistance to PCN (Kort *et al*, 1977) and non-specific resistant potato brought down PCN by 98 to 99 per cent and increased yields by 65 per cent in Nilgiri hills (Krishna Prasad, 2000). These crop rotations and inter-cropping of potato with French bean or wheat reduced nematode infestation by 35 per cent, while, nutritive value of the soil increased (Manorama *et al*, 2003). Exposing soil to summer heat, fallowing, mulching, trap-cropping, addition of organic amendments to increase the activity of antagonistic organisms in the soil are other methods of nematode management in potato.

In wheat, infested with the cereal cyst nematode, *Heterodera avenae* causing 'molya' disease in parts of northern India, crop rotation with nonhost crops like carrot, radish, turnip and mustard has proved beneficial as there was 50 per cent reduction in the eggs and larval content in the cyst population. The "mustard effect" is attributed to glucosinolate compounds contained in brassica residues. Toxicity is attributed to enzymatically induced breakdown products of glucosinolates, a large class of compounds known as isothiocyanates and nitriles that suppress nematodes by interfering with their reproductive cycle. These glucosinolate breakdown products are similar to the chemical fumigant VAPAM (metam sodium), which degrades in soil to methyl isothiocyanate. Glucosinolate compounds are also responsible for the pungent flavors and odors of mustards and radish (Brown and Morra, 1997).

Sesame when added into rotation with cotton, peanuts, and soybeans, nematode levels are reduced and yields significantly increased among those crops in fields previously planted in sesame. Sesame may be an effective rotation crop to control peanut root knot nematode (*Meloidogyne arenaria*) and *Meloidogyne incognita*. Sesame rotation is not effective, however, for another root knot nematode, *Meloidogyne javanica* (Starr and Black, 1995). The best rotation to control the root-knot nematode in potatoes involves planting a summer non-host crop, followed by a winter cover crop (rapeseed) incorporated as a green manure. Non-host crops include supersweet corn, pepper, lima bean, turnip, cowpea, muskmelon, watermelon, squash, rapeseed, canola, mustard, and sudangrass (Ingham, 1990).

In a crop rotation program, it is advantageous to be able to plan several years in advance. If the expected rate of population increase is known, one can forecast what the population will be at harvest. This combined with knowing the expected rate of decline will allow forecasting when the most profitable crop can be planted once again and so forecast if the rotation chosen will be economical. One could also consider other possible options such as harvesting the crop early when nematode numbers might be lower which might then permit a shorter rotation period.

Crop rotation programs function because, under nonhost conditions, nematodes will starve to death. The time required for starvation is different for different nematodes and for many has not been studied. The economics of a program in which a profitable host crop can be grown every other year, for example, is different from one in which a host can only be grown every 8 years.

9.4.2 Time of Planting

Nematode population levels fluctuate both with season of the year and with cropping cycles. Control of some nematodes can be obtained by manipulating sowing dates to avoid periods of peak nematode activity. Van Gundy *et al* (1974) found that wheat (*Triticum aestivum*) yields are affected by *P. thornei* when the crop is sown in late October. Late planted deep-water rice crop escapes ufra disease caused by *Ditylenchus angustus* (McGeachie and Rahman, 1983). Similarly, wheat crop if grown in the second week of December using late sown varieties will escape the early threat from the cereal cyst nematode, *Heterodea avenae* as the nematode starts haching in the first week of November under Indian conditions.

For each nematode species, there is a minimum temperature at which the nematodes are able to penetrate plant roots. If planting is to take place at a time of year when temperatures are below this infection minimum, or when soils are cooler, less damage can be expected than when soil temperatures are optimum. Nematode damage can be reduced by growing crops at periods that are not favorable for nematode growth (Bridge, 1996). Low temperatures increase nematode life cycle duration and reduce reproduction and hatching; thus, sensitive vegetable species have to be cropped in the coldest period tolerable.

The soil temperatures over the period of a year relative to the minimum temperature at which *M. incognita* is able to penetrate plant roots demonstrates how altering the date of planting by only a few weeks can reduce nematode damage. One main drawback to utilizing this technique can be in situations in which date of planting is driven by market values. It is generally more profitable to sell a particular crop when it is in short supply than when it is plentiful. Growers will try to time harvests to coincide with times they think it will be most profitable to market the crop. This timing may make it necessary to plant at times when nematodes are active. In some situations, altering the time of harvest can significantly affect the amount of nematode damage experienced. For this technique to be utilized successfully, there needs to be some ability to predict when nematode populations will reach levels at which the economics in a particular field will tip in favor of an early harvest.

9.4.3 Flooding

When considering the potential for flooding to control nematodes, one must consider that soil dwelling nematodes are aquatic organisms living in the film of moisture that lines soil pore spaces. There is also evidence in the literature that nematodes are capable of anaerobic respiration. Flooding the soil for seven to nine months kills nematodes by reducing the amount of oxygen available for respiration and increasing concentrations of naturally occurring substances such as organic acids, methane, and hydrogen sulfide that are toxic to nematodes. However, it may take two years to kill all the nematode egg masses (Yepsen, 1984). Flooding works best if both soil and air temperatures remain warm. An alternative to continuous flooding is several cycles of flooding (minimum two weeks) alternating with drying and disking. (MacGuidwin, 1993) But note that insufficient or poorly managed flooding can make matters worse, as water is also an excellent means of nematode dispersal.

In situations where flooding is being conducted for other reasons, such as the leaching of salts, or the management of waterfowl, nematode reductions can also be realized. For other nematodes in other locations, short term flooding has been utilized to stimulate nematode egg hatch. When followed by fallow, the nematode population may then decline more rapidly.

9.4.4 Fallow

The basic principle behind utilizing fallow to manage nematodes is that nematodes will starve to death in the absence of a host crop. Fallow is not successful in practice because the importance of weeds as hosts is not taken seriously. The length of time that roots remaining in the ground are able to serve as hosts for nematodes in the absence of above ground plant growth is often overlooked. Efforts to reduce populations of stem and bulb nematode on alfalfa, garlic, and daffodils, and root-knot nematode on potatoes have been confounded by the presence of "volunteers" surviving around the edges of fields or as "weeds" within the field. For fallow to be effective, soil needs to be weed free.There is conflicting evidence in the literature as to whether dry or wet fallow is of greater usefulness. The particular nematodes of interest, soil type, climatic conditions, and other variables are likely to all play a role. Fallow can be costly practice because the land is out of production and soil conservation - wind and water erosion can occur; soil structure may deteriorate.

Research indicates that soil disturbance (*e.g.* plowing or roto-tilling) in the field during fallow will increase the rate of nematode decline (as well as add to the cost of the fallow procedure). Hot, dry climates offer an opportunity to employ fallow in the rotation. This consists in plowing the soil two or three times, the nematodes thus being turned up to the heat of the sun, where they perish from desiccation.

Generally, 6-18 months of fallow are required for most plant-parasitic species as it causes depletion of metabolic reserves and there is considerable effect of increasing biological activity in the soil. Although in some cases, only two to three months fallow are required, even this length of time is out of synch with current agricultural practices in some areas. The length of time that a field must be in nonhost crops or fallow can be predicted by knowing the nematode population at harvest and the rate of population decline or percent yearly decline expected, although this can vary from field to field. Fields that are left fallow but kept weed-free for one to two years usually have an 80 to 90 per cent peryear reduction in root-knot populations (Sasser, 1990). This host-free period can be achieved in one season, rather than two years, by disking every ten days all summer. Such disking, however, while having the added advantage of reducing perennial weeds, is expensive in terms of fuel costs, possible erosion, and loss of organic matter through oxidation (Ingham,1996).

There are many cases of successful reductions in nematode damage by the use of planted fallows, mainly with non-host pasture grasses, such as their use to manage rootknot on tobacco (Shepherd and Barker, 1990), which provide some immediate return to the farmer. There is also at present considerable interest in the use of leguminous plants (the 'improved fallows') which are designed to maintain or increase the productivity of the land under intensive systems. These improved fallows can give effective management of some nematodes but, conversely certain leguminous trees and bushes are themselves severely damaged by nematodes, thus preventing their successful introduction. In comparison, the non-productivity of bare fallow is the main factor that works against it being taken up by the farmers.

9.4.5 Weed Control

Any nematode pest management program, including rotation programs, should involve good field management, particularly weed control. Weeds are potential hosts for many of the important common nematodes, and can act as reservoirs of infection and as hosts that maintain or even build up nematode populations. Weed control is especially important in rotation programs because a program's success depends on the absence of host roots to prevent nematode feeding and reproduction. Therefore, the presence of weed hosts will negate the benefits of growing a resistant or nonhost crop plant. Cyst nematodes have hosts among weeds of the families Chenopodiaceae and Cruciferae, and presence of such weeds as mustard can nullify rotation programs for the cyst nematode. Similarly, root knot nematodes have weed hosts representing many plant families, especially broad leaf plants. An example of common weed hosts of root knot nematodes are the Solanaceous nightshade plants, which allow good root knot nematode reproduction in many vegetable and field crops, such as tomatoes and cotton. Noling and Gilreath (2002) considered that controlling *Amaranthus* spp. was essential for limiting a nematode population, because that species is a very good host for root-knot nematodes. Schroeder *et al.* (1993) showed that when weeds were not controlled in fallows, nematode population levels increased. Kutywayo and Been (2006) and Rich *et al.* (2009) indicated that inadequate weed control may even counteract nematode control strategies, like fallows and resistant crops.

9.4.6 Burning Straw and Field Sanitation

Good field sanitation practices are becoming more important pest management tools due to the loss of pesticides with fewer new products coming in, and the reduction in efficacy due to resistance problems. Field sanitation practices should be an important aspect of production from field preparation through harvest and beyond. Open field burning is a widely adopted cultural practice which was established as a technique for field sanitation in the late 1940's. At that time in USA, some 50,000 acres of perennial ryegrass grown for seed became infested with blind seed fungus which markedly reduced viable seed production. It was found that open field burning of crop residues after harvest was an effective and economic means for destroying the fungus in infected seeds which wintered in crop residues and at the soil surface (Hardison, 1964). Field burning serves a threefold function of field sanitation, residue removal, and residue utilization or disposal. Physical removal of grass residue substitutes for only one of these functions. It requires physical collection and transfer of residue from the field.

Immediately after harvesting, foliar nematodes such as *D. angustus* that overwinter in rice straw can be eliminated by burning the straw and other plant remains. For effective control, complete burning of all the straw is essential, but this is not always possible, as subsistence farmers normally remove some straw for use as livestock feed, bedding, and thatching. In such cases there is only a partial burning of the remaining straw leaves and infested plant material. Burning of the maximum quantity of straw should be recommended in all the ricegrowing areas where *D. angustus* occurs. The removal of infested plant material from the soil and its destruction by burning soon after harvest reduces nematode populations and prevents further spread. This method can make a significant contribution towards control of the root nematodes on a wide range of crops. Uprooting of infected plants after harvest, and then exposing their roots to solar heat or burning the rootknot infectred root stubbles have similar lethal effect on nematodes. Bridge (1996) advised uprooting plants after each harvest and exposing the roots to sun radiation to kill nematodes in root tissues. This has become a common practice for some tropical crops, but its efficiency has rarely been quantified. Barker and Koenning (1998) considered that taking this precaution could reduce *Meloidogyne* populations by 90 per cent compared to leaving residual roots in the soil.

Root-knot nematodes can be easily spread by human activities that provide communication between contaminated and healthy areas; for example, the transport of infested soil, plant debris or water. At the farm level, experts have recommended cleaning all agricultural machines and tools to avoid transporting nematodes with the soil (Mateille *et al.*, 2005; Djian-Caporalino *et al.*, 2009). In protected crops, most nematode damage appears to occur at the entrance of the greenhouse. Those observations have led to the use of airlocks fitted with foot or wheel baths. Hugo and Malan (2010) reviewed many situations with dissemination of nematodes (especially *Meloidogyne* spp.) through irrigation water and pointed out the difficulties for controlling the phenomenon.

9.4.7 Use of Organic Amendments

Many different types of amendments and composted materials have been applied to soil to suppress populations of plant parasitic nematode and improve crop yield and plant health. Animal manures, poultry litter, and disk-incorporated cover crop residues are typical examples of soil amendments used in agriculture to improve soil quality and as a means for enhancing biocontrol potential of soil. Some amendments which contain chitin and inorganic fertilizers that release ammoniacal nitrogen into soil suppress nematode populations directly and enhance the selective growth of microbial antagonists of nematodes. More recently, composted municipal wastes and sludges have been used to amend soil to improve soil fertility, organic matter content, water holding capacity, nutrient retention and cation exchange capacity.

Suppression of soilborne pathogens via the incorporation or simple mulching of composted amendments is reputedly based on enhanced microbial activity and increased numbers of antagonists generated by decomposition of the amendment in soil. Soils with a diversity of beneficial microorganisms are more suppressive to pathogens than soils with little or no biological diversity. Other possible mechanisms for pathogen suppression by composts include direct inhibition of the pathogen or reduced infectivity of the organisms into the plant host. A population increase of beneficial organisms in soil appears to be the direct result of environmental changes brought about by the amendments after addition to soil. This suggests that to sustain soil suppressiveness, amendments must be periodically reapplied to maintain the soil environment conducive to antagonists.

Application of chemical fertilizers to soil and incorporation of various organic amendments influence nematode populations. Their application may increase the growth and yield of nematode-infected plants by improving the soil structure and increasing the water-holding capacity, supplying mineral nutrients to deficient soils and providing substrate on which parasites of nematodes may propagate. Application of nitrogen fertilizers increases root growth in cereals and diminishes the symptoms of cereal cyst nematode (*Heterodera avenae*) infection in wheat. In Japan, control of rice root nematode (*Hirschmanniella* sp) is obtained by application of calcium silicate and compost. In India, application of fertilizers has been found to improve the growth of rice (*Oryza sativa*) and increases the number of nematodes, although populations of *Hirshmanniella oryzae* are lowered in fields by application of urea (Mathur and Prasad, 1971). The yield of irrigated wheat infected with *Pratylenchus thornei* is improved by the application of nitrogen, when the nematode population is near threshold level. At greater than threshold levels application of nitrogen does not compensate for nematode damage and yield is reduced. Application of lime and ash leads to an increase in the bacterial feeding nematode populations, probably due to an increase in soil pH. Populations of *Acrobeloides* sp, *Protorhabditis* sp, *Rhabditis* sp, and other bacterial feeding nematodes increase in limed soils and the effect is visible months after application of lime. These nematodes may play a role in organic matter decomposition. Oilcakes of linseed (*Linum usitatissimum*), mustard (*Brassica* sp), groundnut (*Arachis hypogaea*), castor (*Ricinus communis*), *mahua* (*Madhuca indica*), and neem (*Azadirachta indica*) are all capable of reducing the number of galls on the roots of the plants when they are incorporated into soil infested with the root-knot nematodes (*Meloidogyne* spp). These amendments reduce the numbers of nematodes in plant roots. Mustard and neem oilcakes, applied before planting, reduce populations of *H. oryzae* and improve the growth of paddy rice (Mathur and Prasad, 1973).

From time immemorial, Indian farmers have used the neem tree (*Azadirachta indica*) for its pesticidal, antifungal, and antifeedant properties. All parts of neem (leaf, flower, bark, gum, root, fruit, seed, seed kernel and seed coat) have nematicidal potential (Mojumder and Mishra, 1993). In research trials, potting soil amended with plant parts from the neem and 'drek' tree (*Melia azadirach*) inhibited root-knot nematode development on tomatoe (Siddiqui and Alam, 2001). Neem cake, made from crushed neem seeds, provides nitrogen in a slow-release form in addition to protecting plants against parasitic nematodes. Neem cake can be mixed with fertilizers such as compost and seaweed. Recommended rates are 180 to 360 lbs. per acre or 2 lbs. per 100 to 160 sq. ft. (Anon, 1998). Neem cake is toxic to plant-parasitic nematodes and is not as detrimental to beneficial free-living soil organisms (Riga and Lazarovits, 2001). In greenhouse trials, 1 percent neem cake caused a 67 to 90 percent reduction in the number of lesion (*Pratylenchus penetrans*) and root-knot (*Meloidogyne hapla*) nematodes in tomato roots grown in three different soils. In the field, 1 percent neem cake reduced the number of lesion nematodes by 23 percent in corn roots and 70 percent in soil reduced the number of lesion nematodes by 23 percent in corn roots and 70 percent in soil around roots (Abbasi *et al.*, 2005). Investigating the influence of soil applications of neem-based (Achook and Nimbidin) and fertilizer (urea and cattle manure) treatments, separately and in combination on free-living and plant-pathogenic nematode populations infecting pigeon pea, the results showed that all the neem based products along with fertilizers significantly decreased the number of plant-parasitic nematodes (Abuzar and Haseeb, 2009). A field experiment using dry powder of leaves of *Calotropis gigantea, Azadirachta indica, Tagetes erecta* and *Citrullus lanatus* for the management of *M. incognita* infecting tomato showed significant reduction in root knot index and increase in plant growth parameters (Sarvanapriya and Sivakumar, 2005). Goswami and Vijaylakshmi (1981) have also reported the efficacy of *Eclipta alba, Azadirachta indica* and *Dhatru metch* in reducing the root knot galls and increasing growth of tomato against *M. incognita*. The use of neem seed powder and neem based formulations *viz.* Achook,neemark, neemgold, nimbecidine and fieldmarshal as seed soaking @ 5 and 10 per cent against *Meloidogyne incognita, Heterodera cajani* and *Rotylenchulus reniformis* was found to be an efficient method to reduce the nematode population in soil and to improve the plant growth of pigeon pea (Das and Mishra, 2000). Seed treatment with neem seed powder @ 5 per cent w/w + latex of *Calotropis procera* @ 1 per cent v/w against root-knot nematode and wilt complex in chickpea proved to bean effective approach (Mishra *et al.*, 2005). Combined application of neem seed powder as seed treatment @ 5 per cent w/w and chemicals (carbofuran/phorate @ 0.75 kg/ha) provided good reduction in the population of *Heterodera avenae* infesting wheat (Pankaj *et al.*, 2003).

By amending the soil with organic materials, there is an attempt to improve soil structure, soil water rentention, improve plant nutrition, add beneficial microbes, try to stimulate growth of nematophagus fungi and produce nematicidal break down products (Bridge, 1996; Oka, 2010). There is abundant evidence that this group of substances loosely classified as soil amendments, natural products, or organic amendments, can have an affect on nematode populations or stimulate plant growth in spite of the presence of plant parasitic nematodes. There is less evidence that they perform as consistently as traditional chemical nematicides or provide the same degree of nematode control when tested side by side in replicated field trials. In spite of this, there are probably abundant situations in which such products could be utilized in agricultural situations. Many previous reviews have focused on the use of organic amendments to control plant-parasitic nematodes (Rodríguez-Kábana, 1986; D'Addabbo, 1995; Akhtar and Malik, 2000; Oka, 2010; Thoden *et al.*, 2011). Farm manure trials have frequently involved poultry or cattle litter. Poultry litter appeared to be an appropriate choice (Gamliel and Stapleton, 1993), especially when combined with sorghum cover crop (Everts *et al.*, 2006), but it may be phytotoxic at high dosages (Kaplan and Noe, 1993).

Many, perhaps hundreds, of substances have been applied to soil in attempts to minimize problems from plant parasitic nematodes. In some cases the substance has been a waste product (*e.g.* coffee grounds, newsprint, crab shells, quinoa bran) from a manufacturing process. In other cases, a crop is grown specifically for this function (*e.g.* marigolds, vetch and sesame). In still others, microbial organisms or mixtures of microbials have been selected and reproduced in a fermentation process.

The mode of action of most of these natural substances is unclear (as is the mode of action of many chemical nematicides) and because of the variability within agricultural soils, activity could be expected to vary from one situation to another. Some products claim to be inoculating soils with beneficial microbials originally derived from soils, and multiplied in fermentation media, or with toxins produced by these organisms (*e.g.* killed microbials). Some products are marketed to reinoculate soils with beneficial organisms after these have been removed by chemical treatments.

Some substances such as those containing chitin (*e.g.* crab and shrimp shells) are thought to stimulate the growth of nematophagous fungi which utilize chitin for food. Chitin has been shown to be present in the eggshells of nematodes and this is thought to be the target of the nematophagous fungi. The addition of organic matter to soils is known to improve soil structure, aid in water retention, and provide nutrients. Improving soil conditions reduces plant stress which in turn can make the stress caused by plant parasitic nematodes less severe or apparent.

As some plant based products (*e.g.* marigolds, brassicas, and sesame) decompose in the soil, they release chemicals thought to be nematicidal. Brassicas, for example, have been shown to release a compound similar to the active ingredient in metam-sodium containing (*e.g.* Vapam) products. Kimenju *et al.* (2004) reported that application of organic amendments stimulated the activity of natural antagonists of plant parasitic nematodes. However, the available reports do not mention the contribution of nematode destroying fungi in the reduction of plant parasitic nematodes yet they are known to destroy nematodes in the soil. *In vitro* experiments have shown that nematode destroying fungi increase in numbers or activity when organic substrates are incorporated into the soil (Gomes *et al.*, 2000; Timm *et al.*, 2001). In a related study, Jaffee (2006) reported that alfalfa (*Medicago sativa* L.) leaves enhanced the populations of *Dactylellina candidum* but the study did not mention other nematode destroying fungi.

Addition of organic amendments into the soil also resulted in an increase of bacterial and fungal feeding nematodes and reduction of plant parasitic nematodes. Specifically there was a 225, 96 and 62 per cent increase in bacterial feeding nematodes and 391, 96 and 74 per cent increase in fungal feeding nematodes in soil amended with chicken manure alone, combination of chicken and cow manure alone in that order. Numbers of plant-parasitic nematodes were 92 per cent lower in soil treated with chicken manure compared to the control. It has been demonstrated that application of organic soil amendments resulted in changes in nematode community structure by increasing the abundance of free-living nematode populations and suppressing plant parasitic nematodes. Application of soil amendments is becoming a conventional practice that helps in the control of nematodes and other soil-borne diseases. Oil seed cakes of neem/margosa (*Azadirachta indica*), castor (*Ricinum communis*), mustard (*Brassica compestris*), rocket salad/duan (*Eruca sativa*) were found to be highly effective in reducing the multiplication of nematodes and consequently plant growth and bulk density of woody stem of pigeonpea increased significantly. The multiplication rate of nematodes was less in presence of *Paecilomyces lilacinus* along with the above amendments (Suhail, 2003). In a greenhouse trial, 1 per cent neem cake (mass/mass soil) caused a 67 per cent -90 per cent reduction in the number of lesion (*Pratylenchus penetrans*) and root-knot (*Meloidogyne hapla*) nematodes in tomato roots (Abassi *et al.*, 2005). Dhawan and Kaushal (1988) showed reduction in hatching incidence of *H avenae* juveniles by using neem emulsion. Kaushal (1993) showed that neem powder and neem based chemicals were effective against *H avenae.*

In an another study, Akhtar and Mehmood (1997) have found that the application of poultry and cattle manures, a neem based urea coating agent Nimin and another neem based biopstcide, Suneem G significantly reduced populations of *Meloidogyne incognita* and other plant parasitic nematodes and increased the population of microbivorous nematodes. Comprehensive studies like those of Koenning *et al.* (2003) have revealed the nematicidal potential of organic products used as soil amendments. When incorporated into the soil, organic substrates undergo biologically mediated decomposition to release NH_4^+, formaldehyde, phenols and volatile fatty acids, among other compounds (Wang *et al.*, 2004; Viaene and Abawi, 1998; Rich and Rahi, 1995). The involvement of soil micro-organisms in nematode control in amended soils has been confirmed by the fact that soil irradiation disrupts the nematicidal effect of these amendments (Kaskavalci, 2007). It has been established that application of organic substrates leads to build-up of micro-organisms which serve as food substrates for free-living nematodes hence their build-up. Populations of free-living nematodes such as bacteriovores, fungivores and predacious have been shown to rapidly increase following the addition of organic amendments (Akhtar and Malik, 2000; Jaffee, 2002). In addition, free-living nematodes may accelerate the decomposition of soil organic matter and increase mineralization of nitrogen and phosphorous thus triggering a chain reaction that favours their build-up (Widmer and Abawi, 2000; Kimenju *et al.*, 2004). Mian and Rodríguez-Kábana (1982) reported that the nematode management potential of an organic amendment is directly related to its nitrogen (N) content. Soil amendments with low carbon: nitrogen (C/N) ratios (*e.g.*, animal manures, oilcakes, and green manures) exhibit high nematicidal activity (Lazarovits *et al.*, 2001; Oka, 2010). Yucel *et al.* (2002) were categorical that organic amendments that have high nitrogen content and release ammonia upon decomposition are more effective in nematode suppression. Organic amendments such as distillery sludge, neem cake, poultry manure and vermicompost significantly reduced the populations of *H. oryzicola* and other nematodes in the soil and roots of banana in Tamil Nadu, India (Sundararaju *et al.*, 2002).

The increase in crop vigour may partly be attributed to reduced plant-parasitic nematode populations, but nutrients availability cannot be ignored. In an earlier study, decrease in populations of parasitic nematodes has been associated with increased in crop yield (Zolda, 2006). In turn, the decline in plant parasitic nematodes in this study could probably be attributed to the high number of nematode destroying fungi. The association between nematode-trapping fungi, organic matter, plant growth and nematodes community is complex. It would be difficult to conclude that the increase of nematode destroying fungi would lead to automatically reduction of plant parasitic nematodes hence health plants. Akhtar and Malik (2000) suggested that free-living nematodes reproduce rapidly when presented with organic substrate and play an important role in recycling of plant nutrients making them available to plants. Such organic substrates again have been seen to support high numbers of nematode destroying fungi and reduced populations of plant parasitic nematodes. Therefore, low numbers of plant parasitic nematodes coupled with high numbers of nematode destroying fungi and high nutrient levels in the soil has a positive effect to plant growth. It would be impossible to attribute the performance of the crop to either of them. In order to access the contribution of nematode destroying fungi, there is need to address the quantification and efficacy methods (Jaffee, 2006). More study should be devoted to the correlation between populations of nematode destroying fungi, plant parasitic nematodes and the actual plant yield.

Decomposed dry leaves of *Cynodon dactylon, Datura stramonium, Eichhomia crassipes, Emex spinosus, Ricinus communis* and *Sisymbrium irio* when mixed with soil at the rate of 1, 3, 5 and 10 g/kg soil and compared their nematicidal potential with carbofuran as a standard against the root-knot nematode, *M. incognita* infecting tomato resulted in lower *M. incognita* populations in the soil and root galling were significantly suppressed. The highest reduction was noticeable with the plants grown in *Sisymbrium irio* amended soil followed by *Datura stramonium* and *Emex spinosus* (Radvan *et al.*, 2006). Although the efficacy of these products under controlled conditions is commonly recognized, results in field conditions are rather inconsistent (Abawi and Widmer, 2000). For example, some experiments showed no significant effect of compost on nematode control (Szczech *et al.*, 1993; McSorley *et al.*, 1997). Thoden *et al.* (2011) even reviewed several studies in which root-knot nematode populations were increased after the application of organic amendments.

Because of the variability in agricultural soils, and the suspected mode of action of many soil amendments, it is likely that growers and workers will need to develop different programs for particular crops and sites. In such instances, it is important to try to leave some replicated areas untreated in order to be able to observe the success of the treatment program.

9.4.8 Trap Crops

Trap cropping is a technique useful for removing a portion of the population of a sedentary endoparasitic nematode such as root-knot (*Meloidogyne* sp.) or cyst nematode (*Heterodera* sp.). A host for the nematode of interest is planted, allowed to grow for a time, and killed prior to the development of egg bearing adults. It is important to destroy the roots of the crop so they do not continue to support nematode development. The technique is not useful for migratory ectoparasitic or endoparasitic nematodes because the juveniles and adults are able to move to other roots and continue to feed and develop.

Plants with nematicidal properties, when intercropped with nematode susceptible crops or grown as cover crop, can effectively reduce populations of nematodes and improve yields. Their acceptability and usefulness depends on their economic or other value to the farmers. Crops such as sesame (*Sesamum indicum*), mustard, and asparagus (*Asparagus officinalis*) which have local use or market can be readily recommended. Some weeds that have nematicidal root exudates can be encouraged as cover crops if the farmer can derive some benefit from their presence. *Sphenoclea zeylanica*, a common weed in India, suppresses populations of *Hirschmanniella* spp in rice fields within two months (Mohandas *et al.*, 1981). It decomposes quickly when incorporated into the soil and can be used as a green manure crop before rice.

9.4.9 Tillage Practices

Tillage reduces the aggregate size and changes the pore volume, thus altering the drainage, aeration, and soil chemistry due to accelerated oxidative breakdown of soil organic matter. Continuous cultivation of land leads to decline in soil organic matter content and degradation of soil structure. The reduced or conservation tillage practices conserve antagonists of nematodes to a greater extent than conventional tillage practices. The nematode community is more diverse in stable natural ecosystems than in agricultural systems because tillage disturbs the soil as well as soil microfauna and flora. Lenz and Eisenbeis (2000) observed that various tillage treatments (with a cultivator or a two-layer plow) affected both the structural (taxonomic) and functional (trophic group, life strategy) characteristics of nematode communities; the density of plant-parasitic nematodes was reduced after tillage, and the populations of bacteriovorous and fungivorous nematodes was increased (Freckman and Ettema, 1993).

Intensively managed ecosystems tend to support a greater relative abundance of plant-feeding nematodes than the non-agricultural ecosystems. The rate of reproduction of plant feeders is high in soil under deep chiseling and low in no tillage soil whereas the rate of reproduction of useful nematodes such as bacterial feeders is high in no tillage soil or shallow tillage soil. Most cereal growers sow their crops in a conventionally prepared seedbed following a period of fallow. There is an increasing interest in minimum tillage practice to preserve soil structure, control soil erosion, reduce the turn around time, and reduce non-productive costs. Wheat sown by direct drilling is less severely damaged by cereal cyst nematode, *H.avenae* than when it is traditionally sown.

9.4.10 Solarization

Soil solarization is a nonchemical technique in which transparent polyethylene tarps are laid over moist soil for a 6 to 12 week period to heat noncropped soils to temperatures lethal to nematodes and other soil-borne pathogens. Soil solarization, a method of pasteurization, is well documented as an appropriate technology but the economics of purchasing and applying plastic restrict its use to high-value crops.This is a method which is environmentally friendly, poses no toxic threat to humans and livestock, which can be easily applied, is suitable for small and large farms, nurseries and glasshouses and is particularly suitable for the tropics (Gaur and Perry, 1991). It is consistently effective only where summers are predictably sunny and warm. The basic technique entails laying clear plastic over tilled, moistened soil for approximately six to eight weeks. Solar heat is trapped by the plastic, raising the soil temperature. Irrigation before covering soil with polyethylene improves the effect of solarization in reducing the nematode densities. Agroclimatic factors such as rain during solarization, solar radiation intensity, and sunshine hours influence the effect of solarization. The residual effects of solarization, however, does not last more than a crop season.This method of control is practical only with suitable climatic conditions, and only against nematodes which parasitize shallow rooted crops, principally annuals. It is an environment-friendly approach without any toxic residual effects. Steaming the soil suppresses nematodes in a manner similar to solarization. There are prototype steam machines capable of performing field applications, but steaming is probable economical only for greenhouse operations or small plantings of high-value crops.

The incorporation of poultry litter prior to solarization, or use of a second layer of clear plastic, can reduce effective solarization time to 30 days (Brown *et al.*, 1989; Stevens *et al.*, 1990). Brassica residues are also known to increase the solarization effect, in a process known as biofumigation. The plastic holds in the gaseous breakdown products of the brassica crop (or food processing wastes), thereby increasing the fumigation-like effect (Gamliel and Stapleton, 1993). Large-scale field experiments using cabbage residues with solarization obtained results comparable to solarization combined with methyl bromide (Chellami *et al.*, 1997). Soil solarization reduces the population densities of nematodes parasitic to chickpea (*Cicer arietinum*) and pigeonpea (*Cajanus cajan*) (Sharma and Nene, 1990). Large-scale field experiments using cabbage residues with solarization was found comparable to solarization combined with methyl bromide. (Chellami *et al.*, 1997)

In an experiment in USSR by Khurramov (1990) covering greenhouses with polyethylene film and thereby raising the soil temperature by some 7-14°C, virtually eliminated natural populations of *Meloidogyne incognita, M. acrita, M. javanica* and *M. arenaria* in the root zone of tomatoes of a susceptible variety and increased yields which were comparable with aldicarb. All fruits in the controls were damaged. Comparable reductions were obtained with plastic coverings in a field experiment. In another experiment, Ijoyah and Koutatouka (2009) found that the number of root-galls from lettuce roots produced under solarized mulched plot was significantly (P=0.05) reduced by 72.0 and 72.6 per cent, respectively, in the year 2004 and 2005 compared to that obtained from the non-mulched plot. Similarly, the number of leaves, leaf length, leaf width, biological weight, economic weight and yield of lettuce were significantly (P=0.05) increased by 15.6, 40.0, 24.0, 4.7, 8.5 and 19.8 per cent respectively in the year 2004 and by 8.0, 40.0, 28.5, 7.7, 9.9 and 15.4 per cent respectively in the year 2005. Candido *et al.* (2008) reported that a single solarization treatment significantly increased yields by 116 per cent, and strongly reduced nematode infestation by 99 per cent of infested plants and of 98 per cent of the root gall index in the following melon crop. It also suppressed annual weed emergence three years later. Plant yields from two- and three-year solarized soil were always higher than nonsolarized control for tomato and melon. Further, two- and three-year solarization treatments almost completely suppressed the infestation of the *M. javanica* in tomato, and reduced the nematode effect in melon by 86 per cent and 79 per cent, respectively. Repeated solarization treatments also resulted in a high reduction of emergence of most weed species in all crop cycles.

A study by the USDA's Agricultural Research Service (Kaplan, 1991) found that red plastic mulch suppresses root nematode damage to tomatoes because the light reflection keeps more of the plant's growth above ground. The plant's energy goes into developing fruit and foliage; rather than roots. Nematodes feed on roots. The far-red light reflection to the above-ground plant draws away nutrients from the roots and the nematodes. Fewer roots mean less food for nematodes. Less food means fewer nematodes.The research team planted tomatoes in sterilized soil, mulched them with red or black plastic, and inoculated the roots with nematodes. Plants inoculated with 200,000 nematode eggs and mulched with black plastic produced 8 pounds of tomatoes, while those mulched with red plastic produced 17 pounds.

Deep summer ploughings along with fallowing (by keeping the fields free of plants and weeds) done in the hottest months will keep the nematodes exposed to the sun heat and starved, killing most of the population. It should be repeated at every 10-15 days interval. The months of May and June are ideal for the summer ploughings (north Indian conditions) to be carried out.

9.5 Plant Resistance

Host resistance to a pest/pathogen has always been considered a very attractive method to fight against disease causing agent. It can be one of the most useful, economical and effective means of managing nematodes in all farming situations. However, nematode resistance is found in relatively few crops and is not available for many crop-nematode combinations. Once developed a resistant variety is normally able to maintain of being unaccepted to the pathogen. There is a wide range of plant defences against pests and pathogens. Immune plants do not allow nematode attack including initial root invasion. Resistant or nonhost plants may be invaded by nematodes and may show damage but chemical or physical unsuitability of the plant will prevent population increases. The nematode penetrates but fails to further develop in its life cycle. Susceptible plants allow normal nematode reproduction and may or may not tolerate nematode attack. Cucurbits and a whole lot of vegetables if infected in the initial stages of growth are highly susceptible and may suffer huge damage. Tolerant host plants are able to withstand nematode attack. The concept of tolerance to the pest is a preferred alternative to absolute resistance. The possibility of finding tolerant varieties for a crop which is a known host of a particular nematode should not be overlooked. This type of research can be conducted utilizing a combination of greenhouse, microplot and field research. Intolerant host plants are more likely to be damaged by nematode attack.

A resistant cultivar is more effective against sedentary endoparasitic species such as root-knot and cyst nematodes than against ectoparasitic species. Root-knot and cyst nematodes spend most of their life cycle within the root, relying on specialized cells for feeding. Upon entering the roots of resistant cultivars, these nematodes become trapped as the feeding cells necessary for their survival fail to develop. By using resistant varieties, the suppression of nematode reproduction leads to lower nematode densities levels with obvious long term benefits in an agricultural system. A number of reasons are responsible for the absence of resistant breeding 1) the unavailability of a resistance source 2) the difficulties of transferring resistance to commercially acceptable cultivars 3) problems are localised and resistant breeding has a low priority 4) acceptable effective, immediate alternative management methods are available like nematicides (Cook and Evans, 1987). Food crops with a low commercial value would also be considered as a low priority for resistant breeding.

There are many examples of crop varieties bred for resistance against one or two *Meloidogyne* species. But in U.S.A., Nemaguard rootstock is resistant to all root knot nematode species currently present there. The reistance mechanism of Nemaguard and other resistant cultivars does not restrict attack and penetration by root knot. However, the resistance mechanism is effective in halting nematode reproduction. Massive attack of young Nemaguard roots can cause visible growth reduction in the first year's growth, but if nematode reproduction is limited by controlling weed hosts, the resistance mechanism can be highly effective in the field. Some cultivars resistant to root knot nematode exist for some field and vegetable crops but they can be damaged by heavy infestation. Thompsom Seedless grape sometimes can be heavily galled by root knot nematodes, but once established it usually suffers light-to-moderate galling only. Apparently root knot nematodes have difficulty penetrating roots of the grapevine. Once inside, development is slowed, but relatively high egg production occurs. Field corn and wheat are also hosts for root knot nematodes, but are tolerant to damage and can yield well under moderate-to-heavy infection.

Some of the resistance varieties exhibiting resistance or tolerance reaction to root-knot nematodes are (Khan, 2008):

- ☆ *Tomato*: SL-120, Hisar Lalit, PNR-7, Hisar N-1, Hisar N-2, Hisar N-3, NT-3, NT-8 NT-12, Ronita, Patriot, PAU-5, Mangla and Karnataka Hybrids.
- ☆ *Brinjal*: Giant of Banaras, Black beauty, Gola, Gachha Baigan, Pbr-91-2, IC-95-13, HOE-101, Red Wonder.
- ☆ *Okra*: Kanki local green, Harichickni, Vaishali Badher
- ☆ *Cowpea*: Barasati mutant, 82-IB, C-152, IHR-29-5, GAU-1
- ☆ *Chilli*: Pusa Jawala, CAP-63, CA-2057, Sindhuri, NP-46 A, Mohini, SP-26, P-6-3, K-235
- ☆ *Ridge gourd*: Panipati, Meerut special
- ☆ *Pea*: B-58, C-50
- ☆ *Potato*: Kufri Dewa
- ☆ *Pumpkin*: Dasna, Jaipuri
- ☆ *Water melon*: Shahjanpuri

High degree of PCN resistance available in *S. vernei* (clone 62-33-3) has been used for developing a PCN resistant variety *'Kufri Swarna'* at the Nilgiris (CPRI, 1986). This variety offers better management of PCN as it is resistant to pathotypes Ro 1 and Pa 2 (Maximum occurrence). Continuous cultivation of resistant varieties helps build up other pathotypes (Krishna Prasad, 1996). Therefore, it has to be alternated often with a susceptible potato variety. South Indian hills, which grow potato throughout the year, are also endemic to the late blight (LB) pathogen, *Phytopthora infestans* (Krishna Prasad and Latha, 1999), and hence, potato varieties should have combined resistance to PCN and LB. Potato breeding and evaluation resulted in developing 21 promising advance hybrids with combined resistance to PCN and LB (Krishna Prasad *et al*, 2001; Joseph *et al*, 2003). Non-host like *Solanum sisymbrifolium*, proved capable of inducing large hatching of potato cyst nematode and is fully resistant to potato cyst nematode can effectively be used against the nematode.

There is relevance in considering both natural genes for resistance (R-genes) and a wide range of molecules that are either pre-formed or occur after nematodes invade roots. The latter may help underpin biotechnology aimed at disrupting root migratory endoparasites. This latter area has received much less attention at the molecular level than characterisation of R-genes that are effective against sedentary endoparasites such as *Meloidogyne* spp.

9.5.1 Natural Resistance Genes (R-genes)

The identification of R-genes particularly Mi-gene displaying resistance to rootknot nematodes and its linkage to an acid phosphatase gene is a major breakthrough (Rick and Forbes, 1974) and gives a high degree of specificity for a particular pathogen. This specificity stems from a so-called "gene-for-gene" interaction whereby a component of the pathogen, encoded by an avirulence gene, interacts either directly or indirectly with a corresponding R-gene in the plant. The presence of both the avirulence (avr) and the R-gene product results in an incompatible interaction. The plant recognises the pathogen and elicits a suitable defensive response. If the pathogen does not possess an avr product, or it possesses that is non-functional due to mutational or other change, there will be no interaction with the R-gene product. Such a pathogen will be able to infect a plant.

Several R-genes are targeted against nematodes. The first described was Hs1pro-1 from a wild species of beet that confers resistance to the cyst nematode *Heterodera schachtii* (Cai *et al.*, 1997). Subsequently two genes for resistance to *Globodera* spp have been described (van der Vossen *et al.*, 2000). The resistance genes in tomato to *Meloidogyne* are of greatest interest as this nematode attacks majority of food crops. These include the Mi2 through Mi8 genes (all from *Lycopersicon*) and the Me and N genes from pepper. The Mi-2 gene of tomato confers resistance against *Meloidogyne incognita* and some other *Meloidogyne* species such as *M. javanica* (Milligan *et al.*, 1998) but not to *M. hapla* or to *M. enterolobii* (Sikora and Fernandez, 2005). It also confers resistance to the peach potato aphid *Macrosiphum euphorbiae* (Rossi *et al.*, 1998). This is the first example of a single R-gene conferring resistance to pathogens that differ so remarkably in the nature of their feeding from the host plant. The Mi-2 gene encodes a putative protein of 1257 amino acids that belongs to the family of nucleotide binding domain-LRR proteins and has an N-terminal region that encodes a potential leucine zipper domain, perhaps involved in protein dimerization. Some other R-genes are similar to Mi but so far they have only been described from Solanaceous plants (Hwang and Williamson, 2003). In many cases, however, these genes become ineffective at higher temperatures.

9.6 Biological Control

Biological control can be defined as the use of natural enemies to reduce the damage caused by a pest population. Stirking (1991) said that the biological control of plant parasitic nematodes is the reduction of nematode populations which is accomplished through the action of living organisms other than nematode resistant plants, which occurs naturally or through the manipulation of the environment or the introduction of antagonists. It differs from 'natural control' which occurs much of the time, natural enemies keeping populations of potential pests in check without intervention. Biological control, on the other hand, requires intervention, rather than simply letting nature take its course. Biological control is an approach that fits into an overall pest management program, and represents an alternative to continued reliance on pesticides. Biocontrol as an integral part of management is an attractive option for plant parasitic nematodes that should be pursued besides the cultural practices of crop rotation and organic amendment to include the use of microorganisms isolated, cultured and packaged in the tropics for tropical farmers (Agbenin, 2011).

In last few years, some developments have occurred which have had significant effects on the prospects and opportunities for the biological control of plant-parasitic nematodes. First, several nematicides have been withdrawn from the market because of health and environmental problems associated with their production and use (Thomason, 1987). As a result of this and increasing public concern over the use of pesticides in food production, there has been increased interest in the development of alternative methods of control, including the use of biological agents. Second, it has been demonstrated in several soils that nematophagous fungi and bacteria increase under some perennial crops and under those grown in monocultures, and so they may control some nematode pests, including cyst and root-knot nematodes (Stirling, 1991). Such nematode-suppressive soils have been reported from around the world and include some of the best documented cases of effective biological control of nematode pests. Finally, a number of commercial products based on nematophagous fungi and bacteria have been developed, but all so far have had only limited success.

While applying biological agents the following characteristics should be kept in mind: they should be host specific, their parasitism/preying ability should always be lethal, be easily manipulated in the laboratory, amenable for mass production, can easily be disseminated with standard equipment, they should have capacity for establishment and recycling, should be able to provide control for extended period and should not be harmful to environment. Additionally they should have good shelf-life. These are characteristics that any worker may consider when developing biological control agents. The idea here is that the more desirable characteristics a candidate organism has, the more likely it will be able to be developed into a viable commercial product. There are pro and con arguments to host specificity. Generally, it would be desirable to have an organism that was only active against one or more plant parasitic nematodes. This would reduce the chance of killing nontarget beneficial organisms. However, if only one of several damaging organisms is killed by a particular agent, costs may increase if more than one agent is used. Killing the target organism is considered more desirable than merely altering behavior or other sublethal effects from which recovery might be possible. Many candidate organisms have proven difficult to produce on artificial media. If an organism must be reared on its host, this raises the question of how it can be utilized without accidentally introducing the host into a new environment. Organisms which can be reared in fermentation facilities have been the most attractive to commercial entities. Producing a formulated product that can be applied with equipment currently in use by growers will ensure easier acceptance. Commercialization will likely be more acceptable if a natural product can be produced without needing to be reared in or on another organism.

Microbial pesticides are generally slow-acting, lacking persistence and systemic ability in the field. They are readily inactivated by environmental factors such as sunlight, rain and wind and, therefore, remain infective for a short while (Thacker, 2002). There may be less incentive for commercialization of a product that only needs to be applied a single time. One should not assume that just because a product is natural or biological that it is safe to use in all situations. For example, some natural insecticides such as pyrethrins can be harmful to fish. Many biological insecticides contain proteins and other substances to which some could become allergic, as is possible with products containing shellfish.

Over millions of years, balanced relationships have evolved in nature between parasites/predators and hosts. Scientists studying natural systems may find it difficult to change this balance in favor of increased parasitism. A host specific organism runs the risk of becoming extinct if it kills off all of its hosts. This has at times been cited as a reason why "natural" biological control will never be practical for nematodes. Although it may not be feasible to effect control with a single organism, combinations of organisms could be developed or organisms utilized in combination with cultural controls which might provide control comparable to chemical nematicides.

9.6.1 Botanicals against Nematodes

There has been a search for alternative nematode control methods or less toxic nematicides. One way of searching for such nematicidal compounds is to screen naturally occurring compounds in plants. Several such compounds, *e.g.*, alkaloids, phenols, sesquiterpenes, diterpenes, polyacetylenes, and thienyl derivatives, have nematicidal activity (Gommers, 1981). For example, á-terthienyl from *Tagetes* spp. is a highly effective nematicidal compound. Natarajan *et al.* (2006) have reported that whole *T. erecta* plant extracts were more efficacious than stem extracts although both were more effective than root extracts from 40-day old plants for controlling root knot nematode, *M. incognita* infecting tomato. *T. erecta* is capable of suppressing a wide range (up to 14 genera) of nematode pests. *T. erecta* lowered levels of *Radopholus similis, Helicotylenchus multicinctus* and *Hoplolaimus indicus* nematodes when intercropped with a highly susceptible banana crop (Wang *et al.*, 2007). Recently Osei *et al.* (2011) have demonstrated the potential of antagnostic plants, *Mucuna pruriens* and *Tithonia diversifolia* (known as tree marigold or Maxcican) in controlling plant parasitic nematodes, *Meloidogyne* spp. and *Pratylenchus brachyurus* by 93 and 95 per cent and 86 and 87 per cent over the control treatment respectively.

Other compounds with nematicidal activity have been isolated from plants, mainly from the family Asteraceae. Many allelopathic compounds in their native or processed forms have potential for development as viable components of plant-parasitic nematode management strategies. Allelochemicals are defined as plant metabolites or their product that are released intithe environment through volatization, exudation from roots, leaching from plants or plant residues and decomposition of compounds. As such they are classified as biopesticides which are defined as being derived from natural materials such as plants and and microorganisms that include both biochemical and microbial pesticides. They possess differing levels of activity against a wide range of plant-parasitic nematodes. In general, these compounds are less toxic to nontarget species, and less persistent in soil than chemical nematicides. Operative mechanisms for plant-parasitic nematode control with allelopathic compounds include nematicidal activity, nematostatic activity, and nematode behavior modification.

Allelochemicals are sometimes produced in large quantities in plant material or as agricultural waste, making the use of rotation crops, cover crops, and organic amendments effective means for production and/or distribution of the active compounds (Burelle and Kabana, 2006). These compounds affect the behavior of other organisms in the plant's environment. For example, sudangrass (and sorghum) contain a chemical, dhurrin that degrades into hydrogen cyanide, which is a powerful nematicide (Luna, 1993; Forge, *et al*, 1995; Wider and Abawi, 2000). Some other plants produce allelochemicals that function as nematode-antagonistic compounds, such as polythienyls, glucosinolates, cyanogenic glycosides, alkaloids, lipids, terpenoids, steroids, triterpenoids, and phenolics, among compounds from these plants *e.g.*, castor bean, chrysanthemum, partridge pea, velvetbean, sesame, jackbean, crotalaria, sorghum-sudan, indigo, tephrosia are exuded during the growing season or released during green manure decomposition. Sunn hemp, a tropical legume, is popular nematode-suppressive cover crops that produce the allelochemicals known as monocrotaline (Chitwood, 2002; Grossman, 1988; Hackney and Dickerson, 1975; Quarles, 1993; Wang *et al.*, 2002; Williams and Williams, 1990a, 1990b, 1993). Some cover crops have exhibited nematode-suppressive characteristics equivalent to aldicarb (Grossman, 1990). Another North American native known as "Indian Blanket" or "Blanket Flower" (*Gaillardia pulchella*) was effective in controlling root knot nematode (*Meloidogyne incognita*) on sweet potato. Tissue extracts of Indian Blanket were lethal to various plant-parasitic nematodes but were innocuous to free-living nematodes. Root exudates of Indian Blanket were lethal to mobile juveniles of *M. incognita* and were inhibitory to the hatch of eggs at concentrations of 250 parts per million or higher. Indian Blanket could be used to manage root knot nematode as a rotation crop, a co-planted crop, or a soil amendment to control root-knot nematode (Tsay *et al.*, 2004). Essential oils from various plants have shown promise as potential sources for new nematicides. Most of these plants are aromatic and culinary herbs that contain the nematicidal compounds carvacrol and thymol. At very low concentrations (1000 micrograms per liter, or.001 gm per liter, or.0038 gm per gal, or 0.38 gm per 100 gal) several oils immobilized juvenile root-knot nematodes and some also reduced hatching of eggs. The essential oils from the following plants ranked the highest for nematicidal activity: caraway, fennel, applemint, spearmint, Syrian oregano, and oregano (Oka *et al.*, 2000). The toxicity of the essential oil from wormwood or Sweet Annie (*Artemisia annua*) leaves was evaluated in vitro against second-stage juveniles (J2) of the root knot nematode (*Meloidogyne incognita*) and pre-adults of the reniform nematode (*Rotylenchulus reniformis*). Complete mortality (100 percent) of both nematodes was found in 500 and 250 parts per million concentrations of the essential oil and gradually decreased with lower concentrations (Shakil *et al.*, 2004).

To determine the effects of *Bridelia micrantha, Mallotus oppositifolius, Hunteria umbellata* and *Citrus medica* leaf extracts on growth and root-knot infection caused by *M. incognita* on cacao seedlings, Okeniyi *et al.* (2010) found that water extract of all the test plants significantly inhibited egg hatching of nematode and caused 100 per cent mortality of the second juveniles of *M. incognita* in vitro after 12h of exposure.

Undiluted crude leaf extracts of *H. umbellata* and *M. oppositifolius* exhibited 100 per cent inhibition of egg hatch and larva mortality, while undiluted leaf extracts of *B. micrantha* and *C. medica* exhibited 92 and 93.2 per cent inhibition of egg hatch and 62.1 and 73 per cent larval mortality respectively. Egg inhibition and larval mortality decreased with increase in dilution of all the extracts. Juvenile mortality increased corresponding to an increased time of exposure. The leaf extracts of individual plant significantly enhanced the growth of cacao seedlings in the presence of the nematode in the nursery when compared to the control ($p<0.05$). There was a significant increase in plant height (68.2, 70.2, 65.9 and 65.7 for *B. micrantha, M. oppositifolius, H. umbellate* and *C. medica* respectively), shoot weight and root weight of the seedlings treated with all the leaf extracts even at the lowest concentration of 10 per cent compared to the rest. Extracts of plants such as *Azadirachta indica, Acacia alatta, Borrelia* spp., *Acalypha cilliata* and garlic bulbs have been identified to have nematicidal properties (Egunjobi and Onayemi 1981; Agbenin *et al.*, 2005; Rotimi and Moen 2002). In India extracts of *Calendula officinalis, Enhydra fluctans* and *Solanum khasianum*, chilli pepper and garlic were all effective against the root-knot nematode (Goswami and Vijayalakshmi 1986; Sukul, 1992). Plant extracts are applied either as soil drench, root dip or foliar spray.

9.6.2 Predators of Nematodes

The soil dwelling organisms such as tardigrades, turbellarians, enchytraeids, insects, mites and nematodes have been shown to provide some control of plant parasitic nematodes in natural and agricultural environments.

9.6.3 Bacterial Parasites of Nematodes

Bacterial parasites are mainly *Pasteuria penetrans, Bacillus thuringiensis* and *Streptomyces avermtilus*. Of the three bacteria which have been most intensively studied in connection with nematodes, only *P. penetrans* is a true parasite. The others produce toxins or metabolites which have been found to affect nematodes. Other dominant bacterial genera like *Pseudomonas fluorescens, Bacillus subtilis, B. sphaericus, Corynebacterium, Rhizobium* and *Xanthomonas* which are able to antagonise plant parasitic nematodes especially the root knot nematodes (Oliveira *et al.*, 2007). *Pseudomonas fluorescens* also provide effective control of root-knot nematodes on vegetable crops (Haseeb and Kumar, 2006; Krishnaveni and Subramanian, 2004; Stalin *et al.*, 2007). *P. fluorescens* has been found to be associated with enhanced PR activity and accumulation of phenols while managing *Meloidogyne* populations (Anita *et al.*, 2004). Nematode control by rhizosphere bacteria may be achieved through direct antagonism by producing toxins, enzymes, interference with plant-nematode recognition, competition for nutrients and through inductment of systemic resistance.

9.6.3.1 *Pasteuria penetrans*

The *Pasteuria penetrans* is the primary bacterial parasite which has been studied and was first described by Thorne (1940) as a protozoan. A number of different strains have been isolated and tested. In spite of extensive work, these bacteria still cannot be reared on artificial media and this has greatly hindered commercial development. Spores of these bacteria adhere to the cuticle of a nematode and a penetration tube develops which penetrates the nematode. Once gaining access, the bacteria reproduces inside of the nematode utilizing the cuticle as protection. Two million spores have been shown to be formed within a single nematode. Typically, the nematode survives and feeds following infection but reproduction is greatly reduced. The fungus, *Paecilomyces lilacinus,* is effective against root knot nematodes on potato (Jatala, 1985) and was extensively used for RKN management in other crops under the International *Meloidogyne* Project (Sasser, 1989), while, the endomychorhhizal fungi, *Glomus fasciculatus* and *G. mossae,* offer possibilities for nematode management, especially RKN, in India (Krishna Prasad, 1993). Two bioagents, *Paecilomyces lilacinus* and *Pochonia chlamydosporia,* in talc formulations, were tested for PCN management under field conditions, with and without the nematicide (Krishna Prasad and Nagesh, 2007). Both these organisms were able to reduce cysts and propagules by 46 to 49 per cent while increasing marketable-potato yields by 40 to 70 per cent. Similarly, *Pseudomonas flourencens,* in combination with *Paecilomyces lilacinus,* brought about substantial yield increase upto 70 per cent while reducing PCN (Malavika *et al,* 2008). The rapidity with which many of these nematode biocontrol agents have been used in the Indian agriculture for management of several parasitic nematodes like RKN, in vegetables (Rao *et al,* 1997, 1998; Rao, 2004), ornamentals (Rao *et al,* 2003) and reniform and burrowing nematodes in cotton (Shivakumar *et al,* 2004) and banana (Karuna *et al,* 2004) has shown that indigenously isolated and multiplied biocontrol agents will become an important component of nematode management in potato as well (Walia, 2004, Krishna Prasad, 2007).

9.6.3.2 *Bacillus thuringienesis*

A number of strains of *Bacillus thuringiensis* (popularly called B. T.) are utilized in the biological control of insects. When eaten by insects, delta endotoxins that were produced by the bacteria in fermentation are activated and result in death. Although not yet commercially developed, patents have been issued for strains of *B. thuringiensis* which have activity against nematodes.

A second group of toxins produced by some *B. thuringiensis* strains is called β-exotoxin (or beta exotoxin). Although first tested against insects, because the mode of action is to interfere with RNA transcription, they have been found to have activity against many forms of life including nematodes. Again, these toxins have not been commercially developed for nematode management. At the conclusion of the fermentation process, if the medium is centrifuged, the delta endotoxins are found in the centrifugate while B-exotoxins are found in the supernatant.

9.6.3.3 *Streptomyces avermitilis*

Avermectins are a class of macrocyclic lactones produced by *Streptomyces avermitilis,* a member of the actinomycete family, that are commonly used in treating gastrointestinal helminthic parasites in domestic animals. *Abamectin* is the common name assigned to the avermectins. It is active against insects at very low dosages, and formulations containing avermectin are utilized against insects on a number of crops in U.S.A. Avermectin has also become widely used to control nematode parasites of animals. Although activity has been shown against plant parasitic nematodes, unfortunately, their utility as soil-applied nematicides has been limited due to expense, low water solubility, and their tendency to adsorb to soil particles (Bull *et al.*, 1984). However, application of certain avermectins (abamectin) as a seed dressing has shown promise for nematode suppression in cotton seedlings (Monfort *et al.*, 2006).

In another study, an isolate of the actinomycete, *Streptomyces* sp. CMU-MH021 produced secondary metabolite, Fervenulin, a low molecular weight compound at the concentration of 250 µg/ml showed killing effect on second-stage juveniles of *M. incognita* up to 100 per cent after incubation for 96 h. and also inhibited egg hatch (Ruanpanun *et al.*, 2011).

9.6.4 Nematophagus Fungi

9.6.4.1 Nematode-trapping Fungi

Nematophagous fungi have been found in all regions of the world, from the tropics to Antarctica. They have been reported from agricultural, garden and forest soils, and are especially abundant in soils rich in organic material. There is a growing evidence to suggest that the parasitic habit of nematode-trapping fungi has evolved among cellulolytic or lignolytic fungi as a response to nutrient deciencies in nitrogen-limiting habitats (Barron, 1992). In soil environments with a high carbon:nitrogen ratio, nematodes might serve as an important source of nitrogen during growth on carbohydrate containing substrates. A fossil of nematodes parasitized by nematophagous fungi has been dated to 22.5–26 million years ago (Jansson and Poinar, 1986) and can be used as a reference to estimate the divergence time of trapping devices. If the fossil is used in time calibration, the predatory fungi would be well established in the Tertiary period. However, *Orbilia fimicola*, a predatory fungus, was estimated to be first derived from its ascomycete ancestor (nonpredatory) at about >900 million years ago (Pandovan *et al.*, 2005), and the time scale is much older than the fossil record. Several fungi have been identified and classified according to their nematophagous properties. They include trappers, endoparasites, egg-parasites and toxin producers (Liu *et al.*, 2009). The nematode-trapping fungi are a group of soil-living hyphomycetes that capture nematodes using specialized structures called traps. The morphological structure of the traps differs depending on the species and can be divided into four major categories: adhesive nets, adhesive knobs, adhesive branches, and constricting rings. The species with adhesive cells capture nematodes using extracellular polymers that accumulate at the site of attachment between the fungus and the nematode (Tunlid *et al.*, 1992). Such polymers are not found on the cell surface on the constructing rings. These cells can rapidly swell when a nematode is passing through the ring and thereby physically ensnare the nematode.

Over 200 species of fungi (distributed in Zygomycota, Basidiomycota, and Ascomycota) use special structures to capture free-living nematodes in the soil (Li *et al.*, 2000; Yang *et al.*, 2007; Sanyal, 2000; Masoomeh *et al.*, 2004). The fungi dier in their saprophytic/parasitic ability. While many of the trap-forming and egg-parasitic fungi can survive in soil saprophytically, the endoparasites are mostly more dependent on nematodes as nutrient (obligate parasites). The ability to capture nematodes is connected with a specic developmental phase of the fungal mycelium. The trapping (predatory) fungi have developed sophisticated hyphal structures, such as hyphal nets, knobs, branches or rings, in which nematodes are captured by adhesion or mechanically. The endoparasites, on the other hand, attack nematodes with their spores, which either adhere to the surface of nematodes or are swallowed by them.

Nematode destroying fungi are natural enemies of plant parasitic nematodes (Nordbring-Hertz *et al.*, 2002). The most widespread predatory fungi are in the family of Orbiliaceae, Ascomycota (Pfister, 1997, Rubner, 1996). *A. oligospora, A. conoides, A. musiformis and A. superb* form three-dimensional adhesive nets, whereas *A. dactyloides* uses constricting rings to capture nematodes mechanically by the swelling of the ring cells. Adhesive branches and adhesive knobs appear in the genus *Monacrosporium. M. haptotylum* (*Dactylaria candida*) produces both adhesive knobs and nonconstricting rings. Predacious fungus of nematodes, has been very useful in understanding the relationship between nematophagous fungi and their nematode hosts. *Arthrobotrys oligospora* is by far the most common nematode-trapping fungus with the characteristic ability of forming adhesive trapping nets once in contact with nematodes. It is considered as a model organism for understanding interaction between fungi and nematodes. The fungi which form rings have been shown to form more rings in the presence of nematodes than in their absence. Within a few hours of close contact with nematodes, the sparse mycelia of these fungi will differentiate spontaneously into functional structures (traps). The mycelial traps then adhere to, penetrate, kill, and digest the nematodes' contents (Barren, 1977). The adhesive network (*Monacrosporium* spp.), the most widely distributed trap, is formed by an erect lateral branch growing from a vegetative hypha, curving to fuse with the parent hypha and developing more loops exterior to the original loop or on the parent hypha.

Many network-forming species do not form a network spontaneously; they are more saprophytic than other nematode-trapping fungi. Formation of network-trapping devices is induced by the presence of nematodes or substances of animal origin known as *nemin*, a morphogenic substance causing trap formation by predaceous fungi. Like carnivorous plants, predatory fungi have the ability to capture and to absorb nutrients from their prey, a fascinating example of convergent evolution. At the genetic level there are basically three mutually non-exclusive explanations that can account for the origin of parasitism in nematode-trapping fungi: (i) Parasitism of nematodes is a result of selection on regulatory genes. Thus, closely related parasitic and non-parasitic species have basically the same set of genes, and adaptation to the nematode parasitic habit is due to differences in regulation of gene expression. (ii) Parasitism is due to the presence of novel genes. Novel genes can be acquired through gene duplications or horizontal gene transfer (HGT) (iii) Parasitism is due to the loss of specific genes. It is also being established that different sets of genes are expressed during development of traps as well as during infection when compared to the vegetatively growing mycelium.

9.6.4.2 Other Fungi against Nematode

Egg-parasitic fungi include *Paecilomyces, Pochonia* and *Verticilium* genera. *Paecilomyces lilacinus* and *Pochonia chlamydosporia* are probably the most effective egg parasites. *Paecilomyces lilacinus* protects the root system against diseases caused by plant parasitic nematodes, specifically root-knot nematodes (*Meloidogyne* spp.), cyst nematodes (*Heterodera* and *Globodera* species), reniform nematode (*Rotylenchulus reniformis*), banana nematodes (*Radopholus similis*) and citrus nematode (*Tylenchulus semipenetrans*). *Paecilomyces lilacinus* has been proven to successfully control root-knot nematodes, *M. javanica* and *M. incognita* on tomato, egg-plant and other vegetable crops (Verdejo-Lucas *et al.*, 2003; Goswami and Mittal, 2004; van Damme *et al.*, 2005; Goswami *et al.*, 2006; Haseeb and Kumar, 2006; Kumar *et al.*, 2009). This fungus colonizes the root surface and is strongly parasitic to eggs and egg-masses of plant parastic nematodes. Its parasitization can destroy upto 90 per cent of eggs and 75-80 per cent of egg-masses of nematodes. *Paecilomyces lilacinus* formulations have been homologated in many countries for vegetables and other crops, including coffee and banana. However, *P. lilacinus* appears to be more suited to tropical conditions (Krishnamoorthi and Kumar, 2007) and acid soils close to pH 6 (Krishnamoorthi and Kumar, 2008) than to temperate or cold conditions. *Pochonia chlamydosporia* prefers mild climate and soil conditions (Atkins *et al.*, 2003), where it occurs naturally (Bent *et al.*, 2008). The cereal cyst nematode, *Heterodera avenae* population often declines within 5-6 years where monoculture of wheat crop is practiced. This decrease in nematode populations is mainly caused by two parasitic fungi, *Nematophthora gynophila* and *Verticillium chlamydosporium*, which attack the developing female on the root surface; in suppressive soils 95 to 97 percent of the females and eggs are destroyed (Kerry *et al.*, 1982). Thus, the natural control of cereal-cyst nematode in a range of soils is predictable and effective, but slow acting.

The edible oyster mushrooms species can also control nematode populations when come in contact with them (Kwock *et al.*, 1992; Thorn and Tsueneda, 1993; Heydari *et al.*, 2006). It was observed that *Pleurotus ostreatus* either alone or in combination with nematode or in combination with nematode and *P. tuberregium* proved more effective in causing larval mortality of *Meloidogyne incognita* infesting soybean resulting in increased plant growth and nodulation than the combined effect of *P. tuberregium* plus nematode (Okorie *et al.*, 2011).

Additionaliy, *Verticillium lecanii* acts against the cyst nematodes *H. glycines*, *Heterodera schachtii*, and *Globodera pallida* (Gintis *et al.*, 1983; Hanssler, 1990; Hanssler and Hermanns, 1981; Meyer *et al.*, 1990; Meyer and Huettel, 1996; Meyer and Meyer, 1995, 1996; Uziel and Sikora, 1992). Research has been conducted on the use of this fungus as a biocontrol agent for *H. glycines*. In laboratory and greenhouse studies, strains of *V. lecanii* reduced SCN populations, with decreases in cyst numbers of up to 85 per cent compared with untreated controls (Meyer *et al.*, 1990; Meyer and Huettel, 1996; Meyer and Meyer, 1995, 1996).

Several microbial pathogens have been developed into commercial formulations against nematodes. These include the bacteria *Pasteuria penetrans* (formerly known as *Bacillus penetrans*), *Bacillus thuringiensis* (available in insecticidal formulations) and *Burkholderia cepacia*. Nematicidal fungi include *Trichoderma harzianum*, *T. hamatum*, *T.koengi*, *Hirsutella rhossiliensis*, *Hirsutella minnesotensis*, *Verticillium chlamydosporum*, *Arthrobotrys dactyloides*, and *Paceilomyces lilacinus*. Affokpon *et al*. (2011) demonstrated significant inhibition of nematode (*M.incognita*) reproduction, suppression of root galling and an increase of tomato yield compared with the non-fungal control treatments. *Trichoderma asperellum* suppressed second stage juvenile (J_2) densities in roots by up to 80 per cent ; *T. brevicompactum* suppressed egg production by as much as 86 per cent. Tomato yields were improved by over 30 per cent following the application of these biocontrol agents, especially *T. asperellum*. Another fungus, *Myrothecium verrucaria*, found to be highly effective in the control of nematodes (Anon, 1997), is available in a commercial formulation, DiTera™, from Abbott Laboratories. Circle One, Inc. offers a combination of several mycorrhizal fungal spores in a nematode-control product called Prosper-Nema™. Stein Microbial products offer the bacterium *Burkholderia cepacia* in a product called Deny™ and Blue Circle™. Rincon-Vitova offers a product called Activate™ whose active ingredient is the bacterium *Bacillus chitinosporus* (Quarles, 2005).

9.7 Entomopathogenic Nematode Against Plant Parasitic Nematodes

Entomopathogenic nematodes, in general, do not have the intrinsic abilities required for routine, effective eld use as biocontrol agents but various direct or indirect effects of their applications may occur within the soil community. They may be fed upon directly by nematophagous mesoarthropods such as mites and collembolans (Epsky *et al.*, 1988; Wilson and Gaugler, 2004), nematode trapping fungi (*e.g.*, Koppenhoffer *et al.*, 1996; Kaya and Koppenhoffer, 1999) or predacious nematodes. These interactions could potentially indirectly inuence the population densities of other species of prey or competing predators. Moreover, the subsequent death and decomposition of large numbers of entomopathogenic nematodes may indirectly affect population densities of bacterivorous and fungivorous soil organisms. Grewal *et al.* (1999) observed direct repellency and allelopathic suppression of plant-parasitic nematode juveniles by metabolites of the EPN symbiotic bacteria *Xenorhabdus* sp., released upon decay of the respective nematode-infected cadavers. However, these secondary metabolites produced by the entomopathogenic nematodes bacterial symbionts and released into the soil from the insect cadaver have a very short-term anti-microbial activity (Chen *et al.*, 1994). Shapiro *et al.* (2006) have precisely demonstrated that entomopathogenic nematodes can also be used as biological control agents to control plant-parasitic nematodes infesting different crops in the fields and greenhouses. To control plant- parasitic nematodes, entomopathogenic nematodes can be applied using standard spraying equipments used for application of chemical pesticides. Entomopathogenic nematodes are generally applied against plant-parasitic nematodes at the rate of 1 billion infective juveniles per acre but this rate can vary with both entomopathogenic nematode and plant- parasitic nematode species. However, Bird and Bird (1986) reported for the first time the efficacy of EPNs against plant parasitic nematodes and proved that *S.glasseri* suppressed *M. incognita* on tomato seedlings in agar. Other examples are, *S. carpocapsae* can reduce the population of potato cyst nematodes (*Globodera rostochiensis*) and foliar nematode *Aphelenchoides fragariae*. *Steinernema riobrave* can reduce the population of stunt nematodes (*Tylenchorynchus* spp.), lance nematodes (*Hoplolaimus* spp.), root-knot nematodes (*Meloidogyne* spp.). It can also reduce the population of sting nematodes (*Belonolaimus longocaudatus*). *Steinernema feltiae* can inhibit hatching in root-knot nematode eggs and infection by hatched infective juveniles of root-knot nematodes (*Meloidogyne* spp.). *Steinernema glaseri* reduced egg masses of root-knot nematodes (*Meloidogyne* spp.). *Heterorhabditis bacteriophora* can reduce the population of ring nematodes (*Mesocriconema* spp., *Criconemoides* spp.), stunt nematodes (*Tylenchorynchus* spp.) and the population of lesion nematodes (*Pratylenchus pratensis*) (Smitley *et al.*, 1992). *H. baujardi* can inhibit hatching root-knot nematode eggs and infection by hatched infective juveniles of root-knot nematodes (*Meloidogyne mayaguensis*) according to Shapiro *et al.* (2006). Somashekhar *et al.* (2002) reported that *Heterorhabditis* sp. reduced the total number of plant parasitic nematodes in turf grass in Ohio.

The insect-attacking nematode *Steinernema riobravis* can provide root-knot nematode control comparable to that achieved with chemical nematicides (Grossman, 1997). Although the exact mechanisms of control are not known, researchers (Jagdale *et al.*, 2002) hypothesize that an allelochemical is involved (perhaps manufactured by symbiotic bacteria that live within *S. riobrave*) that repels plant parasitic nematodes. Recent research measured the effect of beneficial nematodes on root-knot nematodes (*Meloidogyne* species) infecting tomatoes and peanuts. In the laboratory, peanut seedlings treated with the beneficial nematodes *Steinernema feltiae* and *S. riobrave* showed resistance to pest nematode. On peanuts, pre-and post-infestation applications of *S. feltiae* suppressed *M. hapla* penetration but not egg production. Only pre-infestation applications of *S. riobrave* suppressed *M. hapla*. The tomatoes were infested with *Meloidogyne incognita* eggs and treated with *Steinernema glaseri* or *Heterorhabditis megidis* applied at the same times as the tomato treatments. The low rate of *S. glaseri* suppressed *M. incognita* penetration into tomato roots and the high rate of *S. glaseri* reduced egg production (Pérez and Lewis, 2004).

To make entomopathogenic nematodes more successful, realistic strategies through genetic engineering, IPM programs, and new delivery systems and/or training programs to overcome their inherent cost, formulation instability and limited field efficacy toward certain insects are needed.

9.8 Suppressive Soils

A suppressive soil can be loosely defined as one which should have a nematode problem but doesn't. It is thought that one or more biological organisms in the soil are suppressing nematode populations. Trying to document that a soil is suppressive is no easy task. There are indications in the literature that crops grown under monoculture for many years in nematode infested fields will have high nematode populations for several years which will then fall and remain at low levels. It is thought that some biological organisms are functioning to maintain nematode populations at a low level. For example, if a wheat field is found infested with cereal cyst nematode, *Heterodera avenae*, the population will increase in the next four to five years but after that it is found to decline geometrically even without resorting to any control measure. It is also not uncommon in chemical control field trials for chemically treated plots to have significantly higher nematode populations at harvest than untreated controls. At times, this has been interpreted as indicating the field is naturally suppressing nematode populations in the controls. In reality, the difference is likely due to the chemically treated plots having healthier root systems which are able to support larger populations than can be supported by the roots of the untreated controls.

9.9 Physical Techniques

Physical methods comprise a relatively small but important group of methods that are typically utilized either alone or in combination with cultural and chemical management techniques. Application of heat is widely used method for the control of pests and pathogens and is also the most important of physical methods. Other physical methods like irradiation, plasmolysis or electricity are lethal to nematodes but they are seldom applied and moreover their efficacy is also not well determined.

Each nematode has different maximum and minimum thresholds of temperature for activity, infection and growth and different maximum levels for survival. Nematodes on exposed surfaces are easily reached by heat which can be lethal to them. Hot water treatments like drenching and soakings are conducted on planting stocks to insure that only nematode free materials are planted. But precaution has to be taken here not to harm the living plant material. So to develop satisfactory treatment, a thermal-time-death curves must be determined for all the stages of nematodes present in the infected plant material. The plant parts also tested for tolerance to temperature time combinations that kill the nematodes. Various temperature-time combinations for killing nematodes in living plant tissues have been recommended which varies from 4 hrs at 43.5°C in for treatment of onion bulbs in water with 0.5 per cent formalin to 3 minutes at 53°C for controlling root-knot nematode infested grape seedlings. At times, it is necessary to add a chemical to the hot water treatment bath to either enhance the effectiveness against nematodes or to minimize problems from fungal or bacterial infestations. Dry heat is sometimes used to disinfest soil by providing it at a desired temperature to a continuous flow of soil. Heat generated by decomposition of compost heaps is enough to destroy plant parasitic nematodes. Some control of plant parasitic nematode can also results from the biological agents present in the compost. Steam sterilization is most often used to treat batches of soil to be used in pots or raised beds. There have also been attempts to develop equipment to treat entire fields. Reuven *et al.* (2005) applied steam at 100 °C for 1 h with a negative pressure system. They observed a moderate decrease in root galling on flower crops (carnation), but largely insufficient to limit yield decrease. On the contrary, nematode population decrease was higher in Dutch experiments, where steam at 160 °C was blown in the soil until 25-30 cm (Runia and Greenberger, 2005). These authors also compared steam and hot air application treatment, for which temperature was sublethal and therefore less efficient thnt steam itself. Moreover as steaming indifferently kills micro-organisms (including non-pathogen ones), it is likely to reduce natural biocontrol processes (McSorley *et al.*, 2006). The efficiency of steaming depends on soil preparation. The soil must achieve high porosity to allow deep penetration of the steam (20 cm or more).

Electricity is known to influence the direction of nematodes and may be practical to devise some methods to ward off nematodes from plants or soil. Electrical energy applied at radio wave frequency killed potato cyst nematodes present in the bags. Effects of irradiation have been discouraging but the ultraviolet light of longer wave length filtered through earth atmosphere may be harmful to exposed nematodes.

9.9.1 Other Physical Methods

Removal of infected plants can be utilized in cases where nematodes are not widespread in a field. If infested plants manifest distinctive symptoms early in the growing season, they could be removed to prevent nematodes from not only infesting surrounding plants but also prevent them from reproducing further. Daffodils infested with stem and bulb nematode (*Ditylenchus dipsaci*), for example, have distinctive symptoms on the leaves called spikkles consisting of raised bumps and yellowing. This same nematode infection also results in distinctive symptoms on alfalfa, consisting of shortening of the internodes. An aware grower could remove infested plants from a field which would limit spread to other plants. Similarly infestation by earcockle nematode, *Anguina tritici* show early symptoms of basal stem swelling, prostrate leaves and excessive tillering, such plants can easily be removed. Mechanical seed cleaning can also be resorted to get rid of cockles of *Anguina tritici.*

Sugar, which increases the osmotic concentration of soil has been found to kill nematods within 24 hrs when applied at 1-5 percent of soil by weight but 10-50 tons of sugar per acre would be required. Ultrasonics are ineffective in killing nematodes in soil but may be effective in water.

In many cases, we really don't know the cost of alternatives. How much does it cost to fallow a field for six months or a year? This would include the cost of weed control which would typically need to be done several times during the year, and in some cases, the cost of erosion control. If one is only going to grow a crop every other year, then the crop grown will need to yield the grower more than twice the rate of return of one grown every year utilizing chemical control. If one examines the cost of organic produce in their local supermarket, it is obvious that costs are higher, but not usually twice that of produce grown with the use of chemicals. This would indicate that growers are not receiving twice the value for the growing of organic crops. In many cases, the use of a cultural alternative will have a lower efficacy than the use of a chemical method. The differences would depend on both the cultural method and the chemical method. This means that one would have to utilize more than one cultural method, or perhaps a combination of cultural and physical methods, in order to achieve control comparable to a chemical.

9.10 Chemical Control

There is a tendency to think that current efforts to develop nonchemical techniques are a new direction in nematode management. A historical perspective, however, shows that early nematologists put considerable effort into developing similar practices prior to the development of the first chemical nematicides. In retrospect, the development of chemicals was both a blessing and a curse to the science of nematology. It permitted nematologists to demonstrate dramatic differences in crop growth following a chemical treatment. For example, prior to fumigation, the potato crop was devastated by the golden cyst nematode. The next year after plants had been removed and the area fumigated with D-D, a profitable crop could be grown. This is not the case in U.S.A. or whole of Europe but has been witnessed in Nilgiris hills of Tamil Nadu against golden nematode of potato and also against root knot infecting potato and cereal cyst nematode infecting wheat in India. For the management of plant parasitic nematodes, chemicals mostly fumigants, have extensively been used throughout the world to demonstrate the damage caused by the cyst and rootknot nematodes. Because of the lack of personnel and resources devoted to nonchemical management techniques, now that the tide of public opinion has turned against the utilization of chemicals, nematodes are increasingly being recognized as a difficult pest to control with nonchemical methods. The many chemicals mentioned so glowingly for the control of important plant parasitic nematodes are no longer available for use because of concerns for groundwater contamination, air pollution, and/or potential for carcinogenicity.

A review of the various cultural management techniques that have been developed leads one to ponder why if so many techniques are available, does nematode management rely so heavily on the use of chemicals. One reason is that when growers compare the cost of cultural alternatives to chemicals, the cost of chemical control is often much lower than the cost of alternatives. The comparative economics of nematode management often works against the implementation of nonchemical techniques. Prevention, eradication and protection have been possible with use of nematicides. Soil fumigants made it possible to demonstrate that the elimination of plant parasitic nematodes frequently removed certain symptoms formerly attributed to fungi and bacteria. The nematicidal properties of chloropicrin were discovered in 1919 but it was the work of Godfrey *et al.* (1934) that revealed the great potential of this chemical as nematicide. This led to development of agricultural machinery to make field fumigation feasible. The fumigants, actually move through air in soil pores for variable distances related to their volatility and then dissolve into thin film of water surrounding soil particles which contain nematodes. Metam-sodium is the least volatile fumigant and, on its own, will only move about four inches through the soil. For greatest effectiveness, all the chemicals belonging to different groups must be moved through the soil either in irrigation water or by mechanical tillage. Because of the differences in volatility, one will often hear that the true fumigants are better nematicides than the nonfumigants. In reality, if any of the products listed contact nematicide at the proper concentration, they will result in death. The perceived variability in nematicidal efficacy results from the difficulty in application of the nonfumigant products.

Land preparation is an important aspect for successful use of nematicides and is similar for both fumigants and nonfumigants. Fumigants can also reduce levels of beneficial bacteria. This can slow the rate of decomposition in soils and in some cases result in unexpected phytotoxicity until normal levels are re-established. Nematicides do not typically erradicate nematodes. When used properly, they can reduce nematode popoulations below detectable levels allowing a crop to be planted and a healthy root system to develop before populations increase once again to damaging levels.

9.10.1 Use of Carbamates and Organophosphates

The mode of action of carbamates and organophosphates on nematodes is thought to be similar to the action against insects which is to inhibit acetylcholinesterase at nerve synapses. With respect to nematodes, these products are often called nematistats because at low concentrations, they inhibit nematode activity without killing them, and the effects are reversible if nematodes are placed in water. Nematistat is a term coined by nematologists to describe effects which have been observed when nematodes are exposed to low concentrations of organophosphates and carbamates. These products do not always kill nematodes, but at times appear to disorient, paralyze, or confuse nematodes and so to prevent infestation of roots. If the product is present for a long enough period of time, the nematodes will eventually starve to death. In practical application, the product decomposes in soil over a period of several weeks. This can result in nematodes regaining their abilities to feed on and penetrate roots. In annual cropping situations, the goal is to achieve control for a period of several weeks so that a healthy root system develops to carry the crop through to harvest. In perennial cropping situations, repeated applications can be timed to maximize the nematistatic effect. These compounds are active at dosages of 2 to 10 kg a.i./ha which are smaller than the 200 to 300 litres/ha required for treatment with liquid fumigants. Most of the early formulations of these products were as granules that, when incorporated in the top 10 cm of soil, release the active ingredient, which is spread through the field by rainfall or irrigation.

The organophosphates work by tying up or inhibiting certain important enzymes of the nervous system, namely *cholinesterase* (ChE). The enzyme is said to be phosphorylated when it becomes attached to to the phosphorous moiety of the insecticide, a binding that is irreversible. This inhibition results in the accumulation of acetylcholine (ACh) at the neuromuscular junctions or synapses, causing rapid twitching of voluntary muscles and finally paralysis. The first of these products to be used were the organophosphates thionazin and dichlofenthion, VC-13, which were marketed for nematode control in turf and ornamentals. In the ensuing years, numerous organophosphates like phorate (thimet), Dasanit (fensulfothion) etc. were registered as nematicides.

Carbamates inhibit cholinesterase (ChE) as organophosphates do, and they behave in almost identical manner in biological systems, but with two main differences. Some carbamates are potent inhibitors of aliesterase (miscellaneous aliphatic esterase whose exact functions are not known), and their selectivity is sometimes more pronounced against the ChE of different species. Second, ChE inhibition by carbmates is reversible.Carbofuran (Furadane) and Aldicarb (Temik) are the two carbamates generally in use throughout the world.

Biorational approaches to nematode control have received only limited attention. Recently, a protein from *Erwinia amylovora* (harpin) with the reported ability to elicit a resistant response in certain plant species (Wei and Beer, 1996) has been suggested as a means to enhance crop growth, development, and pest resistance. Studies in cotton indicate that reproduction of *Meloidogyne incognita* may be slightly lower in harpin-treated plants, but yield responses have been disappointing (Bednarz *et al.*, 2002). A second interesting biorational approach involving genetic engineering of the cotton plant has been suggested (Atkinson *et al.*, 2003). Aldicarb in the soil solution at rates that are usually applied results in cholinesterase inhibition leading to paralysis and ultimately nematode mortality (Spurr, 1982). However, at much lower concentrations, aldicarb acts as a chemoreception inhibitor in nematodes (Winter *et al.*, 2002). Certain peptides have been shown to elicit this same response, and development of plants that produce these proteins in roots has been suggested as a way to suppress nematode infection.

A material with nematicidal properties, sodium methyl dithiocarbamate (metam sodium), was reported in 1956 (Lear, 1956). Metam sodium (Vapam®) releases methyl isothiocyanate as it degrades, a chemical that is active against nematodes as well as certain soil borne fungal pathogens and weeds.

Massive chemical control attempt was also made under the Indo-German Nilgiris Development Project during 1971-75 to check potato cyst nematode. The treatment was made mandatory under the Tamil Nadu Pest Act 1971 and all the infested fields at that time in Nilgiris were treated with 30 kg ai/ha of Fensulfothion (Dasanit 10 per cent G) in the first year, followed by 15 kg ai/ha in the subsequent years. Nearly 1000 tons of 10 per cent Fensulfothion was used treating about 3100 hectares of annual cropped area, over a period of five years (Seshadri, 1978). Thus, each hectare, on average, received 325 kgs of the pesticide. At present, application of Carbofuran at 2 kg ai/ha is recommended to keep nematodes at below the economic threshold levels of less than 10 eggs or larvae per 100 ml of soil at the time of planting potato (Krishna Prasad, 2000). This will help reduce the initial nematode inoculum in soil and also reduce build-up of pathotypes that multipliy on the resistant cv. 'Kufri Swarna' (Krishna Prasad, 2007).

9.10.2 Fumigants Against Nematodes

The fumigants are small, volatile, organic molecules that become gases at temperatures above 40°F. They are usually heavier than air and commonly contain one or more of the halogens (Cl, Br, or F). Most of them are highly penetrating, reaching through large masses of material. They are used to kill insects, insect eggs, nematodes, and certain microorganisms in buildings, warehouses, grain elevators, soils, and greenhouses and in packaged products such as dried fruits, beans, grain, and breakfast cereals. Methyl bromide is the most heavily used of the fumigants but now it is being phased out. Fumigants, as a group, are narcotics, that is, they are more physical than physiological. The fumigants are liposoluble (fat soluble); they have common symptomology; their effects are reversible; and their activity is altered very little by structural changes in their molecules. As narcotics, they induce narcosis, sleep, or unconsciousness. Liposolubility appears to be an important factor in their action, since these narcotics lodge in lipid-containing tissues including their nervous system.

In 1941, a practical method for using another volatile biocide (methyl bromide) was reported (Taylor and McBeth, 1941). Methyl bromide was extremely effective across a range of soil borne organisms including nematodes, fungal pathogens, and many weeds, but it was highly toxic and rather expensive to use, limiting its use to only the highest value crops. Within the next few years, however, two new chemicals that were less expensive and much easier to apply were discovered. The first of these was a product obtained from the Shell Chemical Corporation known as D-D (a mixture of 1, 3-dichloropropene and 1,2-dichloropropane) (Carter, 1943). The second was a related material, ethylene dibromide (EDB) that was marketed by both Shell and Dow Chemical Companies (Christie, 1945). Both materials were much less volatile than methyl bromide and were highly effective against nematodes. In addition to providing an effective means of managing nematode damage to various crops, these materials were perhaps most important in demonstrating to the general public the role of nematodes as plant pathogens (Johnson and Feldmesser, 1987).

In 1954, 1, 2-dibromo-3-chloropropane (DBCP) was reported as having nematicidal properties (McBeth, 1954; Raski, 1954). This material provided a considerable improvement over previously used nematicides because it was effective at relatively low application rates (making it less expensive to use), and was much less phytotoxic and less volatile than either D-D or EDB. DBCP launched a new era of nematicide use because it could be applied either before or after planting on an array of crops. The introduction of DBCP into the market place as Nemagon® (Shell Chemical Co.) or Fumazone® (Dow Chemical Co.) solidified the concept of nematicide application as a primary strategy in nematode management in agronomic crops including cotton. D-D at the rate of 300l/ha and DBCP at 45l/ha appeared to be the optimum dose for effective reduction in nematode population as well as yield response in wheat and barley (Swarup *et al.*, 1976). Unfortunately, environmental and human health concerns resulted in its elimination from the marketplace in the mid-1970s.

Initially, DD, EDB, MBr, DBCP, telone (a mixture of DD and EDB) were used for controlling root knot and potato cyst nematodes. Qiao *et al.* (2011) indicated that methyl bromide was generally superior to the treatments involving 1, 3-D and avermectin, which in turn were superior to the control, for improving cucumber yield and to control root knot nematode. Experience has shown that neither root-knot nor potato cyst nematodes can be eradicated once these get established in a locality (Sethi and Gaur, 1986; Hamid and Alam, 1998). Their similarities in population dynamics, biology and life cycle facilitate integrated management of root knot and potato cyst nematodes in potato. Hence, these nematodes need to be managed by combining several plant-protection strategies (Kamra and Dhawan, 1998). Root knot nematodes in the north Indian hills are being managed by application of Carbofuran 3G at 1 or 2 kg ai/ha at potato planting (Deshraj, 1983); two-year crop rotation with wheat or maize and seed certification after inspection for root knots on seed tubers (Krishna Prasad, 1986, 1993). This has helped keep juveniles of root-knot nematodes at less than the threshold level of 10 to 20 larvae /100 g of soil sample. Carbofuran is equally effective against cyst nematodes as Kaushal and Seshadri (1986) has reported that the peneteration of juveniles of *H. avenae* was adversely affected by carbofuran @ 2 and 4 kg ai /ha causing marked reduction in cyst formation in wheat.

9.10.3 Hazards of Nematicidal Use

Once we make use of chemicals for the management of nematodes, many associated problems crop up during the course of implementing these methods. The main problems are of public health and safety of workers, environmental pollution, groundwater contamination, damage to nontarget organisms and the problem of chemical residues in the food. This summarizes potential problems from the use or misuse of nematicides. Nematicides which have been found in groundwater in various situations include 1, 2-dichloropropane (one of the components of D-D mixture), Vydate (oxamyl), Temik (aldicarb), and DBCP (dibromochloropropane). Telone II (1, 3-dichloropropene) has been detected at parts per billion in several parts of the world which resulted in suspension of use for several years. Because of acute toxicity, the organophosphates and carbamates are considered to be more likely to cause harm to nontarget organisms and endangered species than are the fumigant nematicides. The carbamates and organophosphates also have greater potential for absorption by roots and thus to leave residues in food than do the fumigants. On the other hand, several of the fumigant nematicides have been implicated as potential carcinogens. (DBCP).

Fumigants are designed to volatilize and thus may dissipate into the air where they can be decomposed by sunlight. Within soil, fumigants are also decomposed by hydrolysis. In soil, organophosphates and carbamates are typically decomposed by microbial activity. The zone of greatest microbial activity surrounds roots and decreases with increasing depths in soil. To be used successfully, nematicides must be able to dissolve in the water lining soil pores in order to contact nematodes. This same water solubility, though, can result in their leaching below the depth of microbial activity. Once past this zone, degradation proceeds very slowly, if at all, resulting in the potential for groundwater contamination.

There have been a number of reports of "loss of activity" of nematicides following repeated application. Nematicides are more difficult to move successfully through fine textured soils (high percent silt or clay) and soils with high levels of organic matter and reports of control failures are more frequent in these types of soils than in sandy soils with low amounts of organic matter. Unfortunately, those conditions which contribute to the success of a nematicide application also contribute to the potential for leaching to groundwater. Greater caution needs to be exercised when utilizing nematicides in areas with shallow groundwater, high rainfall, or frequent irrigation as these factors increase the potential for groundwater contamination. Application procedures for some products and crops call for delaying irrigation following application in order to maintain the product within the root zone. Low soil temperatures delay the rate of nematicide decomposition by microbial organisms and allow more time for leaching to occur.

Long-term exposures of plant-parasitic nematodes, *Xiphinelna index*, *Meloidogyne incognita,* and *Pratylenchus vulnus* to subnematicidal stresses of carbofuran, oxamyl and phenamiphos can alter their responses to subsequent nematicidal treatments. Responses vary not only with the different nematicide used but with the different nematode species. Subnematicidal stressing with carbofuran and phenamiphos reduced reproductive potential in *X. index,* while phenamiphos stressing reduced reproductive potential in *M. incognita.* However, carbofuran stressing appeared to increase, reproductive potential in *P. vulnus.* Cross-susceptibility was observed in carbofuran and oxamyl stressing, which made *X. index* more susceptible to phenamiphos, while carbofuran stressing increased the susceptibility of *P. vulnus* to oxamyl. Long-term subnematicidal stressing of all three nematodes with the three nematicides appeared to induce resistance and cross-resistance phenomena in the form of indifference and increased populations in response to nematicidal treatments (Yamashita *et al.*, 1986).

9.11 Integrated Management

The most successful approaches to nematode control rely on integrated pest management strategies (IPM). IPM combines management options to maintain nematode densities below economic threshold levels. IPM techniques can still be difficult to implement against a pathogen as aggressive and resilient as root-knot nematodes. Nevertheless, a combination of management tactics/tools, including cultural practices (rotations with nonhost crops and cover crops that favor the build-up of nematode antagonists), resistant cultivars, and chemical soil treatments, if necessary, generally provide acceptable control of root-knot nematodes. The extent of this success, however, is dependent upon having accurate damage threshold densities and available and readily acceptable resistant cultivars.

Integration of different management practices is considered as better option to an individual approach. Integration of soil solarisation (for six weeks), Vasicular arbuscular mycorrhiza fungus (VAM), *Glomus fasciculatum* inoculation and seed treatment with carbosulfan (@ 3 per cent w/w) is highly effective in reducing population of *M incognita* and *Fusarium oxysporum* and significantly increasing chickpea grain yield (Rao and Krishnappa, 1995). Seed treatment with carbendazim (0.25 per cent w/w) together with carbosulfan (3 per cent w/w) is effective in reducing the *Fusarium-Meloidogyne* wilt complex and increasing the yields. Gamliel and Stapleton (1993) and Oka *et al.* (2007) found that the combination of solarization and organic amendment reduced nematode populations and galling indices in conditions that were not effectively controlled by solarization or organic amendment alone. Another approach is to organize the actions of different techniques by combining short and long-term effects (Roberts, 1993). The short term aim should be to limit plant infestation by temporarily reducing nematode numbers in the soil before planting and reducing their infectivity before growing with resistant or tolerant cultivars. The long term aim should be to reduce the multiplication rate of nematodes on each crop, even partially; this will have beneficial effects on the succeeding crops, and therefore, in the long-term.

9.12 The New Alternatives for Nematode Management

An alternative method to engineer broad spectrum resistance which is not only active against sedentary nematodes but also to many damaging species of migratory nematodes employs effective antinematode genes. The key opportunities expressing anti-nematode genes occur during (a) invasion, (b) feeding site initiation, (c) early feeding of the parasite, (d) feeding and digestion of pre-adults, (e) feeding and digestion of the adult female, and (f) its reproductive development offer opportunities for engineering resistance. Many of these opportunities are common to endoparasitic nematodes irrespective of their precise mode of feeding. Specific expression of genes that affect plant cells may hinder invasion (i) prevents feeding site initiation, (ii) or full development of the feeding site into a plant transfer system (iii). Other nematodes that modify plant cells offer similar opportunities for engineered resistance. Anti-nematode gene is defined as one that produces a peptide/protein that is toxic, damaging or inhibitory to nematodes but is not deleterious or significantly less so to the host plant or to the animals/humans that will eventually feed on the plant or its products. Several classes of potential anti-nematode genes encoding lectins, enzymes and enzyme inhibitors are being evaluated for their ability to confer broad-spectrum nematode resistance.

Antibodies produced in plants (plantibodies) are being investigated as a means of conferring resistance to pests and pathogens. Antibodies against nematode esophageal gland secretions may be utilized to disrupt either feeding site formation or feeding behavior (Hussey, 1989) via expression of functional antibodies in plants (Hiatt *et al.*, 1989). Plantibodies that selectively neutralize such enzymes could disrupt this aspect of host parasite relationship. The monoclonal antibody specific to stylet secretions of *M. incognita* were expressed in tobacco (Baum *et al.*, 1996). Unfortunately, despite the antibodies binding to the nematode pharyngeal glands and stylet secretions, no reduction in the ability of the nematode to parasitize the tobacco roots was seen. Although it is not entirely clear why the plantibodies failed to disrupt giant cell formation by inhibiting stylet secretion action, one feasible explanation is that the plantibodies accumulated apoplastically, whilst nematode stylet secretion was delivered to the cytoplasm. Plantibodies have also been generated that recognize secretory-excretory proteins on the cuticle surface and the amphids of *Globodera pallida* and *G. rostochiensis*. The plantibodies were shown to affect nematode movement and resulted in delayed root penetration. However, these effects were temporary as turnover of the secreted proteins resulted in loss of the bound antibody (Fioretti *et al.*, 2002 and Sharon *et al*,. 2002).

9.12.1 Targeting the Nematode Feeding Site

In the case of endoparasitic nematodes, such as species of root-knot and cyst nematodes, an alternative strategy to targeting the nematode directly is to disrupt the feeding site that these nematodes initiate within the plant. The aim of this approach is to destroy the nutrient supply by expressing nematicidal gene product resulting in feeding cell death or site metabolic attenuation. Once the larvae have set up a feeding site it becomes sessile and its continued existence and development are entirely dependent on the feeding cells. Therefore, killing the feeding cell would result in the death of the nematode and thus confer resistance in the host plant. This could be achieved theoretically by expressing a cell death gene specifically within the feeding site upon its induction. Primarily, this approach would require a promoter that would drive the cell death system specifically in the nematode feeding cell; the value of such a promoter is self-evident. Disruption of feeding structures can be achieved by (1) tissue specific expression of a toxic protein and (2) down regulation of a component necessary for development or maintenance of feeding cells by anti-sense inhibition or co-suppression.

9.12.2 Protease Inhibitors Against Nematodes

A number of biotechnological strategies have been described to perturb the nematode plant interaction. The best characterized approach is that which delivers proteinase inhibitors from the plant to disrupt nematode feeding. Protein inhibitors of a range of proteinase classes are widely expressed in plants where they are often induced by wounding and herbivory. The potential of plant PIs as anti-nematode effectors was first explored using the serine PI cowpea trypsin inhibitor (CpTI). CpTI expressed in transgenic potato influenced the sexual fate of newly established *G. pallida* (Hepher and Atkinson, 1992) and, as a result, the population was biased toward a predominance of the much smaller and less damaging males. No reduction in the fecundity of established females of *G. pallida* occurred, although CpTI did reduce the fecundity of females of *M. incognita* without influencing their sexual fate (Hepher and Atkinson, 1992; Urwin *et al.*, 1998). Transgenic expression of the sweet potato serine PI, sporamin, inhibited growth and development of female *H. schachtii* (beet cyst nematodes) parasitizing sugar-beet hairy roots (Cai *et al.*, 2003). In this case, the severity of the effect was clearly correlated to the level of trypsin-inhibitory activity detected in the transformed root lines. Inhibitory activity of a potato serine PI (PIN2) expressed in transgenic wheat also showed a positive correlation with plant growth and yield following infestation with the cereal cyst nematode *Heterodera avenae* (Vishnudasan *et al.*, 2005).

The potential of cysteine proteinase inhibitors (cystatins) to provide nematode resistance has been explored in more depth. Initially the efficacy of a rice cystatin (Oc-I) was enhanced by protein engineering, using crystallographic data, to produce a PI with greater inhibitory activity. Expression of the engineered variant in root cultures conferred higher amounts of resistance against the potato cyst nematode *G. pallida* than the unaltered molecule (Urwin *et al.*, 1995). In the first demonstration of a transgenic technology working against both major groups of economically important nematodes (root-knot and cyst), the modified cystatin was expressed in transgenic *Arabidopsis* plants and uptake of the cystatin was correlated with loss of nematode cysteine proteinase activity (Urwin *et al.*, 1997). No females of either *H. schachtii* or *M. incognita* developed to egg production during growth on the cystatin-expressing plants. The same plants also showed resistance against the reniform nematode *R. reniformis*, with higher levels of PI expression again correlated with reduced reproductive success. The number of female nematodes was reduced by 35 per cent and the fecundity of each by 69 per cent, corresponding to 81 per cent resistance in terms of reduced overall reproductive success, a common method of expressing nematode resistance (Urwin *et al.*, 2000).

The environmental biosafety of the approach has been addressed in a range of studies. Cystatin-expressing potato plants do not harm nontarget insect aerial feeders (Cowgill *et al.*, 2002, 2004), their natural parasitoid enemies (Cowgill and Atkinson, 2003) or the soil microbial community (Cowgill *et al.*, 2002). Gene flow has been shown to occur to wild relatives growing near potato crops in the main centre of biodiversity for potato, the central Andes. The use of the male sterile cultivar, engineered to provide nematode resistance, would allow the benefits of biotechnology by eliminating the possibility of this route of gene flow (Celis *et al.*, 2004). A *prima facie* case has been established for the food safety of bioengineered cystatin potato plants. The engineered rice cystatin is not toxic and is rapidly digested in simulated gastric fluid, providing a margin of exposure in excess of 2000 fold (Atkinson *et al.*, 2004). A number of phytocystatin genes have been isolated from various plant species, including potato (Turra *et al.*, 2009) and taro (Yang and Yeh, 2005), which have been shown to be involved in plant defense systems against biotic or abiotic stresses.

Marban-Mendoza*et al.* (1987) provided the first indication that lectins may have anti-nematode activity when application of concanavalin A lectin (Con A) to the roots of tomato plants reduced the amount of galling inflicted by *M. incognita* by 75 per cent compared with control plants. Constitutive expression of a snowdrop lectin (GNA) directed by the CaMV35S promoter has been used to assess anti-nematode activity in oilseed rape, potato and *Arabidopsis* (Burrows *et al.*, 1997, 1998; Ripoll *et al.*, 2003). Challenge of GNA-expressing potato plants with *G. pallida*and the migratory nematode *Pratylenchulus bolivianus* produced a significant reduction in the number of mature females recovered from roots when compared with control plants, with greater amounts of resistance obtained against *P. bolivianus* (Burrows *et al.*,1998). Assays performed with *Arabidopsis* plants in tissue culture also showed that GNA confers a protective effect against *M. incognita* (Ripoll *et al.*, 2003). The number of galls formed in the roots of lectin-expressing plants was reduced by up to 50 per cent compared with wild-type, although the fitness of progeny hatched from females reared on the transgenic plants was not compromised. The lectin may exert its effect either through interaction with glycoproteins present in the intestine or through binding to glycoproteins present externally on the nematode. Interaction with glycoproteins associated with chemoreception, on the amphid sensory organs themselves or in their secretions, could disrupt sensory perception and compromise the nematode's ability to establish feeding sites.

9.12.3 Disruption of Chemoreception

Location and invasion of host roots by plant-parasitic nematodes requires input from chemoreceptive neurons, responding to distinct chemical cues. Phage display has led to the isolation of two different peptides that either bind and inhibit acetylcholinesterase or bind to nematode nicotinic acetylcholine receptors, two targets in the cholinergic nervous system of nematodes. Both peptides inhibited cyst nematode chemoreception, blocking their response to an attractant, at concentrations from 1 nm and are likely to undergo retrograde transport along certain chemoreceptive neurons in order to exert their effect (Winter *et al.*, 2002). The strategic development of this approach provided a prototype potato plant expressing a secreted form of the peptide that inhibits acetylcholinesterase. Its presence in potato roots and the near rhizosphere disorientated the invading cyst-nematode *G. pallida*, causing a 52 per cent reduction in the number of female nematodes that developed in the roots. This effect was achieved with ultra-low conentrations of the peptide in soil water that were equivalent to 1 per cent of the acetylcholinesterase inhibition level allowed in USA drinking water (Liu *et al.*, 2005).

9.13 RNA Interference for Nematode Management

Developments in the application of RNA interference (RNAi) for gene silencing in plant-parasitic nematodes have recently culminated in the demonstration that plants expressing double-stranded RNA (dsRNA) targeting a nematode gene display resistance to infection (Gheysen and Vanholme, 2007; Lilley *et al.*, 2007). RNA interference (RNAi) represents a major breakthrough in the application of functional genomics for plant-parasitic nematode control. RNAi-induced suppression of numerous genes essential for nematode development, reproduction or parasitism has been demonstrated, highlighting the considerable potential for using this strategy to control damaging nematode pest populations (Li *et al.*, 2011). Gene silencing triggered by dsRNA was first demonstrated for *C. elegans* (Fire *et al.*,1998) and the underlying mechanism of RNAi has subsequently been studied in depth for this free-living nematode. The impact of this work was recognized in 2006 by the award of the Nobel Prize in Physiology or Medicine to Andrew Fire and Craig Mello. A similar phenomenon had previously been described for plants as post-transcriptional gene silencing (Jorgensen *et al.*, 1996; Waterhouse *et al.*, 1998). The detailed molecular mechanisms and various component proteins involved in RNAi are still being elucidated for *C. elegans* (Pak and Fire, 2007; Sijen *et al.*, 2007; Winston *et al.*, 2007). Plants can be designed to produce dsRNAs which can then silence specific essential genes contributing to development and growth or parasitism of the nematode. On feeding these engineered plants, the nematode would ingest these dsRNA, or its siRNAs, from the plant cytoplasm, and once inside the nematode the RNAi process would inactivate the gene targeted by the dsRNA and stop further development or parasitism of the nematodes. The utility of RNAi for functional analysis of plant-parasitic nematode genes was first reported in 2002. Genes encoding a cysteine proteinase, C-type lectin and major sperm protein of cyst nematodes were targeted by soaking infective J2s in a solution of dsRNA with the addition of the neurochemical octopamine to stimulate ingestion (Urwin *et al.*, 2002). Careful target selection plays a key role in the biotechnological application of RNAi towards the control of plant parasitic nematodes. Two kinds of genes were considered as potential candidate targets, the one kind is essential and conserved genes, of which *C. elegans* orthologs has lethal or highly damaging RNAi effect (Bakhetia *et al.*, 2005). The other kind of targets were parasitism genes, which are expressed in the esophageal gland cells of plant parasitic nematodes, encode proteins that are secreted into host root. These proteins include cell-wall-modifying enzymes, multiple regulators of host cell cycle and metabolism, suppressors of host defense and mimics of plant molecules (Niu *et al.*, 2010). A number of genes, expressed in a range of different tissues and cell types, have now been successfully targeted for silencing in both cyst and root-knot nematodes. More recently, nematode resistance has been demonstrated for plants expressing hairpin constructs targeting plant-parasitic nematode genes (Huang *et al.*, 2006; Steeves *et al.*, 2006; Yadav *et al.*, 2006). The first study demonstrated silencing by RNAi delivered from host tobacco plants of *Meloidogyne* genes encoding a splicing factor and a component of a chromatin remodelling complex (Yadav *et al.*, 2006). The target sequences were selected for their presumed essential role and homology to *C. elegans* genes with lethal knockout phenotypes and robust RNAi data. Plants expressing hairpin constructs for each of the two sequences displayed > 95 per cent resistance to *M. incognita*. The few nematodes that formed galls appeared developmentally compromised and lacked detectable transcript for the targeted genes (Yadav *et al.*, 2006). Soybean roots expressing inverted repeat constructs of three genes Hg-rps-3a, Hg-rps-4 and Hgspk-1 involved in different aspects of mRNA metabolism displayed 81 per cent –88 per cent reductions in numbers of *H. glycines* cysts (Klink *et al.*, 2009). RNAi of the *H. glycines* Prp-17 gene which encodes a mRNA splicing factor resulted in significant reductions in cysts D g root tissue (53 per cent reduction) and eggs D g root tissue (79 per cent reduction) in the transgenic soybean plants inoculated with *H. glycines,* which supported the important role of Prp-17 gene in *C. elegans* (Li *et al.,* 2010).

9.14 Quarantine Procedures

Perhaps the most important principles for preventing nematode problems are-quarantine *i.e.* keeping the nematodes at large from the unaffected areas, use of certified planting material free from nematode infection,check suspect material of any infection before planting, nematodes may be present in manure which is generally not suspected to be contaminated by the nematodes, various equipments to be used in the field should free from nematode infection and one should avoid using irrigation water passing through infected fields to the non-infected fields. The rapid growth in world population and expansion of global trade are making sustainable crop and pest management systems and related pest quarantines increasingly important. Although nematode reproduction and survival are dependent upon a suitable soil type, presence of a host plant, temperature, and soil moisture (Nelson and Boag,1996 and Trudgill,1991), increased shipment of equipment and agricultural products will likely result in new infestations of diverse nematode species in many regions of the world.

The chronic nature of nematode problems should indicate that measures to minimize the spread of nematodes should be a high priority. Unfortunately, this is often not the case. There is a general lack of awareness of nematodes among the general public because of their hidden nature. This can easily result in the unintentional dispatch of nematodes to many unaffected areas. Growers who do not have nematode problems are often unaware of their existence. This can result in unwarranted movement of nematodes in soil, irrigation water, on equipment and on planting stock. On farms, nurseries, and in government departments, there is a continuous transfer of personnels and consequently there is a continuing need for education about the importance of prevention in the management of nematodes.

A successful and well run quarantine program will result in greater precautions on the part of those shipping plants into the state because of the financial losses that will occur if infested plants are found. Nematode sampling and processing is time consuming and therefore expensive but any slackening of vigilence can relatively quickly result in the infestation of new areas. For example, one acre of infested nursery stock can result in the infestation of more than 100 acres of farm land. Successful attempts to erradicate nematodes are rare, or perhaps even nonexistent. It is only due to the strict quarantine measures that India has been able to restrict potato cyst nematode in few localities in its southern parts although there are few reports of it being present in Shimla hills of Himachal Pradesh (Ganguly, 2011).

Planting material especially seed is the most prevalent form of exchanged or traded agricultural commodity and seed-transmitted pests or soil as contaminant of seeds and packing material are the major source of transmission of various pests and diseases including nematodes to different parts of the world. In India, the examination of samples (of non-Indian origin) revealed the presence of plant nematodes of quarantine importance, *viz., Anguina tritici, Aphelenchoides besseyi, Aphelenchoides arachidis, Ditylenchus angustus, D.destructor, D.dipsaci, Heterodera schachtii, Pratylenchus crenatus, P.penetrans* and *Rhadinaphelenchus cocophilus.* All of these species except

Anguina tritici, *Aphelenchoides besseyi* and *P. penetrans* are not reported to occur in India.

It is expected with the unprecedented movement of plant-derived commodities around the globe, climate change, the continuing growth of the human population and consequent changes in land use and agricultural practices, and new threats to agricultural production from either undescribed or recently distributed species of plant-parasitic nematodes will continue to occur and be an ongoing concern. There is a list of the top 15 plant-parasitic nematode species regulated by twenty or more countries in international quarantine legislation in 2000. *Globodera rostochiensis* is at the top of the list with 106 countries, and despite its status as a quarantine pest new incidents continue to be found (See 9.15). There is a need for standardized diagnostic protocols for plant protection agencies that are rapid and accurate to limit the movement of regulated species. Diagnostic protocols are being developed by the European and Mediterranean Plant Protection Organisation with quarantine status in the European Union and these protocols provide a resource for the international community of diagnostics that are standardized and comparable between laboratories.

9.15 Top 15 Regulated Nematode Species

Nematode Species	*No. of Countries Regulating in 2000*
Globodera rostochoensis	106
Aphelenchoides besseyi	70
Ditylenchus dipsaci	58
Radolpholus similis	55
Globodera pallida	55
Ditylenchus destructor	53
Heterodera glycines	56
Aphelenchoides fragariae	47
Bursaphelenchus xylophilus	46
Xiphenema index	42
Nacobbus aberrans	38
Xiphinema americanum	30
Anguina tritici	24
Heterodera schachtii	22
Bursaphelenchus cocophilus	21

(http://nematode.unl.edu/regnemas.htm)

(Lehman, 2002)

Analyses done by many workers have found that most techniques listed above had a partial effect on nematode control. Currently, these alternative techniques are difficult to promote in intensive agriculture, because farmers can and do compare it to chemical efficiency. Moreover, there is great variability in efficiency among studies, due to several factors. Those that depend on practical modalities (*e.g.*, organic amendment rates, maximum temperatures achieved in solarization, etc.) have been highlighted above. Others include soil and climate conditions. For example, clay soils offer poor conditions for the development of nematodes (Mateille *et al.*, 1995; Barker and Koenning, 1998). Consequently, the effect of any particular technique probably depends on the ratio between clay and sand in the soil, which is not always reported. Understanding the variability among studies is difficult, because, several techniques may contribute to the same process, and a single technique may contribute to several different processes. As an example of the first case, killing nematodes can be achieved by different techniques, alone or concurrently solarization or steaming (thermal effects), biocontrol or nematicidal products (nematicidal oil-cakes, chemicals, natural or not). Conversely, green manure affects multiple processes; it may have a biocidal effect once buried, it may break the biological cycle of the nematodes if a non-host or resistant species is chosen and it enhances the competition by providing new organisms and feeding those present in soil. However, green manure may also have a negative effect on nematode levels by enabling root-galling during cropping. For example, sorghum, generally considered as non-host, can increase root-gall occurrences in heavily infested soils. Therefore, the resultant effects will depend on the balance between the intensity of these contradictory processes. It can be noted that while many techniques contribute to killing nematodes, fewer are available for alternative ways of control.

Combining techniques does not necessarily lead to synergistic effects, and complex interactions can occur. Therefore, it may be appropriate to rethink the whole system instead of trying to control it with a single action. Along these lines, we advocate systemic agronomic research (Lewis *et al.*, 1997), which aims to rebuild cropping systems as a whole and formulate cropping systems that naturally limit the increase of pathogens, rather than sticking to the therapeutic paradigm. Identifying the most promising combinations is the key. To date, few operational propositions have been made that efficiently control nematodes in vegetable production. Most rely on advisory services and local experimentation. For example, Melton (1995) in North Carolina, built an efficient cropping system based on host resistance, crop rotation and residual root destruction immediately after harvest, and cover cropping. Some variability in the effects of different techniques (*e.g.* organic amendment) may arise from differences in micro-organism competition. Thus, despite quite good efficiency under controlled conditions, biocontrol may not operate in the field. The added micro-organisms may not survive in field soil, they may be unable to adequately reproduce, or the competition may be too strong. In light of the world population increase and given that current means of control through pesticides pose serious environmental risks, a holistic approach is required to design different management systems according to the goals of production, nematode management and other services, like environmental preservation. Outlined in this proposal are examples of how genomics approaches have been and are being used to identify novel gene targets in the nematode and how these are used to identify nematode-specific chemistries that inhibit nematode growth and development.

References

Anon. (1998). Plasma Neem Cake. Plasma Power Web site.

Abbasi, P. A., Riga, E., Conn, K. L. and G. Lazarovits (2005). Effect of neem cake soil amendment on reduction of damping-off severity and population densities of plant-parasitic nematodes and soil borne plant pathogens. Can. J. Plant Path. 27: 38 - 45

Abawi, G.S. and Barker, K.R. (1984). Effects of cultivar, soil temperature, and population levels of *Meloidogyne incognita* on root necrosis and *Fusarium* wilt of tomatoes. Phytopathology 74: 433-8.

Abawi, G.S. and Widmer, T.L. (2000). Impact of soil health management practices on soilborne pathogens, nematodes and root diseases of vegetable crops. Appl. Soil Ecol. 15: 37-47.

Abebe, E., Tesfamariam, M., and William, K. Thomas (2011). A critique of current methods in nematode taxonomy. African Journal of Biotechnology 10 (3) : 312-323,

Abdel-Rahman, F.and Maggenti, A.R. (1987). *Meloidogyne californiensis* n. sp. (Nemata: Meloidogyninae), parasitic on bulrush, *Scirpus robustus* Pursh. J. Nematol. 19: 207-217.

Abrantes, LM.de O. and Santos, S.N. de A. (1991) *Meloidogyne lusitanica* n. sp. (Nematoda: Meloidogynidae), a root-knot nematode parasitizing olive tree (*Olea europaea* L.). J. Nematol. 23: 210-224.

Abdel-Momen, and S.M., Starr, J.L. (1998). *Meloidogyne javanica-Rhizoctonia solani* disease complex of peanut. Fundamentals of Applied Nematology 21: 611–16.

Abdul Hamid, W. and Alam, M.M. (1998). Strategies for the control of plant parasitic nematodes with special reference to IPM. pp. 216-225. **In:** Recent advances in plant nematology. Ed. P. C. Trivedi, CBS publishers and distributors, New Delhi - 110002, pp. 277.

Aboul-Eid, H.Z. and Ghorab, A.I. (1974). Pathological effects of *Heterodera cajani* on cow pea. P1. Dis. Reptr. 58: 1130.

Aboul-Eid, H.Z. and Ghorab, A.I. (1981). The occurrence of *Heterodera zeae* in maize fields in Egypt. Egyptian J. Phytopath. 13: 51-61.

Abuzar, Syed and Akhtar Haseeb (2009). Bio-Management of Plant-Parasitic Nematodes in Pigeon Pea Field Crop Using Neem-Based Products and Manurial Treatments. World Applied Sciences Journal 7: 881-884.

Adams, P.J.M and Tyler, S. (1980). Hopping locomotion in a nematode: functional anatomy of the caudal gland apparatus of *Theristus caudasaliens* sp. n. Journal of Morphology 164: 265-285.

Adamson, M. (1989). Constraints in the evolution of life histories in zooparasitic nematoda. *In: Current Concepts in Parasitology* (R C. Ko, ed.), pp. 221-253. Hong Kong University Press, Hong Kong.

Affokpon, A., D. L. Coyne, C. C. Htay, R. D. Agbèdè, L. Lawouin and J. Coosemans. (2011). Biocontrol potential of native *Trichoderma* isolates against root-knot nematodes in West African vegetable production systems. Soil Biology and Biochemistry 43:600-608

Agatha, S. and Strüder-Kypke, M.C. (2007). Phylogeny of the order Choreotrichida (Ciliophora, Spirotricha, *Oligotrichea*) as inferred from morphology, ultrastructure, ontogenesis, and SSrRNA gene sequences. Eur. J. Parasitol. 43: 37-63.

Agbenin, N.O. (2011): Biological control of plant parasitic nematodes: prospects and challenges for the poor Africa farmer. Plant Protect. Sci., 47: 62–67.

Agbenin, N.O., Emechebe, A.M., Marley, P.S. and Akpa, A.D. (2005): Evaluation of nematicidal action of some botanicals on *Meloidogyne incognita in vivo* and *in vitro*. Journal of Agriculture and Rural Development in the Tropics and Subtropics, 106: 29–39.

Akhtar, M. and Mahmood, I. (1997). Impact of organic and inorganic management and plant based products on plant paraditic and microbivorous nematode communities. Nematol. Medit. 25: 21-23.

Akhtar, M. and A. Malik, (2000). Roles of organic soil amendments and soil organisms in the biological control of plant-parasitic nematodes: A review. Bioresour. Technol., 74: 35-47.

Alavez,S., Maithili C. Vantipalli, David J. S. Zucker, Ida M. Klang and Gordon J. Lithgow (2011). Amyloid-binding compounds maintain protein homeostasis during ageing and extend lifespan. Nature, 2011; DOI:10.1038/nature09873

Al-Banna, L.,Williamson,V.M. and Gardner,S. L. (1997). Phylogenetic analysis of nematodes of the genus *Pratylenchus* using nuclear 26s rDNA. Mol. Phylogen Evol. 7: 94-102.

Al Banna, L., and Gardner, S. L. (1996). Nematode diversity of native species of *Vitis* in California. Can. J. Zool. 74: 971-982.

Aldrovandus, U. (1623). De animalibus insectis libri septum. fol. Franeofurti.

Allen, M.W., Hart, W. H. and Boghott, K.V. (1970). Crop rotation controls barley root knot nematode at Tulelake. California Agriculture, 24:4-5.

Amin, A.W. (1993). A new local race of the root-knot nematode *Meloidogyne thamesi* Chitwood in Chitwood, Specht and Havis, 1952 in Hungary. Opuscula Zoologica (Budapest) 26: 3-8.

Anderson, R C. (1992). Nematode Parasites of Vertebrates: Their Development and Transmission. CAB International, Wallingford, UK.

Andersen, S., and Andersen, K. (1982). Suggestions for determination and terminology of pathotypes and genes for resistance in cyst-forming nematodes, especially *Heterodera avenae.* EPPO (Eur. Mediterr. Plant Prot. Organ.) Bull. 12: 379-386.

Anderson, R.V., Coleman, D.C., Cole, C.V.,and Elliott, E.T. (1981). Effect of nematodes *Acrobeloides* sp. and *Mesodiplogaster lheritieri* on substrate utilization and nitrogen and phosphorus mineralization in soil. Ecology 62: 549–555.

Andrássy, I. (1976). Evolution as a basis for the systematization of nematodes. Pitman Publishing Ltd., London.

Andres, M.F., Romero, M.D., Montes, M.J. and Delibes, A. (2001). Genetic relationships and isozyme variability in the *Heterodera avenae* complex determined by isoelectrofocusing. Plant Pathology 50: 270-279.

Anju Kamra and Dhawan, S.C. (1998). Nematode management. Pp. 171-190. In: Recent advances in plant Nematology. Ed. P.C. Trivedi, CBS publishers and distributors, New Delhi -110002, pp.277

Anwar,A. and M.A. Khan (2002). Studies on the interaction between *M. incognita* and *F.solani* on soybean. Ann. Pl. Prot. Sci.15: 189-194.

Araki, M. (1992). Description of *Meloidogyne ichinohei* n. sp. (Nematoda: Meloidogynidae) from *Iris laevigata* in Japan. Japanese Journal of Nematology 22: 11-20.

Aristotle (384-322 B.C.) (1910). *Historia animalium.* Translated by D'Arcy Wentworth Thompson. In: Works. J. A. Smith and W. D. Ross, eds. Vol. IV. Garrison Morton.

Armstrong, M.R., Blok, V.C., and Phillips, M.S. (2000). A multipartite mitochondrial genome in the potato cyst nematode *Globodera pallida*. Genetics 154:181–192

Atilano, R.A., Menge, J.A. and Vangundy, S.D. (1981). Interaction of *Meloidogyne arenaria* and *Glomus fasciculatum* in grape. J. Nematol. **13**: 52-57.

Atkinson, G.F. (1889). Nematode root galls. Report of agricultural experiment station, Alabama Polytechnic Institute, Auburn, Alabama. Pp176-226.

Atkinson, G.F. (1892). Some diseases of cotton. Bulletin No. 41, Alabama Polytechnical Institute Agricultural Experiment Station, 64-65.

Atkinson, H.J., P.E. Urwin, and M.J. McPherson (2003). Engineering plants for nematode resistance. Annual Review of Phytopathology 41: 615-639.

Atkinson, H.J., Johnston, K.A. and Robbins, M. (2004). Prima facie evidence that a phytocystatin for transgenic plant resistance to nematodes is not a toxic risk in the human diet. Journal of Nutrition 134: 431–434.

Babatola, J.O. (1983a). Pathogenicity of *Heterodera sacchari* on rice. Nematol. medit., 11: 21.

Babatola, J.O. (1983b). Rice cultivars and *Heterodera sacchari*. Nematol. medit., 11: 103.

Babatola, J.O. (1984). Rice nematode problems in Nigeria; their occurrence, distribution and pathogenesis. Tropical Pest management 30: 256-265.

Baermann, G. (1917). Eine einfache Methode zur Auffindung von *Ankylostomum* (nematoden) Larven in Erdproben. Geneesk. Tijdschr. Ned-Indië 57: 131-137.

Back, M.A., Haydock, P.P.J., Jenkinson, P. (2002). Disease complexes involving plant parasitic nematodes and soilborne pathogens. Plant Pathology 51: 683–97.

Bhadury, P., Austen, M.C., Bilton, D.T., Lambshead, P.J.D., Rogers, A.D., Smerdon, G.R. (2007). Exploitation of archived marine nematodes — a hot lysis DNA extraction protocol for molecular studies. Zoologica Scripta 36:93–98.

Bagyaraja, D. J., A. Manjunath, and D. D. R. Reddy. 1979. Interaction of vesicular-arbuscular mycorrhizae with root-knot nematodes in tomato. Plant Soil 51: 397-403.

Baird, S.M., and Bernard, E.C. (1984). Nematode population and communidynamics in soybean- wheat cropping and tillage regimes. J. Nematol. 16: 379-386.

Baird, R.E., J.R. Rich, G.A. Herzog, S.I. Utley, S. Brown, L.G. Martin, and B.G. Mullinix. (2000). Management of *Meloidogyne incognita* in cotton with nematicides. Nematologia Mediterranea 28:255-259.

Bajaj, H.K. and D.C. Gupta, 1994. Existence of host races in *H. zeae* Koshy *et al.*, Fundam. Appl. Nematol., 17: 389–390

Bakhetia, M., Charlton, W.L., Urwin, P.E., McPherson, M.J. and Atkinson, H.J. (2005). RNA interference and plant parasitic nematodes. Trends Plant Sci. 10: 362-367.

Baldwin, J.G. and Sasser, J.N. (1979) *Meloidogyne megatyla* n. sp., a root-knot nematode from loblolly pine. J. Nematol. 11:47-56.

Baldwin, J.G., Souza, R.M., and Dolinski, C.M. (2001). Fine structure and phylogenetic significance of a muscular basal bulb in *Basiria gracilis* (Tylenchidae). Nematology 3: 681–688.

Barker, K.R. and Koenning, S.R. (1998). Developing sustainable systems for nematode management. Annu. Rev. Phytopathol. 36: 165-205.

Barron, G.L. (1975). Detachable adhesive knobs in *Dactylaria candida*. Trans. Brit. Mycol. Soc. 65: 311–312

Barron G.L. (1977) The Nematode-destroying Fungi. Guelph, Ontario: Canadian Biological Publications.

Barron, G.L. (2003). Predatory fungi, wood decay, and the carbon cycle. Biodiversity 4: 3–9.

Basnet C.P. and Jayaprakash A. (1984). *Heterodera raskii* n. sp. (Heteroderidae: Tylenchina), a cyst Nematode on grass, from Hyderabad, India. J Nematol. 16 (3): 213-6.

Batten, C.K. and Powell, N.T. (1971). The *Rhizoctonia-Meloidogyne* disease complex in flue-cured tobacco. J. Nematol. 3:164-9.

Baum, T.J., Hussey, R.S., Davis, E.L. (2007). Root-knot and cyst nematode parasitism genes: the molecular basis of plant parasitism. Genet Eng (*N Y*) 28: 17-43.

Baunacke, W. E. (1923). *Arb. Biol. Bund Anst. Land und Forstw.* 11 : 185–288.

Bedding, R. A. (1967). Parasitic and free living cycles in entomopahtogenic nematodes of the genus *Deldenus*, Nature, London 214: 174-175.

Bedding, R.A. (1992). Current status of the biological control of *Sirex noctilio* with the nematode *Deladenus ciricidicola.* Page 303 *in* Proceedings of the 19th International Congress of Entomology, Beijing, China.

Bednarz, C.W., S.N. Brown, J.T. Fanders, T.B. Tankersley, and S.M. Brown. (2002). Effects of foliar applied harpin protein on cotton lint yield, fiber quality, and crop maturity. Communications in Soil Science and Plant Analysis 33: 933-945.

Behrens, E. (1975). *Globodera* Skarbilovich, (1959), an independent genus in the subfamily Heteroderinae Skarbilovich, 1949 (Nematoda: Heteroderidae)]. Vortragstagung zu Aktuellen Problemen der Phytonematologie No. 1, pp.12-26.

Bekal, S., Gauthier, J.P. and Rivoal, R. (1997). Genetic diversity among a complex of cereal cyst nematodes inferred from RFLP analysis of the ribosomal internal transcribedspacer region. Genome 40: 479-486.

Berge, J.B., Dalmasso, A., Person, F., Revoal, R. and Thomas, D. (1981). Isoestérases chez le nematode *Heterodera avenae* I. Polymorphisme chez différentes races françaises. Revue de Nématologie 4: 99-105.

Bergeson, G.B. (1972). Concepts of nematode—Fungus associations in plant disease complexes: A review. Experimental Parasitology 32(2): 301-314

Bergeson, G.B., Van Gundy, S.D. and Thomason, I.J. (1970). Effect of *Meloidogyne javanica* on rhizoshere microflora and *Fusarium* wilt of tomato. Phytopathology 60: 1245-9.

Berkeley, M. J. (1855).Vibrio forming cysts on the roots of cucumbers. Gardener's Chronicle and Agricultural Gazette 14: 220.

Bernard, E.C. and Eisenback, J.D. (1997). *Meloidogyne trifoliophila* n. sp. (Nemata: Meloidogynidae), a parasite of clover from Tennessee. J. Nematol. 29: 43-54.

Bernard, E. C., Handoo, Z. A., Powers, T. O., Donald, P. A.and Heinz, R. D. (2010). *Vittatidera zeaphila* (Nematoda: Heteroderidae), a new genus and species of cyst nematode parasitic on corn (*Zea mays*). J. Nematol. 42 : 139-150

Bessey, E.A. (1911). Root knot nematode and its control. Bull. U.S. Dept. Agric. 217.

Bessey, E.A. and Byars, L.P. (1915). The control of Root Knot. U.S. Dept. Agric. Farmer's Bull. 648: 1-19.

Bhadury, P., Austen, M., Bilton, D., Lambshead, P., Roger,s A. and Smerdon, J. (2008). Evaluation of combined morphological and molecular techniques for marine nematode (*Terschellingia* spp.) identification. Mar. Biol.154: 509-518.

Bhatti, D.S. and Gupta, D.C. (1973). Guar, an additional host of *Heterodera cajani*. Indian J. Nematol. 3: 160.

Bird, A. F. (1958). The adult female cuticle and egg sac of the genus *Meloidogyne* Goeldi, 1887. Nematologica 3: 205 - 212.

Bird, D.M. (2004). Signaling between nematodes and plants. Current Opinion in Plant Biology 7 (4): 372-376

Bird, A.F., Bird, J., Fortuner, R., and Moens, R. (1989). Observations on *Aphelenchoides hylurgi* Massey, 1974 feeding on fungal pathogens of wheat in Australia. Revue de Nematologia 6:285–290.

Bishnoi, S.P. and Bajaj, H.K. (2004). On the species and pathotype of *Heterodera avenae* complex of India. Indian J. Nematol. 34: 147-152.

Bishnoi, S.P., Singh, S., Mehta, S. and Bajaj, H.K. (2004). Isozyme patterns of *Heterodera avenae* and *H. filipjevi* populations of India. Indian J. Nematol. 34: 33-36.

Blaauw, R.H., Brière, J-F., de Jong, R., Benningshof, J.C.J., van Ginkel, A.E., Fraanje, J., Goubitz, K., Schenk, H., Rutjes, F.P.J.T and Hiemstra, H. (2001). Intramolecular photochemical dioxenone- alkene cycloadditions as an approach to the bicyclo hexane moiety of Solanoeclepin A. Journal of Organic Chemistry 66: 233-242.

Blaxter, M.L., De Ley, P., Garey, J.R., Liu, L.X., Scheldeman, P., Vierstraete, A., Vanfleteren, J.R., Mackey, L.Y., Dorris, M., Frisse, L.M., Vida, J.T., and Thomas, W.K. (1998). A molecular evolutionary framework for the phylum Nematoda. Nature 392: 71–75.

Blouin, M., Zuily Fodil Y., Pham-Thi, A.T., Laffray, D., Reversat, G., Pando, A., Tondoh, J., and Levelle, P. (2005). Below ground organisms activities affect plant above ground phenotype inducing plant tolerance to parasites. Ecology letters 8: 202-208.

Bolla, R.I. (1987). Nematodes as model systems for nutritional studies. Pages 424–432 *in* Vistas on nematology: A commemoration of the Twentyfifth Anniversary of the Society of Nematologists (Veech, J.A., and Dickson, D.W., eds.). Hyattsville, Maryland, USA: Society of Nematologists, Inc.

Bongers, T. (1990). The maturity index: an ecological measure of environmental disturbance based on nematode species composition. Oecologia 83: 14-19.

Bongers, T. (1999). The Maturity Index, the evolution of nematode life history traits, adaptive radiation and cpscaling. Plant and Soil 212: 1322.

Bongers, T., Alkemade, R., and Yeates G.W. (1991). Interpretation of disturbance-induced maturity decrease in marine nematode assemblages by means of Maturity Index.Mar. Ecol. Prog.Ser. 76: 135-142.

Bongers, T., and Ferris, H. (1999). Nematode community structure as a bioindicator in environmental monitoring. Trends Ecol. Evol. 14: 224-228.

Borchert Nadine, Christoph Dieterich, Karsten Krug, Wolfgang Schütz, Stephan Jung, Alfred Nordheim, Ralf J. Sommer and Boris Macek (2010). Proteogenomics of *Pristionchus pacificus* reveals distinct proteome structure of nematode models. Genome Res. 20: 837-846

Borellus, P. (1653). Page 240 in: Historiarum, et observationum medicophysicarum, centuria prima, etc. Castris.

Borgonie, G., García-Moyano, A., Litthauer, D., Bert, W., Bester, A., van Heerden, E., Möller, C., Erasmus M. and T. C. Onstott (2011). Nematoda from the terrestrial deep subsurface of South Africa. Nature 474: 79–82.

Bossis, M. and Rivoal, R. (1990). Polymorphisme estérasique chez *Heterodera avenae* Woll.: Variations intra et inter parcellaires. Nematologica 35 (1989): 335-339.

Bossis, M. and Rivoal, R. (1996). Protein variability in cereal cyst nematodes from different geographic regions assessed by two-dimensional gel electrophoresis. Fundamental and Applied Nematology 19: 25-34.

Boucher, G., and Lambshead, P. J. D. (1995). Ecological biodiversity of marine nematodes in samples from temperate, tropical, and deep-sea regions. Conserv. Bio. 9: 594-604.

Bouchet, F. (1995). Recovery of helminth eggs from archeological excavations of the Grand Louvre (Paris, France). Journal of Parasitology 81: 785-787.

Bourne, J.M., Kerry, B.R. and F.A.A.M. De Leij (1996). The importance of the host Plant on the interaction between root-knot nematodes *Meloidogyne* spp. and the nematophagous fungus, *Verticillium chlamydosporium* Goddard. Biocontrol Science and Technology, http://www.informaworld.com/smpp/title~db=all~content=t713409232~tab=issueslist~branches=6 - v66, (4): 539 - 548

Brant, S. V., and Gardner, S. L. (2000). Phylogeny of species of the genus *Litomosoides* (Nemata: Onchocercidae): evidence of rampant host-switching. *J. Parasitol.* 86, 545-554.

Bridge, J. (1996). Nematode management in sustainable and subsistence agriculture. Ann. Rev. Phytopathol. 34: 201-225.

Bridge, J., Luc, M. and Plowright, A. (1990). Nematode parasites of rice. Pages 69–108 *in* Plant parasitic nematodes in subtropical and tropical agriculture (Luc, M., Sikora, R.A., and Bridge, J., eds.). Wallingford, UK: CAB International.

Brooks, D. R., and McLennan, D. A. (1993). *Parascript: Parasites* and the Language of Evolution. Smithsonian Institution Press, Washington, DC.

Brown, R.H. (1987). Control strategies in low value crops. Pages 351–382 *in* Principles and practices of nematode control in crops (Brown, R.H., and Kerry, B.R., eds.) Marrickville, Australia: Academic Press.

Brown, J.E., M.G. Patterson, and M.C. Osborn. (1989). Effects of clear plastic solarization and chicken manure on weed control. p. 76–79. In: Proceedings of the 21st National Agricultural Plastics Congress. Nat. Ag. Plastics Assoc., Peoria, IL.

Brown, Paul D., and Matthew J. Morra. (1997). Control of soil-borne plant pests,using glucosinolate-containing plants. p. 167.215. In: Donald L. Sparks (ed.) Advances in Agronomy. Vol. 61. Academic Press, San Diego, CA.

Brown, R.A. (1981). Nematode diseases. In Economic importance and biology of cereal root diseases in Australia. Report to Plant Pathology Subcommittee of Standing Committee on Agriculture, Australia.Blackwell Publishing Ltd

Buangsuwod, N., Tonboon-E.K. P., Rujirachoon, G., Braun,A.J. and Taylor, A. L. (1971). Nematodes. In : *Rice diseases and pests of Thailand.* English edition, 61-67. Rice Protection Research Centre, Rice Department, Ministry of Agriculture, Thailand.

Bull, D.L., W. Ivie, J.G. MacConnell, V.F. Gruber, C.C. Ku, B.H. Arison, J.M. Stevenson, and W.J.A. Vanden Heuvel (1984). Fate of avermectin B1a in soil and plants. Journal of Agricultural and Food Chemistry 32: 94-102.

Burmeister, H. (1837). Handbuch der Naturgeschichte. Zum Gebrauch bei vorlesungen entworfen.3 2 S Abt.: Zoologie., Pp. 369-858, Berlin.

Burrows, P.R. and Stone, A.R. (1985). *Heterodera glycines. CIH Descriptions of Plant-Parasitic Nematodes* No. 118. CAB International, Wallingford, UK.

Burrows, P.R., Barker, A.D.P., Newell, C.A., and Hamilton, W.D.O. (1998). Plant-derived enzyme inhibitors and lectins for resistance against plant-parasitic nematodes in transgenic crops. Pesticide Science 52: 176–183.

Burrows, P.R. and De Waele, D. (1997). Engineering resistance against plant parasitic nematodes using anti-nematode genes. *In:* Fenoll C,Grundler FMW, Ohl SA, *eds. Cellular and molecular aspects of plant-nematode interactions. Dordrecht, the Netherlands: Kluwer Academic Press*, 217–236.

Butschli, O. (1875). Vorlaufige Mittheilung uber Untersuchungen betreffend die ersten Entwickelungsvorgange im befruchehen Ei von Nematoden und Schnecken. Ztschr. Wiss. Zool. v. 25: 201-213.

Byrne, J., Twomey, U., Maher, N., Devine, K.J. and Jones, P.W. (1998). Detection of hatching inhibitors and hatching factor stimulants for golden potato cyst nematode, *Globodera rostochiensis*, in potato root leachate. Annals of Applied Biology 132: 463-472.

Cai, D.G., Kleine, M., Kifle, S., Harloff, H.J., and Sandal, N.N. (1997). Positional cloning of a gene for nematode resistance in sugar beet. Science 275: 832-34

Cai D, Thurau T, Tian Y, Lange T, Yeh K-W, Jung C. (2003). Sporamin-mediated resistance to beet cyst nematodes (*Heterodera schachtii* Schm.) is dependent on trypsin inhibitory activity in sugar beet (*Beta vulgaris* L.) hairy roots. Plant Molecular Biology **51**: 839–849.

Campos, V, P. Sivapalan, and N. C. Gnanapragasam. (1990). Nematode parasites of coffee, cocoa, and tea. Pp. 387-430 *in* M. Luc, R. A. Sikora, and J. Bridge eds. Plant parasitic nematodes in tropical and subtropical agriculture. Wallingford, UK: CAB International.

Carneiro, R.M.D.G., Carneiro, R.G., Abrantes, I.M.O., Santos, M.S.NA and Almeida, M.R.A. (1996). *Meloidogyne paranaensis* n. sp. (Nemata: Meloidogynidae), a root-knot nematode parasitizing coffee in Brazil. J. Nematol. 28: 177-189.

Carneiro, RM., D.G., Almeida, M.R.A, Gomes, A.C.M. and Hernandez, A. (2005). *Meloidogyne izalcoensis* n. sp. (Nematoda: Meloidogynidae), a root-knot nematode parasitising coffee in EI Salvador. Nematology 7: 819-832.

Carneiro, R.MD.G., Almeida, M.R.A., Cofcewicz, E.T, Magunacelaya, J.C. and Aballay. E. (2007). *Meloidogyne ethiopica*, a major root-knot nematode parasitising *Vitis vinifera* and other crops in Chile. *Nematology* 9: 635-641.

Carter, W. (1943). A promising new soil amendment and disinfectant. Science 97: 383-384.

Carter, W.W. (1975). Effects of soil temperatures and inoculum levels of *Meloidogyne incognita* and *Rhizoctonia solani* on seedling disease of cotton. J. Nematol. 7: 229-33.

Cason, K. M. Thomson, R. S. Hussey and R. W. Roncadori (1983). Interaction of vesicular-arbuscular mycorrhizal fungi and phosphorus with *Meloidogyne incognita* on tomato. J. Nematol. 15(3): 410-417

Castillo, P., Vovlas, N., Subbotin, S. and Troccoli, A. (2003). A new root-knot nematode: *Meloidogyne baetica* n. sp. (Nematoda: Heteroderidae) parasitizing wild olive in Southern Spain. Phytopathology, 93: 1093-1102.

Castillo, P., Vovlas, N., Troccoli, A., Liebanas, 6., Palomares Rivs, J.E. and Landa, B.B. (2009). A new rootknot nematode, *Meloidogyne silvestris* n.sp. (Nematoda: Meloidogynidae), parasitizing European holly in northern Spain. Plant Pathology 58: 606-619.

Celis, C., Scurrah, M., Cowgill, S., Chumbiauca, S., Green, J., Franco, J., Main, G., Kiezebrink, D., Visser, R.G.F. and Atkinson, H.J. (2004). Environmental biosafety and transgenic potato in a centre of diversity for this crop: Nature 432: 222–225.

Celsus, A. C. (53 B.C.-7 A.D.) (1657). De medicina libri octo, ex recognitione Joh. Antonidae von Linden D. and Prof. Med. Pract. Ord.

Charchar, J.M. and Eisenback, J.D. (2002). *Meloidogyne brasilensis* n. sp. (Nematoda: Meloidogynidae), a root-knot nematode parasitising tomato cv. Rossol in Brazil. Nematology 4: 629-643.

Charchar, J.M., Eisenback, J.D. and Hirschmann, H. (1999). *Meloidogyne petuniae* n. sp. (Nemata: Meloidogynidae), a root-knot nematode parasitic on petunia in Brazil. J. Nematol. 31: 81-91.

Charchar, J. M., Eisenback, J. D., Charchar, M. J. and Boiteux, M. E. N. F. (2008a). *Meloidogyne pisi* n. sp. (Nematoda: Meloidogynidae), a root-knot nematode parasitising pea in Brazil. Nematology 10: 479-493.

Charchar, J. M., Eisenback, J. D., Charchar, M. J. and Boiteux, M. E. N. F. (2008b). *Meloidogyne phaseoli* n. sp. (Nematoda: Meloidogynidae), a root-knot nematode parasitising bean in Brazil. Nematology 10: 525-538.

Charles, J S. K and Venkitesan, T. S. (1984). New hosts of *Helerodera oryzicola* Rao and Jayaprakash 1978, in Kerala, India. Indian J. Nematol., 14: 181-182.

Chen, P.S., Peng, D.L. and Zheng, J.W. (1990). The discovery and description of *Meloidogyne fanzhiensis* n. sp. on potato. Journal of Shaxi Agricultural University 10: 55-60.

Chen, Z. X. and Dickson, D.W. (1998). Review of Pasteuria penetrans: biology, ecology, and biological control potential. J. Nematol. 30: 313–340.

Chellami, D.O., S.M. Olson, D. J. Mitchell, I. Secker, and R. McSorley (1997). Adaptation of soil solarization to the integrated management of soilborne pests of tomato under humid conditions. Phytopathology 87: 250-258.

Chilton, N., Gasser R. and Beveridge, I. (1995). Differences in a ribosomal DNA sequence of morphologically indistinguishable species within the Hypodontus macropi complex (Nematoda: Strongyloidea). Parasitol. Int. 25: 647-651.

Chilton, N. B. Gasser, R. B. and Beveridge, I. (1997). Phylogenetic relationships of Australian strongyloid nematodes inferred from ribosomal DNA sequence data. Intnl. J. Parasitol. 27: 481- 494.

Chinnasri, B., N.Tangchitsomkid and Y. Toida (1995). *H. zeae* on maize in Thailand. Japan J. Nematol. 24: 35–38

Chitwood, B. G. (1949). Root knot nematodes, Part1. Revision of the genus, *Meloidogyne* Goeldi, 1887. Proc Helminth. Soc. Wash. 16: 90-104.

Chitwood, David J. (2002). Phytochemical based strategies for nematode control. Annual Review of Phytopathology 40: 221—249.

Chitwood, B. G. and M. B. Chitwood (1933). The characters of a protonematode. Journal of Parasitology 20: 130.

Chitwood, B. G. and M. B. Chitwood (1937). Introduction to Nematology. University Park Press, Baltimore, MD, Monumental Printing.

Chitwood, B. G., and Chitwood, M. B. (1950). Introduction to Nematology.University Park Press, Baltimore, MD, Monumental Printing.

Chitwood, B. G., and Chitwood, M. B. (1974). Introduction to Nematology (Consolidated Edition). University Park Press, Baltimore, MD.

Chitwood, B.G. and M.C.Toung (1960). *Meloidogyne* from Taiwan and New Delhi. Phytopathology 50: 631-632.

Choudhury, B.C. (1981). Root-knot nematode problem in various crop plants in Bangladesh. Pages 142–145 *in* Proceedings of the III International *Meloidogyne* Project Research and Planning Conference on Root-knot Nematodes, *Meloidogyne* spp., Region VI, Jakarta, Indonesia.USA: North Carolina State University and the United States Agency for International Development.

Christian, H., Sikora, R A. and Oerke, E.C. (2011). Influence of different levels of resistance or tolerance in sugar beet cultivars on complex interactions between *Heterodera schachtii* and *Rhizoctonia solani*. Nematology, 13: 319-332

Christie, J.R. (1945). Some preliminary tests to determine the efficacy of certain substances when used as soil fumigants to control the root-knot nematode, *Heterodera marioni* (Cornu)

Goodey. Proceedings of the Helminthological Society of Washington 12:14-19.

Christie, J.R. (1959). Plant nematodes: their bionomics and control. Agricultural Experiment Stations, University of Folrida, Gainsville, FL.

Christie, J.R. and V.G. Perry (1951). A root disease of plants caused by a nematode, of the genus, *Trichodorus*. Science 113: 491-493.

Clapp, J.P., Van Der Stoel, C.D. and Van Der Putten, W.H. (2000). Rapid identification of cyst (*Heterodera* spp., *Globodera* spp.) and root-knot (*Meloidogyne* spp.) nematodes on the basis of ITS2 sequence variation detected by PCRsingle- strand conformational polymorphism (PCR-SSCP) in cultures and field samples. Molecular Ecology 9: 1223-1232.

Cliff, G.M. and Hirschmann, H. (1984) *Meloidogyne microcephala* n. sp. (Meloidogynidae), a root-knot nematode from Thailand. J. Nematol. 16: 183-193.

Cobb, N.A. (1890). *Tylenchus* and root-gall. *Agricultural Gazette of New South Wales* 1: 155-184.

Cobb, N.A. (1914). Nematodes and their relationships. USDA Year Book. pp 457-490.

Cobb, N.A. (1918). Estimating the nema populations of soil. USDA, Bureau of Plant Industry. Agriculture Technical Circular 1:1-48.

Cobb, N. A. (1919). The orders and classes of nemas. Contrib. Sci. Nematol. 8: 213-216.

Cobb, N.A. (1924). The amphids of *Caconema* (*nom. nov.*) and other nemas. Journal of Parasitology 11: 118-120.

Cobb, J. L. S. (1967). The innervation of the ampulla of the tube foot in the starfish *Astropecten irregularis.* Proc Roy Soc London Ser B 168: 91-99

Coetzee, V. (1956). *Meloidogyne acronea,* a new species of root-knot nematode. *Nature, London* 177: 899-900. The *C. elegans* Sequencing Consortium. (1998). Genome sequence of the nematode *C. elegans*: A platform for investigating biology. Science 282: 2012–2018.

Cook, A., Bhadur, P., Debenham, N., Meldal, B., Blaxter, M., Smerdon. G., Austen, M., Lambshead, P. and Rogers, A. (2005). Denaturing gradient gel electrophoresis (DGGE) as a tool for identification of marine nematodes. Mar. Ecol. Prog. Ser. 291: 103-113.

Coomams, A. (1989). Overzicht van de vrijlevende nematofauna vz Belgie. Verh Symp. Invet Belgie pp 43-56.

Coomans, A., Verschuren D., and Vanderhaeghen, R. (1988). The demanian system, traumatic insemination and reproductive strategy in *Oncholaimus oxyuris* Ditlevsen (Nematoda, Oncholaimina). Zoologica Scripta *17*, 15–23.

Cooper, K. M. and Grandisan, G. S. (1986). Interaction of vesicular-arbuscular mycorrhizal fungi and root-knot nematode on cultivars of tomato and white clover susceptible to *Meloidogyne hapla*. Annals of Applied Biology 108: 555-565

Copeland H.F. (1938). The kingdoms of organisms. Quart Rev Biol. 13: 383–420

Correia, F. J. S. and Abrantes, I.M.de O. (2005). Characterization of *Heterodera zeae* Populations from Portugal. J. Nematol. 37(3): 328–335

Cowgill, S.E. and Atkinson, H.J. (2003). A sequential approach to risk assessment of transgenic plants expressing protease inhibitors: effects on nontarget herbivorous insects. Transgenic Research 12: 439–449.

Cowgill, S.E., Bardgett, R.D., Kiezebrink, D.T., and Atkinson, H.J. (2002). The effect of transgenic nematode resistance on nontarget organisms in the potato rhizosphere. Journal Applied Ecology 39: 915–923.

Cowgill, S.E., Danks, C. *and* Atkinson, H.J. (2004). Multitrophic interactions involving genetically modified potatoes, non-target aphids, natural enemies and hyperparasitoids. Molecular Ecology 13: 639–647.

Cowgill, S.E., Wright, C. *and* Atkinson, H.J. (2002). Transgenic potatoes with enhanced levels of nematode resistance do not have altered susceptibility to nontarget aphids. *Molecular Ecolology* 11: 821–827.

Creer, S., Fonseca, V.G., Porazinska, D.L., Giblin-Davis, R.M., Sung, W., Power, D.M., Packer, M., Carvalho, G.M., Blaxter, M., Lambshead, P.J.D. and Thomas, W.K. (2010). Ultrasequencing of the meiofaunal biosphere: practice, pitfalls and promises. Mol. Ecol. 19: 4-20.

Crompton, D. W. T. (1999). How much human helminthiasis is there in the world? J. Parasitol. 85: 397-403.

Cronin, D., Moenne-Loccoz, Y., Fenton, A., Dunne, C., Dowling, D. N., and O'Gara, F. (1997). Role of 2, 4-diacetylphloroglucinol in the interactions of the biocontrol pseudomonad strain F113 with the potato cyst nematode *Globodera rostochiensis*. Applied and Environmental Microbiology 63:1357–1361.

Curran J., Baillie D., Webster J. (1985). Use of genomic DNA restriction fragment length differences to identify nematode species. Parasitology 90: 137-144.

Curran J., McClure M., Webster J. (1986). Genotypic differentiation of *Meloidogyne* populations by detection of restriction fragment length difference in total DNA. J. Nematol. 18: 83-86.

Dackman, C., I. Chet, and B. Nordbring-Hertz. (1989). Fungal parasitism of the cyst nematode *Heterodera schachtii*: Infection and enzyme activity. FEMS Microbiology Ecology 62: 201-208.

Da Ponte, J. J. (1969). *Meloidogyne lordelloi* n. sp., a nematode parasite of *Cereus macrogonus* Salm. Dick. Boletin Cearense da Agronomia 10: 59-63.

Das, Dibakar and Mishra, S.D. (2003). Effect of neem seed powder and neem based formulations for the management of *Meloidogyne incognita, Heterodera cajani* and *Rotylenchulus reniformis* infecting pigeonpea. Ann. PI. Protec. Sci. 11 (1): 110-115

David, Mc., K. Bird, Charles, H., Opperman and Keith, and G. Davies (2003). Interactions between bacteria and plant-parasitic nematodes: now and then. International Journal for Parasitology. 33 (11): 1269-1276

Davis, E.L., Hussey, R.S., Mitchum, M.G.and Baum, T.J. (2008). Parasitism proteins in nematode-plant interactions. Curr. Opin. Plant Biol. 11: 360–366

De Grisse, A. (1961) *Meloidogyne kikuyensis* n. sp., a parasite of kikuyu grass (*Pennisetum clandestinum*) in Kenya. Nematologica 5 (1960): 303-308.

De Coninck, L.A.P. (1965). Systématique des nématodes. In: *Traité de Zoologie: Anatomie, Systématique, Biologie*, Vol. 4, P.P. Grassé, ed. Paris: Masson et Cie., pp. 586–731.

De Deyn, G.B., van Ruijvan, J., Raaijmakers, C.E., Ruiter, P.C. and Vander Putten (2007). Above and below ground insect herbivores differentially alter nematode communities by plant diversity shifts. Oikos 116 : 923-930.

De Ley, P. (2006). A quick tour of nematode diversity and the backbone of nematode phylogeny. WormBook, ed. The *C. elegans* Research Community, Worm Book, doi/10.1895/wormbook.1.41.1, http://www.wormbook.org. (April 29, 2010).

De Ley, P. and Bert, W. (2002). Video capture and editing as a tool for the storage, distribution and illustration of morphological characters of nematodes. J. Nematol. 34: 296–302.

De Ley, P. and Blaxter, M. L. (2002). Systematic position and phylogeny. In: *The Biology of Nematodes*, D.L. Lee, ed., London: Taylor and Francis, pp. 1–30.

De Ley, P. and Blaxter, M. (2004). A new system for Nematoda: combining morphological characters with molecular trees, and translating clades into ranks and taxa. Nematology Monographs and Perspectives 2: 633–653.

De Ley, P., Felix, M., Frisse, L., Nadler, S., Sternberg, P., and Thomas, W. (1999). Molecular and morphological characterization of two reproductively isolated species with mirror-image anatomy (Nematoda: *Cephalobidae*). Nematology 1: 591-612.

De Ley, P. and Mundo-Ocampo, M. (2004). The cultivation of nematodes. In: *Nematology: Advances and Perspectives*, Vol. 1, Chen, Z.X., Chen, S.Y., and Dickson, D.W., eds., Tsinghua: Tsinghua University Press, pp. 541–619.

Den Ouden, H. (1954). Het bieten cysteaaltje en zijn bestrijding. I. Methoden te gebruiken bij het onderzoek naar kunstmatige en natuurlijke lokstoffen. *Meded. Inst. Suikercore.* Bergen-o.-Zoom 24: 101-120.

Derycke, S., Remerie, T., Backeljau, T., Vierstraete, A., Vanfleteren, J., Vincx, M., and Moens T. (2008). Phylogeography of the *Rhabditis* (*Pellioditis*) *marina* species complex: evidence for long-distance dispersal, and for range expansions and restricted gene flow in the northeast Atlantic. Mol. Ecol. 17: 3306-3323.

Deshraj (1983). Potato nematodes and their control. pp. 456- 465: **In:** B.B. Nagaiah (ed.), Potato production, storage and utilization, CPRI, Shimla, 536 p.

Dewang Deng, Allan Zipf, Y. Tilahun, G.C.Sharma, J. Jenkins and K. Lawrence (2008). An improved method for the extraction of nematodes using iodixanol (OptiPrep Ô). African J. Microbiol. Res. 2: 167-170

Dhawan, S.C. and Kaushal, K.K. (1988). Effect of neem emulsuion on the hatching of the cereal cyst nematode, *Heterodera avenae*. Indian J. Nematol. 18: 364.

Dhawan, S.C., Kaushal, K.K. and Srivastava, A.N. (1983). Record of occurrence of *Heterodera sorghi* from Haryana State, India. Indian J. Nematol. 23: 241.

Dobell, C. (1932). Antony Van Leeuwenhoek and His 'Little Animals.' Harcourt, Brace and Co., NY.

Diesing, K. M. (1851). Systema helminthum.2.

Dieterich, C. and Sommer, R.J. (2009). How to become a parasite—lessons from the genomes of nematodes. Trends Genet **25**: 203–209.

Dieterich, C., Clifton, S.W., Schuster, L.N., Chinwalla, A., Delehaunty, K., Dinkelacker, I., Fulton, L., Fulton, R., Godfrey, J. and Minx, P. (2008). The *Pristionchus pacificus*genome provides a unique perspective on nematode lifestyle and parasitism. Nat Genet 40: 1193–1198.

Diggle, M. and Clarke, S. (2004). Pyrosequencing, sequence typing at the speed of light. Mol. Biotechnol. 28: 129-138

Dong, W. B., Brenneman, T. B., Holbrook C. C., P. Timper and A. K. Culbreath (2009). The interaction between *Meloidogyne arenaria* and *Cylindrocladium parasiticum* in runner peanut. Plant Pathology 58: 71–79.

Dorris, M., De Ley, P., and Blaxter, M. L. (1999). Molecular analysis of nematode diversity and the evolution of parasitism. Parasitol. Today 15: 188-193.

Drechsler, C. (1937). Some hypomycetes that prey on free-living terricolous nematodes. Mycologia 29: 447–552.

Drechsler, C. (1950). Several species of Dactylella and Dactylaria that capture free living nematodes. Mycologia 42: 1–79

Dropkin, Victor H. (1980). Introduction to Plant Nematology. John Wiley and Sons, New York, NY. p. 38.44, 242, 246, 256.

Dujardin, F. (1842). Memoire sur les Gordius et lea Mermis. Comet. Rend. Acad. Sci. Paris 15: 117-119.

Dujardin, F. (1845). Histoire naturelle des helminthes ou vers intestinaux. Paris.

Ebsary, B.A. and Eveleigh, E.S. (1983). *Meloidogyne aquatilis* n. sp. (Nematoda: Meloidogynidae) from *Spartina pectinata* with a key to the Canadian species of *Meloidogyne*. J. Nematol. 15: 349-353.

Eisenback, J.D. (1982). Description of the blueberry root-knot nematode, *Meloidogyne carolinensis* n. sp. J. Nematol. 14: 303-317.

Edward, J.C. and Mishra, S.L.(1968). *Heterodera vigni* n.sp. and second stage larvae of *Heterodera* species in Uttar Pradesh, India. Allahabad Fmr. 42 : 155.

Edwards, R., Rodriguez-Brito, B., Wegley, L., Haynes, M., Breitbart, M., Peterson, D., Saar, M., Alexander, S., Alexander, E. and Rohwer, F. (2006). Using pyrosequencing to shed light on deep mine microbial ecology. Genomics 7: 1-12.

Eisenback, J.D. (1993). Interactions between nematodes and root rot fungi. In; Khan, M.W., Nematode Interactions. Chapman and Hall, London, 134-174.

Esbenshade P. and Triantaphyllou A. (1990). Isozyme phenotypes for the identification of *Meloidogyne* species. J. Nematol. 22: 10-15.

Eisenback, J.D.and Triantaphyllou, H. H. (1991). Root-knot Nematodes: *Meloidogyne* species and races. In: Manual of Agricultural Nematology, W. R. Nickle. (Ed). Marcel Dekker, New York. pp 281 – 286.

Egunjobi O.A. and Onayemi S.O. (1981). The efficacy of water extract of neem (*Azadirachta indica*) leaves as a systemic nematicide. Nigeria Journal of Plant Protection **5**: 70–74.

Elbadri, G,, De Ley, P., Waeyenberge, L., Verstraete, A., Moens, M., and Vanfleteren, J. (2002). Intraspecific variation in *Radopholus similis* isolates assessed with restriction fragment length polymorphism and DNA sequencing of the internal transcribed spacer region of the ribosomal RNA cistron. Parasitol. Int. 32: 199-205

Elmiligy, I.A. and de Grisse, A. (1970). Effect of extraction technique and adding fixative to soil before storing on recovery of plant-parasitic nematodes. Nematologica 16: 353-358.

Endo, B.Y. (1964). Penetration and development of *Heterodera glycines* in soybean roots and related anatomical changes. Phytopathology 54:79–88

Ettema, C.H. and Bongers, T. (1993). Characterization of nematode colonization and succession in disturbed soil using the Maturity Index.Biol, fertile. Soils 16: 79-85.

Evans, A. A. F. and Perry, R. N. (1976). Survival strategies in nematodes. *In:* CroJJ, N. A. (Ed.). *The organisation of nematodes.* London and New York, Academic Press: 383-424.

Evans, K. (1993). New approaches for potato cyst nematode management. Nematropica 23: 221-231

Evans, K. and Stone, A.R. (1977). A review of the distribution and biology of the potato cyst nematodes *Globodera rostochiensis* and *G. pallida. PANS*, 23:178-189

Eyualem, A., and Blaxter, M. (2003). Comparison of biological, molecular and morphological methods of species identification in a set of cultured *Panagrolaimus* isolates. J. Nematol. 35: 119-128

Fairbairn, D.J., Cavallaro, A.S., Bernard, M., Mahalinga-Iyer, J., Graham, M.W. and Botella J.R. (2007). Host-delivered RNAi: an effective strategy to silence genes in plant parasitic nematodes. Planta 226: 1525–1533.

Fallon, D.J., Kaya, H.K. and Sipes, B.S. (2006). Enhancing *Steinernema* spp. suppression of *Meloidogyne javanica*. J. Nematology 38: 270-271.

Fargette, M., Phillips, M., Blok, V., Waugh, R., and Trudgill, D. (1996). An RFLP study of relationships between species, populations, and resistancebreaking lines of tropical species of *Meloidogyne*. Fundam. Appl. Nematol.19: 193-200.

Félix, M.A, Ashe, A., Piffaretti, J., Wu, G, and Nuez, I. (2011) Natural and Experimental Infection of *Caenorhabditis* Nematodes by Novel Viruses Related to Nodaviruses. doi:10.1371/journal.pbio.1000586

Fenwick, D.W. (1940). Methods for the recovery and counting of cysts of *Heterodera schachtii* from soil. Journal of Helminthology 18:155–172.

Ferreira, L. F., De Araujo, A. J. G. and Confalonieri, U. E. C. (1983). The finding of helminth eggs in a Brazilian mummy. Transactions Royal Soc. Trop. Med. Hyg. 77: 65-67.

Ferris, H., Venette, R. C. van der Meulen, H. R., and Lau, S. S. (1998). Nitrogen mineralization by bacterial feeding nematodes: verification and measurement. Plant Soil 203: 159-171.

Ferris, V.R., Faghihi, J., Ireholm, A. and Ferris, J.M. (1989). Two-dimensional protein patterns of cereal cyst nematodes. Phytopathology 79: 927-932.

Ferris, V.R., Ferris, J.M., Faghihi, J. and Ireholm, A. (1994). Comparisons of isolates of *Heterodera avenae* using 2-D PAGE protein patterns and ribosomal DNA. J. Nematology 26: 144-151.

Filipjev, I. N. (1934a). Harmful and Useful Nematodes in Rural Economy (Russian). Figs. 1-333.Moskva,Leningrad.

Filipjev, I. N. (1934b). The classification of the freeliving nematodes and their relation to the parasitic nematodes. Smithsonian Misc. Coll. (Publ, 3216) 89: 1-63.

Filipjev, I.N. and Schuurmans Stekhoven, J.H. (1941). Mannual of Agricutural Helminthology. Leiden, E. J. Brill.

Findlay, S., and Tenore, K.R. (1982). Effect of free-living nematode (*Diplolaimella chitwoodi*) on detrital carbon mineralization. Marine Ecology 8: 161–166.

Fire, A., Xu, S., Montgomery, M.K., Kostas, S.A., Driver, S.E. and Mello, C.C. (1998). Potent and specific genetic interference by double-stranded RNA in *Caenorhabditis elegans*. Nature 391: 806–811.

Flood, P.R. (1966). A peculiar mode of muscular innervation in amphioxus. Light and electron microscopic studies of the so-called ventral roots. Journal of Comparative Neurology 126: 181-218.

Fonseca, G. and Derycke S, Moens, T. (2008). Integrative taxonomy in two free-living nematode species complexes. Biol. J. Linn. Soc. 94: 737-753.

Forge, Thomas, A., Russell, E. Ingham, and Diane, Kaufman (1995). Winter cover crops for managing root-lesion nematodes affecting small fruit crops in the Pacific Northwest. Pacific Northwest Sustainable Agriculture. March. p. 3.

Foucher, A., Bongers, T., Noble, L., and Wilson, M. (2004). Assessment of nematode biodiversity using DGGE of 18S rDNA following extraction of nematodes from soil. Soil Biol. Biochem. 36: 2027-2032.

Franklin, M.T. (1957). A review of the genus, *Meloidogyne*. Nematologica 2: 387-397.

Freckman, D.W. (1988). Bacterivorous nematodes and organic matter decomposition. Agriculture, Ecosystems and Environment 24:195–217.

Freckman, D.W. and Ettema, C.H. (1993).Assessing nematode communities in agroecosystemsof varying human interventions.Agric. Ecosystems Environ. 45: 239-261.

Freckman, D.W., Kaplan, D.T.and Van Gundy, S.D. (1977). A comparison of techniques for extraction and study of anhydrobiotic nematodes from dry soils. J. Nematology 9: 176-181.

Freckman, D. W. and Virginia, R. A. (1991). Nematodes in the McMurdo Dry Valleys of southern Victoria Land. Antarctic Journal of the United States, 26: 233–234.

Fresenius, G. (1852). Beitrage zur Mykologie. Heft 1-2. Pp.1-80.

Fry, G. F. and Moore, J. G. (1969): *Enterobius vermicularis*: 10,000 year-old human infection. Science 166 (1): 620.

Fuchs, A. G. (1915). Die Naturgeschichte der Nematoden and einiger anderer Parasiten. 1. Des Ips typographus L., 2. Des Hylobius abietis L. Zool. Jahrb. Jena Abt. Syst. 38: 109-222.

Fuchs, A.G. (1929). Die Parasiten einiger Russel-und Borkenkafern. Z. ParasitKde 2: 248-285.

Fuchs, A.G. (1937). Neue Parasitische und halparasitische Nematoden bei Borkenkafern und einige andere Nematoden. I. Zool. Jb. 70: 291-380.

Fullaondo A., Barrena E., Viribay M., Barrena I., Salazar A. and Ritter E. (1999). Identification of potato cyst nematode species *Globodera rostochiensis* and *G. pallida* by PCR using species specific primer combinations. Nematology 1: 157-163.

Fürst von Lieven, A. (2002). The sister group of the Diplogastrina (Nematoda). Russ. J. Nematol. 10: 127–137.

Fürst von Lieven, A., and Sudhaus, W. (2000). Comparative and functional morphology of the buccal cavity of Diplogastrina (Nematoda) and a first outline of the phylogeny of this taxon. J. Zool. Syst. Evol. Res. 38: 37–63.

Gamliel, A.and Stapleton, J.J. (1993a). Effect of chicken compost or ammonium phosphate and solarization on pathogen control, rhizosphere microorganisms, and lettuce growth. Plant Dis. 77:886-891.

Gamliel, A. and Stapleton, J.J. (1993b). Characterization of antifungal volatile compounds evolved from solarized soil amended with cabbage residues. Phytopathology 83: 899-905.

Ganaie, M.A. and T.A. Khan (2011). Studies on interactive effect of *Meloidogyne incognita* and *Fusarium solani* on *Lycopersicum esculentum* Mill. International J. Botany 7: 205-208.

Gao, X., T. A. Jackson, G. L. Hartman, and T. L. Niblack (2006). Interactions between the soybean cyst nematode and *Fusarium solani* f. sp. *glycines* based on greenhouse factorial experiments. Phytopathology 96 (12): 1409-1415.

Garber R.H., Jorgenson E.C., Smith S. and Hyer A.H. (1979). Interaction of population levels *of Fusarium oxysporum* f. sp. *vasinfectum* and *Meloidogyne incognita* on cotton. J. Nematology 11: 133-137.

Gardner, S. L. (1991). Phyletic coevolution between subterranean rodents of the genus *Ctenomys* (Rodentia: Hystricognathi) and nematodes of the genus *Paraspidodera* (Hetevakoidea: Aspidoderidae) in the neotropics: temporal and evolutionary implications. Zool. J. Linnean Soc. 102: 169-201.

Gaugler, R, and Kaya, H. K. (1990). *Entomopathogenic Nematodes in Biological Control.* CRC Press; Boca Raton, FL.

Gaur, H. S., Perry, R. N. and Beane, J. (1992). Hatching behaviour of six successive generations of the pigeon-pea cyst nematode, *Heterodera cajani* in relation to growth and senescence of cowpea, *Vigna unguiculala.* Nematologica 38 : 190-202.

Gaur, H.S. and Singh, I. (1977). Pigeon-pea cyst nematode, *Heterodera cajani,* associated with the mung crop in the Punjab State. J. Res. Punjab agric. Univ., 14: 509.

Gegenbaur, C. (1859). Grundzuge der verglei Chenden Anatomie. XIV + 606pp. Leipzig.

Gera, W.H. and Roger Cook (2005). An overview of arbuscular mycorrhizal fungi–nematode interactions. Basic and Applied Ecology 6 (6): 489-503

Germani, G., Reversat, G., and Luc, M. (1983). Effect of *Sesbania rostrata* on *Hirschmanniella oryzae* in flooded rice. Journal of Nematology 15: 269–271.

Gheysen, G. and Vanholme, B. (2007). RNAi from plants to nematodes. Trends in Biotechnology **25**: 89–92.

Gill, J.S. and Swarup, G. (1971). On the host range of the cereal cyst nematode, *Heterodera avenae* Woll. 1924, the causal organism of 'Molya' disease of wheat and barley in Rajasthan. Indian J. Nematol., 1:63.

Gintis, B.O., G. Morgan-Jones, and R. Rodriguez- Kabana (1983). Fungi associated with several developmental stages of Heterodera glycines from an Alabama soybean field. Nematropica 12:181-200.

de Goede, R. (1993). Terrestrial nematodes in a changing environmnent. CIP-gegevens Koninkljke Bibliotheek, Den Haag, Wegeningen, The Netherlands.

Goeldi, E.A. (1887). Relatorio sobre a molestia do cafeeiro na provincia do Rio de Janeiro. Archivos do Museo Nacional 8: 7-123.

Goellner, M., Wang, X., Davis, E.L. (2001). Endo-b-1,4-glucanase expression in compatible plant-nematode interactions. Plant Cell 13: 2241–2255.

Goeze, J. A. E. (1782). Versuch einer Naturgeschichte der Eingeweidewur-mer thierischer Korper.

Golden, A.M. and Birchfield, W. (1972). *Heterodera graminophila* n.sp. (Nematoda: Heteroderidae) from grass with a key to closely related species, J. Nematol. 4:147.

Golden, A.M. and Birchfield, W. (1978). *Meloidogyne incognita wartellei* n. subsp. (Meloidogynidae) a root knot nematode on resistant soybeans in Louisiana. J. Nematol. 10: 269-277.

Golden, AM. and Kaplan, D.T (1986). Description of *Meloidogyne christiein.* sp. (Nematoda: Meloidogynidae) from oak with SEM and host-range observations. J. Nematol. 18: 533-540.

Golden, A.M. and Slana, L.J. (1978). *Meloidogyne grahami* n. sp. (Meloidogynidae), a root-knot nematode on resistant tobacco in South Carolina. J. Nematol. 10 : 355-361.

Golden, AM., O'Bannon, J.H., Santo, G.S. and Finley, A.M. (1980). Description and SEM observations of *Meloidogyne chitwoodi* n. sp. (Meloidogynidae), a root-knot nematode on potato in the Pacific Northwest. J. Nematol. 12: 319-327.

Golden, AM., Rose, L.M. and Bird, G.W. (1981). Description of *Meloidogyne nataliei* n. sp. (Nematoda: Meloidogynidae) from grape (*Vitis labrusca*) in Michigan, with SEM observations. J. Nematol. 13: 393-400.

Golden, A.M. and R.M. Mulvey (1982). Morphological and diagnostic features of *H. zeae*, the corn cyst nematode. J. Nematol. 14: 442

Golden, A.M. and Mulvey, R.H. (1983). Redescription of *Heterodera zeae*, the corn cyst nematode, with SEM observations. J. Nematol. 15: 60.

Golden, A.M. and Birchfield, W. (1968). Rice rootknot nematode (*Meloidogyne graminicola*) as a new pest of rice. Pl. Dis. Reptr. 5 : 423.

Golden, A.M. and Ellington, D.M.S. (1972). Redescription of *Heterodera rostochiensis* (Nematoda: Heteroderidae) with a key and notes on closely related species. Proceedings of the Helminthological Society of Washington 39: 64-78.

Golden, J.K., van Gundy, S.D. (1975). A disease complex of okra and tomato involving the nematode, *Meloidogyne incognita* and the soil-inhabiting fungus, *Rhizoctonia solani*. Phytopathology 65: 265–73.

Golden, A. M., J. M. Epps, R. D. Riggs, L. A. Duclos, J. A. Fox, and R. L. Bernard. (1970). Terminology and identity of infraspecific forms of the soybean cyst nematode (*Heterodera glycines*). Plant Disease Reporter 54:544–546.

Goldstein, B., Frisse, L.M., and Thomas, W.K. (1998). Embryonic axis specification in nematodes: evolution of the first step in development. Curr. Biol. *8*, 157–160.

Gomes, A.P.S., M.L. Ramos, R.S. Vasconcellos, J.R. Jensen and M.C.R. Vieira-Bressan. (2000). *In vitro* activity of Brazilian strains of the predatory fungi, *Arthrobotrys* spp. on free-living nematodes and infective larvae of Haemonchus placei. Mem. Inst. Oswaldo Cruz, Rio de Janeiro, 95: 873-876.

Gommers, F. J. (1981). Biochemical interactions between nematodes and plants and the irrelevance to control: A review. Helminthol. Abstr. 50:9-24.

Gommers, F.J., Bakker, J., and Smits, L. (1980). Effects of singlet oxygen generated by the nematicidal compound μ-terthienyl from *Tagetes* on the nematode *Aphelenchus avenae*. Nematologica 6:369–375.

Gomes,V. M., R. M. Souza, V. M.Dias, S. F. da Silveira and C. Dolinski (2011). Guava Decline: A Complex Disease Involving *Meloidogyne mayaguensis* and *Fusarium solani*. J Phytopathol 159:45–50 (2011)

Goodey, J.B. (1963). Soil and Fresh Water Nematodes. (2nd edition, revision of T.Goodey's 1st edition, 1951). London.

Goodey, T. (1933). Plant Parasitic Nematodes and the Diseases they Cause. New York: E. P. Dutton and Company.

Goodey, T. (1951). Soil and Fresh Water Nematodes. London.

Goswami B.K. and Vijayalakshmi K. (1981). Effect of some indigenous plant material and oil cakes amended soil on the growth of tomato and root knot nematode population. Indian Journal of Nematology 11: 121 (Abstr.).

Goswami B.K. and Vijayalakshmi K. (1986): Nematicidal properties of some indigenous plant materials against root-knot nematodes *Meloidogyne incognita* on tomato. Indian Journal of Nematology, **16**: 65–68.

Goswami, B.K.and Mittal, A. (2004). Management of root-knot nematode infecting tomato by *Trichoderma viride* and *Paecilomyces lilacinus*. Indian Phytopathol. 57:235-236.

Goswami, B.K., Pandey, R.K., Rathour, K.S., Bhattacharya, C.and Singh, L. (2006). Integrated application of some compatible biocontrol agents along with mustard oil seed cake and furadan on *Meloidogyne incognita* infecting tomato plants. J. Zhejiang Univ. Sci. B 7: 873-875.

Gould, W. (1747). An Account of English Ants. London.

Goverse, A., Overmars, H., Engelbertink, J., Schots, A., Bakker, J.and Helder, J. (2000) Both induction and morphogenesis of cyst nematode feeding cells are mediated by auxin. Mol. Plant Microbe. Interact. 13: 1121–1129

Grunewald, W., van Noorden, G., Van Isterdael, G., Beeckman, T., Gheysen, G, and Mathesius, U. (2009). Manipulation of auxin transport in plant roots during Rhizobium symbiosis and nematode parasitism. Plant Cell 21: 2553–2562

Grassé, P. P. (1965). Traité de Zoologie. Anatomic, Systématique, Biologic. Némathelminthes (Nématodes) (Nématodes Gordiacés-Rotiféres-Gastrotriches-Kinorhynques). Tome IV. Fascicule II, III.

Greco, N. (1981). Hatching of *Heterodera carotae* and *H. avenae.* Nematologica, 27: 366-371.

Grewal, P.S., Ehlers, R.-U. and Shapiro-Ilan, D.I. (Eds.) (2005). Nematodes as Biocontrol Agents. CABI Publishing, CAB International, Oxon, U.K.,

Griffin, G.D. and Thyr, B.D. (1988) Interaction of *Meloidogyne hapla* and *Fusarium oxysporum f. sp..medicaginis* on alfalfa. Phytopathology 78: 421-5.

Griffiths, B.S. (1990). Approaches to measuring the contributions of nematodes and protozoa to nitrogen mineralization in the rhizosphere. Soil Use and Management 6: 88–90.

Griffiths B., Donn S., Neilson R. and Dniell T. (2006). Molecular sequencing and morphological analysis of a nematode community. Appl. Soil Ecol. 32: 325-337

Grossman, Joel. (1988). Research notes: New directions in nematode control. The IPM Practitioner. February. p. 14.

Grossman, Joel. (1990). New crop rotations foil root-knot nematodes. Common Sense Pest Control. Winter. p. 6.

Gupta, P. and Edward, J.C. (1973). A new record of a cyst forming nematode (*Heterodera chaubattia* n.sp.) from the hills of Uttar Pradesh, Curr. Sci., 42: 618.

Haeckel, E. (1866). Generelle Morphologie der Organismen. Vol. II. Berlin: Georg Reimer

Hackney, R.W., and O.J. Dickerson. (1975). Marigold, castor bean, and chrysanthemum as controls of *Meloidogyne incognita* and *Pratylenchus alleni.* Journal of Nematology. 7: 84-90.

Hadley, N. F. (1994). Water Relations of Terrestrial Arthropods. Academic Press, New York. 356 pp.

Hajibabaei M., Singer G., Hebert P. and Hickey D. (2007). DNA barcoding: how it complements taxonomy, molecular phylogenetics and population genetics. Trends Genet. 23: 167-172.

Hallmann, J., Quadt-Hallmann, A., Miller, W. G., Sikora, R. A., and Lindow, S. E. N (2001). Endophytic colonization of plants by the biocontrol agent *Rhizobium etli* G12 in relation to *Meloidogyne incognita* infection. Phytopathology 91: 415-422.

Handa, D.K., and Yadav, B.D. (1991) Comparative losses in husked, huskless barley and wheat due to *Heterodera avenae* in Rajasthan, India. Current Nematology 2: 99-102.

Handoo, l.A., Huettel, R.N. and Golden, AM. (1994). Description and SEM observations of *Meloidogyne sasseri* n. sp. (Nematoda: Meloidogynidae), parasitizing beachgrasses. Journal of Nematology 25 (1993): 628-641.

Handoo, l.A., Nyczepir, AP, Esmenjaud, D., van der Beek, J.G., Castagnone-Sereno, P., Carta, L.K., Skantar, A.M. and Higgins, J.A. (2004). Morphological, molecular, and differential-host characterization of *Meloidogyne floridensis* n. sp. (Nematoda: Meloidogynidae), a root-knot nematode parasitizing peach in Florida. Journal of Nematology 36: 20-35.

Handoo, l.A., Skantar, A.M., Carta, L.K. and Erbe, E.F. (2005). Morphological and molecular characterization of a new root-knot nematode, *Meloidogyne thailandica* n. sp. (Nematoda: Meloidogynidae), parasitizing ginger (*Zingiber* sp.). Journal of Nematology 37: 343-353.

Hanny Van Megen, Sven van den Elsen, Martijn Holterman, Gerrit Karssen, Paul Mooyman, Tom Bongers, Oleksandr Holovachov, Jaap Bakker and Johannes Helder (2009). A phylogenetic tree of nematodes based on about 1200 full-length small subunit ribosomal DNA sequences. Nematology, 11 (6): 927

Hanssler, G. (1990). Parasitism of *Verticillium lecanii* on cysts of *Heterodera schachtii.* Journal of Plant Diseases and Protection 97: 194—201.

Hanssler, G. and M. Hermanns. (1981). Verticillium lecanii as a parasite on cysts of *Heterodera schachtii.* Zeitschrift fur Pflanzenkrankheiten und Pflanzenschutz 88: 678-681.

Hardison, John R. (1964). "Justification for Burning Grass Fields," in *Proceedings: 24th Annual Oregon Seed Growers Conference*, Corvallis..

Hart, L.P. and Endo, R.M. (1981). The effect of time exposure to inoculum, plant age, root development and root wounding on *Fusarium* yellows of celery. Phytopathology 71: 77-79.

Hasan, N. and Jain, R.K. (1987). Parasitic nematodes and vesicular arbuscular mycorrhizal fungi associated with berseem (*Trifolium alexandrium* L.) in Bundelkhand region. Indian Journal Nematology, **17**: 184-188.

Haseeb, A. and Kumar, V. (2006). Management of *Meloidogyne incognita-Fusarium solani* disease complex in brinjal by biological control agents and organic additives. Ann. Plant Protect. Sci. 14: 519-521.

Hashioka, Y. (1963). Report to the Government of Thailand on blast and other diseases of rice in Thailand. *FAO, EPTA Rep.* 1734, 41 p.

Heald, C.M., Burton, B.C. and Davis, R.M. (1989). Influence of *Glomus intraradices* and soil phosphorus on *Meloidogyne incognita* infecting *Cucumis melo*. Journal of Nematology 21: 69-73.

Hendrix, P.F., Parmelee, R.W., Crossley D.A. Jr., Coleman, D.C., Odum, E.P. and Groffman, P.M. (1986). Detritus food webs in conventional and no tillage agroecosystems. Bioscience 36: 374-380.

Hesling, J.J. (1978). Cyst nematodes: morphology and identification of *Heterodera, Globodera* and *Punctodera*.In: Plant Nematology, J.F. Southey ed., HMSO.

Hewitt, W. B., Raski, D.J. and A. C. Goheen (1958). Nematode vector of soil borne fanleaf virus of grapevines. Phytopathlogy 46: 586-595.

Heydari, R.E., E. Pourjam and Mohammadi E. Golapeh (2006). Antagonistic effect of some species of *Pleurotus* on the rootknot nematode, *Meloidogyne javanica* invitro. Asian Network for Scientific Information. Plant Pathol. J. 5: 173–177

Hippocrates (460-375 B.C.) (1849). Works of Hippocrates, translated by F. Adams. London, "Aphorisms."

Hirsch, A.M., Bauer, W.D., Bird, D.M., Cullimore, J., Tyler, B. and Yoder, J.I. (2003). Molecular signals and receptors: Controlling rhizosphere interactions between plants and other organisms. Ecology 84(4): 858-868.

Hirschmann, H. (1982). *Meloidogyne platani* n. sp. (Meloidogynidae), a root-knot nematode parasitizing American sycamore. Journal of Nematology 14: 84-95.

Hirschmann, H. (1986). *Meloidogyne hispanica* n. sp. (Nematoda: Meloidogynidae), the 'Seville root-knot nematode'. Journal of Nematology 18: 520-532.

Hoberg, E., Monsen, K., Kutz, S., and Blouin, M. (1999). Structure, biodiversity, and historical biogeography of nematode faunas in holarctic ruminants: morphological and molecular diagnoses for *Teladorsagia boreoarcticus* n. sp. (Nematoda: *Ostertagiinae*), a dimorphic cryptic species in muskoxen (*Ovibos moschatus*). J. Parasitol. 85: 910-934.

Hollaway, G., Ophel-Keller, K., Taylor, S., Burns, R. and McKay, A. (2004). Effect of soil water content, sampling method and sample storage on the quantification of root lesion nematode (*Pratylenchus* spp.) by different methods. Australas. Plant Pathol. 32: 73-79.

Hong, R.L. and Sommer, R.J. (2006). *Pristionchus pacificus*: A well-rounded nematode. Bioessays 28: 651–659.

Huang, G.Z., Allen, R., Davis, E.L., Baum, T.J., and Hussey, R.S. (2006). Engineering broad root-knot resistance in transgenic plants by RNAi silencing of a conserved and essential root-knot nematode parasitism gene. Proceedings of the National Academy of Sciences, USA 103: 14302–14306.

Hübschen, J., Kling, L., Ipach, U., Zinkernagel, V., Brown, D. and R. Neilson (2004). Development and validation of species-specific primers that provide a molecular diagnostic for virus-vector longidorid nematodes and related species in German viticulture. *European* Journal of Plant Pathology.110: 883–891.

Huettel, R.N., Dickson, D.W. and D.T. Kalpan (1984). *Radopholus citrophilus* sp.n (Nematode), a sibling species of *Radopholus similis*. Proc Helminthol Soc Wash 51:32-35

Hugot, J.P. (1999). Primates and their pinworm parasites: the Cameron hypothesis revisited. System. Biol. 48: 523-546.

Hunt, H.W., Coleman, D.C., Ingham, E.R., Ingham,R.E., Elliot, E.T., Moore, J.C., Rose, S.L., Reis, C.P.P. and C.R. Morley (1987). The detrital food web in a short grass prairie. Biol. Fertil. Soils 3: 57-68.

Hunt, H.W., Coleman, D.C., Cole, C.V.,Ingham, R.E., Eliott, E.T. and L.E. Woods (1984). Simulation model of a food web with bacteria, amoeba and nematodes in soil. Pages 346–352 *in* Current perspectives in microbial ecology (Klug, M.J., and Reddy, C.A., eds.). Washington, DC, USA: American Society of Microbiology.

Hunt, D.J. and D. Moore (1999). Rhigonematida from New Britain diplopods. 2. The genera *Rhigonema* Cobb, 1898 and *Zalophora* Hunt, 1994 (Rhigonematoidea: Rhigonematidae) with descriptions of three new species. Nematology 1: 225–242.

Hussaini, S.S. (1983). Quantification of root knot nematode (*Meloidogyne* spp.) damage in F.C V. tobacco. Tob. Res. 9: 61-65.

Hussey, R. S. and Grundler, F. M. W. (1998). Nematode parasitism of plants. In: The Physiology and Biochemistry of free-living and plant-parasitic nematodes. Perry, R. N. and Wright, D. J. (Eds), CABI Publishing, UK. pp 213 – 243.

Hussey R.S. and McGuire J.M. (1987). Interactions with other organisms. Pp. 294-320. In: Principles and Practice of Nematode Control in Crops (Brown R.H. and Kerry B.R., eds). Academic Press, Marrickville, Australia.

Hussey, R. S., and R. W. Roncadori. (1978). Interaction of *Pratylenchus brachyurus* and *Gigaspora margarita* on cotton. J. Nematol. 10:16-20.

Hussey, R.S. and Roncadori, R.W. (1982). Vesicular arbuscular mycorrhizae may limit nematode activity and improve plant growth. Plant Disease 66: 9-14.

Huxley, T. H. (1864). Lectures on the elements of comparative anatomy. Xi + 303 pp., 111 figs. London

Hwang, Williamson. (2003). Leucine-rich repeat-mediated intramolecular interactions in nematode recognition and cell death signalling by the tomato resistance protein Mi. Plant Journal 34: 585-593.

Hwang, C.F., Bhakta, A.V., Truesdell, G.M., Pudlo, W.M. and Williamson, V.M. (2000).

Evidence for a role of the N terminus and leucine-rich repeat region of the Mi-gene product in regulation of localized cell death. Plant Cell 12:1319-29

Hyvonen, R. and Persson, T. (1990). Effects of acidification and liming on feeding groups of nematodes in coniferous forest soils. Biol. Fertil. Soils 9: 205-210.

Ibrahim, A.A.M., Al-Hazmi, A.S., Al-Yahya, F.A. and Alderfasi, A.A. (1999). Damage potential and reproduction of *Heterodera avenae* on wheat and barley under Saudi field conditions. Nematology 1(6): 625-630.

Ibrahim, S.K. and Rowe, J.A. (1995). Use of isoelectric focusing and polyacrylamide gel electrophoresis of nonspecific esterase phenotypes for the identi. cation of cyst nematodes

Heterodera species. Fundamental and Applied Nematology 18: 189-196.

Ichinohe, M. (1988). Current research on the major nematode problems in Japan. J. Nematology 20: 184-190.

Ijoyah, M. O. and Koutatouka, M. (2009). Effect of soil solarization using plastic mulch in controlling root-knot nematode (*Meloidogyne* spp.) infestation and yield of lettuce at Anse Boileau, Seychelles. African Journal of Biotechnology, 8 (24): 6787-6790

Inagaki, H. (1977). Soybean nematodes. Technical Bulletin, ASPAC Food and Fertilizer Technology Center, Taiwan No. 31, 17 pp.

Ingham, R.E. (1988). Interactions between nematodes and vesicular-arbuscular mycorrhizae. Agriculture, Ecosystems and Environment, 24: 169-182

Ingham, Elaine (1996). The Soil Foodweb: Its Importance in Ecosystem Health. <http://rain.org:80/~sals/ ingham.html>. 13 p.

Ingham, R.E., J.A. Trofymow, E.R. Ingham, and D.C. Coleman. (1985). Interactions of bacteria, fungi, and their nematode grazers: Effects on nutrient cycling and plant growth. Ecol. Monogr. 55:119-140.

Ingham, R.E. (1990). Biology and control of root-knot nematodes of potato—Research report. Proceedings of the Oregon Potato Conference and Trade Show. p. 109-120.

Ingham, R.E., Trofymow, J.A., Ingham, E.R., and Coleman, D.C. (1985). Interactions of bacteria, fungi and their nematode grazers: effects on nutrient cycling and plant growth. Ecological Monographs 55:119–140.

Inglis, W.G. (1983). An outline classification of the phylum Nematoda. Aust. J. Zool. 31: 243-255.

Ishibashi, N., Kondo, E., Muraoka, M. and Yokoo, T. (1973). Ecological significance of dormancy in plant parasitic nematodes. 1. Ecological difference between eggs in gelatinous matrix and cysts of *Heterodera glycines* Ichinoe. Appl. Ent. Zool., 8: 53-63.

Isreal, P., Rao, Y.S. and Rao, Y.R.V.J. (1963). Investigations on nematodes in rice and rice soils. Oryza 1: 125-127.

Itoh, Y, Ohshima, Y and Ichinohe, M. (1969). A root-knot nematode, *Meloidogyne mali* n. sp., on apple-tree from Japan (Tylenchida: Heteroderidae). Applied Entomology and Zoology, Tokyo 4: 194-202.

Jaffee, B.A. (2002). Soil cages for studying how organic amendments affect nematode-trapping fungi. Applied Soil Ecol., 21: 1-9.

Jaffee, B.A. (2006). Interactions among a soil organic amendment, nematodes and the nematode-trapping fungus *Dactylellina candidum*. Phytopathology, 96: 1388-1396.

Jaffe, H., Hwettel, R.N., Demilo, A. B., Hayes, D.K. and Rebois, R.V. (1989). Isolation and identification of a compound from soybean cyst nematode, *H. glycines* with sex pheromones activity. J. Chem. Ecol. 15:2031-2043.

Jagdale, G.B. and Grewal, P.S. (2008). Influence of the entomopathogenic nematode *Steinernema carpocapsae* in host cadavers or extracts from cadavers on the foliar nematode *Aphelenchoides fragariae* on Hosta. Biological Control 44: 13-23.

Jagdale, G.B., Somasekhar, N., Grewal, P.S., and Klein, M.G. (2002). Suppression of plant parasitic nematodes by application of live and dead entomopathogenic nematodes on Boxwood (*Buxus* spp). Biological Control. 24, 42-49.

Jain, R.K. and Hasan N. (1994). Vesicular-arbuscular mycorrhizal fungi, a potential biocontrol agent for nematode. Vistas in Seed Biology, 11: 49-60.

Jairajpuri, M.S. and Baqari, Q.H. (1992). *Nematode pests of rice.* New Delhi, India, IBH Publishing Co., 66 pp.

Jansson. H.B. and Poinar, G.O. Jr (1986). Some possible fossil nematophagous fungi. Trans Brit Mycol Soc 87:471–474.

Jansson, H. and C.O. Persson (2000). Growth and capture activities of *Nematophagous fungi* in soil visualized by low temperature scanning electron microscopy. Mycologia, 92: 10-15

Jaubert, S., Laffaire, J.B., Abad, P. and Rosso, M.N. (2002). A polygalacturonase of animal origin isolated from the root-knot nematode, *Meloidogyne incognita*. FEBS Lett. 522, 109 –112.

Jayanthi, M., Shankar, G., and Baskaran, P. (1987). Parasitic nematode in white striated plant hopper (WSPH) of rice. International Rice Research Newsletter 12:23.

Jayaprakash, A., and Rao, Y.S. (1984). Cyst nematode, *Heterodera oryzicola* and seedling blight fungus, *Sclerotium rolfsii* disease complex in rice. Indian Journal of Nematology 14:58–59.

Jenkins, L., and H.W. Guengrich. (1959). Chemical dips for control of nematodes on bare root nursery stock. Plant Disease Reporter 43:1095-1097.

Jenkins, W.R. and Coursen, B.W. (1957). The effect of the root-knot nematodes, *Meloidogyne incognita acrita* and *M. hapla* on *Fusarium* wilt of tomato. Plant Disease Reporter, 41:182-86

Jenkins, W. R., and Taylor, D. P. (1967). Introduction. Page 7 in: Plant Nematology. Reinhold Publishing Corporation, New York.

Jepson, S. B. (1983). Identification of *Meloidogyne*: a general assessment and a comparison of male morphology using light rnicroscopy, with a key to 24 species. Revue Nèmatol. 6 (2): 291-309.

Jepson, S.B. (1984). *Meloidogyne kralli* n. sp. (Nematoda: Meloidogynidae) a root-knot nematode parasitizing sedge (*Carex acuta* L.). Revue de Nematologie 6 (1983): 239-245.

Jepson, S.B. (1985). *Meloidogyne chitwoodi.* In: *CIH Descriptions of Plant-parasitic Nematodes,* Set 8, No. 106, Commonwealth Agricultural Bureaux, Farnham Royal, UK.

Jerath, M.L. (1968). *Heterodera sacchari,* a cyst nematode pest of sugarcane new to Nigeria, P1. Dis. Reptr. 52: 237.

Job, S. K. Charles and Thinniyam S. Venkitesan (1995). Biology of banana population of *Heterodera oryzicola* (Nemata : Tylenchina). Fundam. appl. Nematol. 18 (5):493-496.

Jobert, C. (1878). Sur une maladie du cafeier observee au Bresil. C. R. Acad. Sci. Paris 87: 941-943.

Johnson, A.W., and J. Feldmesser (1987). Nematicides – A Historical Review. Pp. 448-454 *In*: J.A. Veech and D.W. Dickson, eds. Vistas on Nematology. Society of Nematologists, Inc. Hyattsville, MD.

Johnson A.W and R.H Littrell (1969). Effect of *Meloidogyne incognita, M. hapla* and *M. javanica* on the severity of *Fusarium* Wilt of Chrysanthemum. J. Nematol.1(2): 122-5.

Johnson, S.R., Ferris, V.R.and Ferris, J.M. (1972) Nematode community structure of forest woodlots: I. Relationships based on the similarity coefficients of nematode species. J Nematol. 4:175-183

Johnson, A.W., Dowler, C.C., and Hauser, E.W. (1975). Crop rotation and herbicide effects on population densities of plant parasitic nematodes. J. Nematol. 7:158–168.

Jonathan, E.I. and G. Rajendran (1998). Interaction of *Meloidogyne incognita* and *Fusarium oxysporum f. cubense* on banana. Nematol. Medit. 26:9-11.

Jones, F. G. W. (1961). The potato root eelworm, *Heterodera rostochiensis* in India. Current Science, 30: 187.

Jones, J.P., Overman, A.J. and Crill, P. (1976) Failure of root-knot nematode to affect Fusarium wilt resistance of tomato. Phytopathology, 66:1339-41.

Jonz, M.G., Riga, E. Mercier, A.J. and Potter, J.W. (2001). Partial isolation of a water soluble pheromone from sugarbeet cyst nematode, *H. schachtii* using a novel bioassay. Nematology 3:55-64.

Jorgensen, R.A., Cluster, P.D., English, J., Que, Q., and Napoli, C.A. (1996). Chalcone synthase cosuppression phenotypes in petunia flowers: comparison of sense vs. antisense constructs and single-copy vs. complex T-DNA sequences. Plant Molecular Biology 31:957–973.

Joseph, T.A. and Krishna Prasad, K.S. (2002). Kufri Giriraj: a high yielding late blight resistant potato variety forNilgiris (Abstr.), National seminar on "Changing Scenario in the Production Systems of Hill Horticultural Crops", February 20-21, pp:196

Joseph., T.A. and Krishna Prasad, K. S. (2005). Adaptibility of potato hybrids with combined resistance to late blight and potato cyst nematodes in Nilgiri hills. J. Ind. Potato Assoc., 32:236

Joseph, T.A., Krishna Prasad, K.S. and Latha, M. (2003). Breeding potato for combined resistance to late blight and potato cyst nematodes in Nilgiris. J. Ind. Potato Assoc., 30: 19-20

Jothi, G. and Babu, R.S.(2000). Interaction effect of VAM and root knot nematode on the growth and nutrient content of brinjal. International Journal of Tropical Plant Disease, 18: 91-100.

Justine, J.L. (2002). Embryology, developmental biology and the genome. In: The Biology of Nematodes, D.L. Lee, ed., London: Taylor and Francis, pp. 121–137.

Kalha, C.S. and Edward, J.C. (1979). *Heterodera vigni* Edward and Misra, 1968, a synonym of *H. cajani* Koshy, 1967, Allahabad Fmr., 50:143.

Kaplan, J. K. (1991). Dress-for-Success Mulch. Ag. Research. 39(9): 10-13.

Kasab, A.S. and Taha, A.H.Y. (1990). Interaction between plant parasitic nematodes, vesicular arbuscular mycorrhiza, Rhizobia and nematicide on Egyptian clover. Annals of Agricultural Sciences, 35: 509-520.

Karczmarek, A., Fudali, S., Lichocka, M., Sobczak, M., Kurek, W., Janakowski, S., Roosien, J., Golinowski, W., Bakker, J., and Goverse, A. (2008). Expression of two functionally distinct plant endo-b-1,4-glucanases is essential for the compatible interaction between potato cyst nematode and its hosts. Mol Plant Microbe Interact 21: 791–798

Karssen, G. (1996). Description of *Meloidogyne fallax n.* sp. (Nematoda: Heteroderidae), a root-knot nematode from The Netherlands. Fundamental and Applied Nematology 19: 593-599.

Karssen, G., van Aelst, A. and van der Putten, W.H. (1998). *Meloidogyne duytsi* n. sp. (Nematoda: Heteroderidae), a root-knot nematode from Dutch coastal foredunes. Fundamental and Applied Nematology 21: 299-306.

Karssen, G., Bolk, R.J., van Aelst, A.C., van den Beld, I., Kox, L.F.F., Korthals, G., Molendijk, L., Zijlstra, C., van Hoof, R. and Cook, R. (2004). Description of

Meloidogyne minor n. sp. (Nematoda: Meloidogynidae), a root-knot nematode associated with yellow patch disease in golf courses. Nematology 6: 59-72.

Kaskavalci, G. (2007). Effects of soil solarization and organic amendment treatments for controlling meloidogyne incognita in tomato cultivars in Western anatolia. Turk. J. Agric. For., 31: 159-167.

Kaushal, K. K. (1993). On the effectiveness of neem products against the cereal cyst nematode, *Heterodera avenae*. Indian J. Nematol. 23: 218-219.

Kaushal, K. K. (1996). A report on *Heterodera spinicauda* from India. Ind. J. Nematol. 26: 272-274.

Kaushal, K. K., Chawla, G., Pankaj, Sirohi, A.and Singh, K. (2008). Report on *Heterodera trifolii* from Kangra, Himachal Pradesh. Indian J Nematol. 38: 261-62.

Kaushal, K.K., Tiwari, S.P. and Uma Rao (2002). *Heterodera glycines* in India – first report. Annals of Plant Protection Sciences 10: 410

Kaushal, K.K and Seshadri, A.R. (1986). Effect of chemical treatment in soilon emergence, penetration and development of *Heterodera avenae* on wheat. Ind. J. Nematol. 16:171-174.

Kaushal, K.K and Swarup, G. (1988). Two new cyst nematode species from India. Indian J. Nematol. 18 : 299-306.

Kellam, M. K., and N. C. Schenck. (980). Interactions between a vesicular-arbuscular mycorrhizal fungus and root-knot nematode on soybean. Phytopathology 70:293-296.

Kerry, B. (1988). Fungal parasites of cyst nematodes. Agriculture, Ecosystems and Environment, 24: 293-305

Kerry, B. R., D. H. Crump, L. A. Mullen (1982). Studies of the cereal cyst-nematode, *Heterodera avenae* under continuous cereals, 1975–1978. II. Fungal parasitism of nematode females and eggs. Annals of Applied Biology, 100 (3): 489 – 499

Khan, A.D. and Husain, S.I. (1965). *Heterodera mothi* n.sp. (Tylenchida: Heteroderidae) parasitizing *Cyperus rotundus* L. in Aligarh, U.P., India. Nematologica, 11: 167.

Khan, M.W. and Müller, J. (1982). Interaction between *Rhizoctonia solani* and *Meloidogyne hapla* on radish in gnotobiotic culture. Libyan Journal of Agriculture, 11: 133-40.

Khan, M.R (2008). Current Options for Managing Nematodes Pest of Crops in India.

Khan, U.V. and Jairajpuri, M.S. (1975). A report on the occurrence of an egg sac in *Heterodera mothi* (Nematoda: Tylenchida), Curr. Sci., 44:590.

Khair, A.M., A.A. Farahat and S.K. Abadir (1989). Studies on the corn cyst nematode (CCN) *H. zeae* in Egypt; IV variation in development and reproduction of four different populations of some corn cultivars. Pakistan J. Nematol., 7: 69–73

Khurramov, Sh. Kh (1990). The use of solarization in control of gall-forming nematodes. Zashchita Rastenii (Moskva) (4): 12-13

Khoury, F.Y. and Alcorn, S.M. (1973). Effect of *Meloidogyne incognita acrita* on the susceptibility of cotton plants to *Verticillium albo-utrum*. Phytopathology, 63:485-90

Kilpatrick, R.A., Gilchrist, L. and Golden, A. M. (1976). Root knot on wheat in Chile. Plant Disease Reporter 60:135.

Kimenju, J.W., D.M. Muiru, N.K. Karanja, M.W. Nyongesa and D.W. Miano (2004). Assessing the role of organic amendments in management of root-knot nematodes on common bean, *Phaseolus vulgaris* L. Trop. Microbiol. Biotechnol., 3: 14-23.

Kirby, M.F., Kirby, M.E., Siddiqi, M.R. and Loff, P.A.A. (1978). Plant parasitic nematodes from Fiji: nematode distributions and host associations. UNDP/FAO Survey agric. Pests and Dis. S. Paficic Reg. - Coord. Comittee Meet., 28 Feb.-2 March, 1978. Rep. Mimeogr. 12 pp.

Kiryanova, E. S.and Ivanova, T. S.(1969). A cyst-forming nematode, *Heterodera cardiolata* n.sp. (Nematoda: Heteroderidae) from Dushanbe, Tadzhikistan. Doklady Akademii Nauk Tadzhikskoi SSR Vol. 12 No. 12 pp. 59-62

Kleynhans, K.P.N. (1986). *Meloidogyne partityla* sp. novo from pecan nut *[Carya illinoensis* (Wangenh.) C. Koch] in the Transvaallowveld (Nematoda: Meloidogynidae). Phytophylactica 18: 103-106.

Kleynhans, K.P.N. (1988). *Meloidogyne vandervegtei* sp. novo from a subtropical coastal forest in Natal (Nemata: Heteroderidae). Phytophylactica 20:263-267.

Klink, V.P., Kim, K.H., Martins, V., Macdonald, M.H., Beard, H.S., Alkharouf, N.W., Lee, S.K., Park, S.C. and Matthews, B.F. (2009). A correlation between host-mediated expression of parasite genes as tandem inverted repeats and abrogation of development of female *Heterodera glycines* cyst formation during infection of Glycine max. Planta 230: 53–71.

Kofoid, C.A. and White, W.A. (1919). A new nematode infection of man. Journal of the American Medical Association 72:567-569.

Koenning, S.R., Overstreet, C., Noling, J.W., Donald, P.A., Becker, J.O., and Fortnum, B.A. (1999). Survey of crop losses in response to phytoparasitic nematodes in the United States for 1994. J. Nematol. 31: 587-618.

Koenning, S.R., K.L. Edmisten, K.R. Barker, D.T. Bowman and D.E. Morrison (2003). Effects of rate and time of application of poultry litter on *Hoplolaimus Columbus* on cotton. Plant Dis., 87:1244-1249.

Kort, J. (1972). Nematode diseases of cereals of temperate climates. In J.M. Webster, ed. Economic nematology, p. 97-126. New York, NY, USA, Academic.

Kort, J., Ross, H., Rumpenhost, H.J. and Stone, A.R. (1977). An interaction scheme for identifying and classifying pathotypes of potato cyst nematodes *Globodera rostochiensis* and *G. pallida*. Nematologica, 23: 333–339.

Koshy, P.K. (1967). A new species of *Heterodera* from India. Indian Phytopath. 20: 272.

Koshy, P. K. and Evans, K. (1986). Hatching from cysts and eggsacs of *Heterodera cruciferae* and effects of temperature on hatching and development of oilseed rape. *Arm. appl. Biol.*, 109: 163-171.

Koshy, P.K. and Swarup, G. (1971a). Investigations on the life history of the pigeon-pea cyst nematode, *Heterodera cajani*, Indian J. Nematol. 1:44.

Koshy, P.K. and Swarup, G. (1971b). On the number of generations of *Heterodera cajani* the pigeon-pea cyst nematode in a year, Indian J. Nematol, 1: 88.

Koshy, P.K. and Swarup, G. (1971c). Distribution of *Heterodera avenue, H. cajani, H. zeae* and *Anguina tritici* in India. Indian J. Nematol., 1: 106.

Koshy, P.K. and Swarup, G. (1971d). Factors affecting emergence of larvae from cysts of *Heterodera cajani* Koshy, 1967. Indian J. Nematol. 1: 209.

Koshy, P.K. and Swarup, G. (1971e). Susceptibility of aerial parts of pigeonpea seedlings to *Heterodera cajani* larvae. Indian J. Nematol. 1:245.

Koshy, P.K. and Swarup, G. (1972). Susceptibility of plants to pigeonpea cyst nematode. Indian J. Nematol., 2: 1.

Koshy, P.K., Swarup, G. and. Sethi, C.L. (1970). *H. zeae* n. sp. (Nematoda: Heteroderidae), a cyst-forming nematode on *Zea mays*. Nematologica, 16: 511–516

Koshy, P.K., Swarup, G.and Sethi, C.L. (1971). Further notes on the pigeonpea cyst nematode *Heterodera cajani*. Nematologica, 16:477.

Krall, E.L. and Krall, Kh. A. (1979). The revision of plant nematodes of the family Heteroderidae (Nematoda, tylenchida) by using a comparative ecological method of studying their phylogeny], and "Printsipy i metodyi zucheniya uzaimootnoshenii mezhdu paraziti- cheskimi nematodami irasteniyamil, Tartu, USSR (1978, publ. 1979). 368

Krishna Prasad, K.S. (2004a). Dominance of *Globodera pallida* over *Globodera rostochiensis* (Abstr.), National Symposium on Paradigms in Nematological Research for Biodynamic Farming, Bangalore, November, 17-19, pp:11-12

Krishna Prasad, K.S. (2004b). Paradigms of potato cyst nematode research in India. Lead paper, National Symposium on Paradigms in Nematological Research for Biodynamic Farming, Bangalore, November, 17- 19. pp: 21-23

Krishna Prasad, K.S. (2006). Potato cyst nematodes and their management in Nilgiris. *Tech. Bull.*, CPRI, Shimla, India, p.20

Krishna Prasad, K.S. (2007a). Dynamics of nematode management in enhancing the potato production. Potato J., **34**:31-33

Krishna Prasad, K.S. (2007b). National perspective of potato nematode management. Lead paper, III National Symposium on Plant Protection in Horticulture: Emerging Trends and Challenges, March 7-9, Bangalore

Krishna Prasad, K.S. (2008). Management of potato nematodes: An overview. J. Hortl. Sci. 3 (2): 89-106.

Krishna Prasad, K.S. and Latha, M. (1999). Potato breeding for combined resistance to cyst nematodes and late blight disease. (Abstr.), Global Conference on Potato, Dec. 6-11, New Delhi, p.171

Krishna Prasad, K.S., Krishnappa, K., Setty, K.G.H., Reddy, B.M.R. and Reddy, H.R.C. (1980) Susceptibility of some cereals to ragi cyst nematode *Heterodera delvii*. Curr. Res., 9 : 114.

Krishna Prasad, K.S. and Nagesh, M. (2007). Evaluation of bioagents for management of potato cyst nematodes. **In:** III National Symposium on Plant Protection in Horticulture: Emerging Trends and Challenges, March 7-9, Bangalore. P. 34

Krishnaveni, M., and Subramanian, S. (2004). Evaluation of biocontrol agents for the management of *Meloidogyne incognita* on cucumber (*Cucumis sativus* L.). Curr. Nematol. 15: 33-37.

Krusberg, L. R. (1988). The corn cyst nematode, *Heterodera zeae*, in the United States. Pp. 171–175 in M. A. Maqbool, A. M. Golden, A. Ghaffar, and L. R. Krusberg, eds. Advances in plant nematology. Proceedings of the U.S.—Pakistan International Workshop on Plant Nematology, Karachi, Pakistan.

Kuhn, J. (1874). Ubers das Vorkommen von Ruben-Nematoden an den Wurzeln der Halmfruchte. Landwirts. Jahrb. 3:47-50.

Kumar, A.C. (1964). A note on the occurrence of *Heterodera cacti* from Mysore, Curr. Sci. 33:534.

Kumar, A.C. (1980). Studies on nematodes in coffee soils of South India. 4. Occurrence of *Heterodera cyperi*. J. Coffee Res., 10:77.

Kumar, V., Haseeb, A., and Sharma, A. (2009). Integrated management of *Meloidogyne incognita- Fusarium solani* disease complex of brinjal cv. Pusa Kranti. Ann. Plant Protect. Sci. 17: 192-194.

Kuhn, J. (1874). Ubers das Vorkommen von Ruben-Nematoden an den Wurzeln der Halmfruchte. Landwirts. Jahrb. 3: 47-50.

Kunert, and H. Lysek. (1987). Mutants of ovicidal fungus *Verticillium chlamydosporium* with chitinolytic activity. Acta Universitatis Palackianae. Otomucensis, Facuhatis Medicinae 116:71-79.

Kutywayo, V. and Been, T.H. (2006). Host status of six major weeds to *Meloidogyne chitwoodi* and *Pratylenchus penetrans*, including a preliminary field survey concerning other weeds. Nematology 8: 647-657.

Kwock, O.C.H., Weisleder D. Plettner and D.T. Wichlow (1992). A nematicidal toxin from *Pleurotus ostreatus* NRRL 3526. J. Chem. Ecol., 18: 127–137

Kyrou, N.C. (1976). Biological notes on *Heterodera avenae* Woll. 1924 studied on wheat in central Macedonia. Annales de l'Institut Phytopathologique Benak New Series 11:187-192.

Labeena, P., Sreenivasa, M. N. and Lingaraju, S. (2002). Interaction effects between arbuscular mycorrhizal fungi and root-knot nematode Meloidogyne incognita on tomato. Indian Journal of Nematology, **32**: 118-120.

Lal, A. and V.K. Mathur (1982). Occurrence of *H. zeae* on *Vetiveria zizanioides*. Indian J. Nematol., 12: 405-407

Lambshead, P. J. D. (1993). Recent developments in marine benthic biodiversity research. *Océanis* 19: 5-24.

Landels, S. (1989). Fumigants and nematicides. Chemical Economics Handbook, Stanford Research Institute International, California

Lankester, E. R. (1877). Notes on the embryology and classification of the animal kingdom; comprising a revision of speculations relative to the origin and significance of the germlayers. Quart. Jour. Micro. Sci. n.s., 17:339454, 17 figs., pl 25 (figs. 120).

Lawton, J. H., Bigtnell, D. E., Bloemers, G. F., Eggleton, P., and Hodda, M. E. (1996). Carbon flux and diversity of nematodes and termites in Cameroon forest soils. Biodiversity Conservation 5: 261-273.

Lazarovits, G., Tenuta, M.and Conn, K.L. (2001). Organic amendments as a disease control strategy for soilborne diseases of high-value agricultural crops. Australas. Plant Pathol. 30: 111-117.

Lee, D.L. (2002). Male and female gametes and fertilisation. In: The Biology of Nematodes, D.L. Lee, ed., London: Taylor and Francis, pp. 73–121.

Lehman, P. S., and Lordello, L. G. E. (1982). *Meloidogyne exigua*, a root-knot nematode of coffee. Nematology Circular No. 88, pp.4. Florida Department of Agriculture and Consumer Services, Division of Plant Industry, Gainesville, FL, USA.

Lenz, R. and Eisenbeis, G. (2000). Short-term effects of different tillage in a sustainable farming system on nematode community structure. Biol. Fertil. Soils 31: 237-244.

Leuckart, R (1865). On the developmental history of nematode worms. Archiv fur Heilkunde. Band ii, pp 195-235.

Lewis, W.J., Van Lenteren, J.C., Phatak, S.C. and Tumlinson, J.H. III (1997). A total systems approach to sustainable pest management. Proc. Natl. Acad. Sci. USA 94, 12243–12248.

Lewis, E.E. and Grewal, P.S. (2005). Interactions with plant-parasitic nematodes. In: Grewal, P.S., Ehlers, R.-U., Shapiro-Ilan, D.I. (Eds.), Nematodes As Biocontrol Agents. CABI Publishing, CAB International, Oxon, U.K., pp. 349-362.

Li-Hui Wei, Qing-Yun Xue, Ben-Qing Wei, Yong-Ming Wang, Shi-Mo Li, Li-Feng Chen and Jian-Hua Guo (2010). Screening of antagonistic bacterial strains against *Meloidogyne incognita* using protease activity. Biocontrol Science and Technology, http://www.informaworld.com/smpp/title~db=all~content=t713409232~tab=issueslist~branches=20 - v2020 (7):739 – 750.

Li, J., Todd, T.C., Lee, J. and Trick, H.N. (2011). Biotechnological application of functional genomics towards plant-parasitic nematode control. Plant Biotechnology Journal: 1-9.

Li, J., Todd, T.C., Oakley, T.R., Lee, J. and Trick, H.N. (2010). Host-derived suppression of nematode reproductive and fitness genes decreases fecundity of *Heterodera glycines* Ichinohe. Planta, 232: 775–785.

Li, S.J. and Yu, Z. (1991). A new species of root-knot nematode (*Meloidogyne actinidiae*) on *Actinidia chinensis* in Henan Province. Acta Agriculturae Universitatis Henanensis 25: 251-253.

Li, T.F., Zhang, K.Q. and Liu, X.Z. (2000).Taxonomy of Nematophagous Fungi. Chinese Scientific and Technological Publications, Beijing.

Li, Y., Hyde, K.D., Jeewon, R., Cai, L., Vijaykrishna, D. and Zhang, K.Q. (2005). Mycologia 97:1034–1046.

Liao, J.L. and Feng, Z. (1995). A root-knot nematode *Meloidogyne hainanensis* sp. novo (Nematoda: Meloidogynidae) parasitizing rice in Hainan, China. *Journal of South China Agricultural University* 16: 34-39.

Liao, J.L. and Feng, Z. (2003). A new species of *Meloidogyne* from China. Russian Journal of Nematology 11: 142 (abstr.).

Liao, J. L., Yang, W. C., Feng, Z. X. and Karssen, G. (2005). Description of *Meloidogyne panyuensis* sp. n. (Nematoda: Meloidogynidae), parasitic on peanut (*Arachis hypogaea* L.) in China. Russian Journal of Nematology 13: 107-114.

Liu, X. Z., Xiang, M.C. and Che, Y.S. (2009). The living strategy of nematophagous fungi. Mycoscience 50: 20-25.

Lilley, C.J., Bakhetia, M., Charlton W.L. and Urwin, P.E. (2007). Recent progress in the development of RNA interference for plant parasitic nematodes. *Molecular Plant Pathology* 8: 701–711.

Linnaeus, Carl von. (1758). Systema naturae regna tria naturae, secundum classes, ordines, genera, species, cum characteribus differentüs, synon-ymis, locis. Editio decima reformats. 1.

Lipscomb D, Platnick N, and Wheeler Q (2003). The intellectual content of taxonomy: a comment on DNA taxonomy. Trends Ecol. Evol. 18: 64-66

Liu, B., Hibbard, J.K., Urwin, P.E. and Atkinson, H.J. (2005). The production of synthetic chemodisruptive peptides *in planta* disrupts the establishment of cyst nematodes. Plant Biotechnology Journal 3: 487–496.

Loos, C.A. (1953). *Meloidogyne brevicauda* n. sp., a cause of root-knot of mature tea in Ceylon. Proceedings of the Helminthological Society of Washington 20: 83-91.

Lopez, R. (1984). *Meloidogyne salasi* sp. n. (Nematoda: Meloidogynidae), a new parasite of rice (*Oryza sativa* L.) from Costa Rica and Panama. Turrialba 34:275-286.

Lopez, R. and Salazar, L. (1989). *Meloidogyne arabicida* sp. n. (Nemata: Heteroderidae) nativo de Costa Rica: un nuevo y severo patogeno del cafeto. Turrialba 39: 313-323.

Lordello, L.G.E. (1956a). *Meloidogyne inornata* sp. n. a serious pest of soybean in the state of Sao Paulo, Brazil (Nematoda, Heteroderidae). Revista Brasil Biologica 16: 65-70.

Lordello, L.G.E. (1956b). Nemat6ides que parasitanica soja na regiao de Bauru. Bragantia 15:55-64.

Lordello, L.G.E. and Zamith, A.P.L. (1960) *Meloidogyne coffeicola* sp. n., a pest of coffee trees in the State of Parana, Brazil (Nematoda: Heteroderidae). Revista Brasil Biologica 20: 375-379.

Lockwood, J.L. (1988). Evolution of concepts associated with soil borne plant pathogens. Annual Review of Phytopathology, 26:93-121.

Logiswaran, G. and Menon, P.P.V. (1969). Preliminary observations on the spread of the golden nematode (Heterodera rostochiensis Woll.) of potatoes in the Niligiris, Madras State, All India Nematol. Symp. New Delhi. p7.

Lopez-Brana, I., Romero, M.D. and Delibes, A. (1996). Analysis of *Heterodera avenae* populations by the random amplified polymorphic DNA technique. Genome 39: 118- 122.

Lorenzen, S. (1981). Entwurf eines phylogenetischen Systems der freilebenden Nematoden. Veröffentlichungen des Institut für Meeresforschungen Bremerhaven, *7*(Suppl.), 472 pp.

Lorenzen, S. (1994). The Phylogenetic Systematics of Freeliving Nematodes. London: The Ray Society.

Luc, M. (1961). Nématodes du genre Heterodera parasites de cultures tropicales en Afrique, C.r. Seanc. Acad. Agric. Fr., 47:940.

Luc, M. (1974). *Heterodera sacchari.* C.I.H. Description P1. Para. Nematodes, Set 4, No. 48.

Luc, M. and Berdon-Brizuela, R.(1961). *Heterodera oryzae* n.sp. (Nematoda: Tylenchoidae) parasite du riz en Côte d'Ivoire. Nematologica 6:272.

Luc, M., Doucet, M.E., Fortuner, R., Castillo, P., Decraemer, W., Lax, P. (2010). Usefulness of morphological data for the study of nematode biodiversity. Nematology 12: 495-504.

Luc, M. and Merny, G. (1963). *Heterodera sacchari* n.sp. (Nematoda: Tylenchoidea) parasite de la canne a sucre au Congo-Brazzaville. Nematologica 9 :31,

Luc, M. and Netscher, C. (1974). Presence of the sugar-beet nematode at Dakar, FAO P1. Protect. Bull.,22:24.

Luc, M. and Taylor, D.P. (1977). *Heterodera oryzae.* C.I.H. Descript. P1. Paras. Nematodes. Set 7. No. 91.

Luebke, R.W. (2007). Nematodes as host resistance models for detection of immunotoxicity. Methods 41:38-47.

Luna, J. (1993). Crop rotation and cover crops suppress nematodes in potatoes. Pacific Northwest Sustainable Agriculture. March. p. 4-5.Chalupova, V., K. Lenhart, J.

Luong, L.T., Platzer, *E.G.*, De Ley, P., and Thomas, W.K. (1999). Morphological and molecular characterization of *Mehdinema alii* (Nematoda: Diplogasterida) from the decorated cricket (*Gryllodes sigillatus*). J. Parasitol. 85: 1053–1064.

Maas, P.WT., Sanders, H. and Dede, J. (1978). *Meloidogyne oryzae* n. sp. (Nematoda, Meloidogynidae) infesting irrigated rice in Surinam (South America). Nematologica 24: 305-311.

Mac Guidwin, A. E., Bird, G. W. and Safir, G. R.(1985). Influence of Glomus fasciculatum on Meloidogyne hapla infecting Allium cepa. Journal of Nematology, 17: 389-395

Madani, M., Subbotin, S., and Moens, M. (2005). Quantitative detection of the potato cyst nematode, *Globodera pallida*, and the beet cyst nematode, *Heterodera schachtii*, using Real-Time PCR with SYBR green I dye. Mol. Cell. Probes, 19: 81-86.

Madulu, J. and Trudgill, D. L. (1994). Influence of temperature on *Meloidogyne javanica*. Nematologica, 40: 230 - 243.

Maheswari, T.U., Sharma, S.B., Reddy, D.D. and M.P. Haware (1997). Interaction of *Fusarium oxysporum* f. sp. *ciceri* and *Meloidogyne javanica* on *Cicer arietinum*. J Nematol.29(1): 117.

Maggenti, A. R. (1963). Comparative morphology in nemic phylogeny. Pp. 273 282 *in* E. C. Dougherty, ed. The lower Metazoa, comparative biology and phylogeny. University of California Press, Berkeley.

Maggenti, A. R. (1981). *General Nematology*. Springer-Verlag, New York, NY.

Maggenti, A. R. (1983). Nematode higher classification as influenced by species and family concepts. Pp. 2540 *in* A. R. Stone, H. M. Platt, and L. F. Khalil, eds. Concepts in nematode systematics. Academic Press Inc., London.

Maggenti, A. R (1991a). Nemata: higher classification. *In*: Manual of Agricultural Nematology, pp. 147-187. Marcel Dekker, Inc., New York, NY.

Maggenti, A. R. (1991b). General nematode morphology. In: Manual of Agricultural Nematology, pp. 3-46. Marcel Dekker, Inc., New York, NY.

Maggenti, A. R. and Allen, M. W. (1960). The origin of the gelatinous matrix in *Meloidogyne*. Proceedings of the Helminthological Society of Washington, 27:4 - 10.

Mai, W. F. and G. S. Abawi (1987). Interactions among root-knot nematodes and Fusarium wilt fungi on host plants. Ann.Rev.Phytopath. 25:317-38

Malakhov, V. V. (1994). Nematodes: Structure Development, Classification and Phylogeny. Smithsonian Institution Press, Washington, DC.

de Man, J. G. (1880). Die einheimischen, frei in der reinen Erde und im siissen Wasser lebenden Nematoden. Tijdschr. ned. Dierk. Vereen. 5: 1-104.

de Man, J.G. (1876). Onderzoekingen over vrij in de aarde levende nematoden. Tijdschr. Ned. Dierk. Ver. 2: 78-196.

de Man, J. G. (1884). Die frei it ïder reinen Erde und im siissen Wasser lebenden Nematoden der niederlandischen Fauna. Eine systematische faunistische Monographie. Leiden, 206 p.

Manser, P.D. (1968). *Meloidogyne graminicola* a cause of root-knot of rice. Pl. Prot. Bull. F.A.O., 16: 11.

Maqbool, M.A. (1981). Occurrence of root knot and cyst nematodes in Pakistan. Nematol. Mediterranea, 9: 211–212

Maqbool, M.A. (1988). Present status of research on plant parasitic nematodes in cereals and food and forage legumes in Pakistan. In M.C. Saxena, R.A. Sikora and J.P. Srivastava, eds. Nematodes parasitic to cereals and legumes in temperate semi-arid regions, p. 173-180. Aleppo, Syria, ICARDA.

Maqbool, M.A. and S. Hashmi (1984). New host records of cyst nematodes *H. zeae* and *H. mothi* from Pakistan. Pakistan J. Nematol., 2: 99–100

Marban-Mendoza, N., Jeyaprakash, A., Jansson, H.B., Damon, R.A. and Zuckerman, B.M. (1987). Control of root-knot nematodes on tomato by lectins. Journal of Nematology **19**: 331–335.

Marcinowski, K. (1909). Parasitisch und semiparasitisch an Pflanzen lebende Nematoden. Arbeiten aus der Kaiserlichen Biologischen Bundesanstalt fUr Land- und Forstwirtschaft, Berlin 7: 1-192.

Marinari-Palmisano, A. and Ambrogioni, L. (2000). *Meloidogyne ulmi* sp. n., a root-knot nematode from elm. Nematologia Mediterranea 28: 279-293.

Masoomeh, S.G., R.A. Mehdi, R.B. Shahrokh, E. Ali and Z. Rasoul (2004). Screening of soil and sheep faecal samples for predacious fungi: Isolation and characterization of the nematode-trapping fungus *Arthrobotrys oligospora*. Iran. Biomed. 8: 135-142.

Mateille, T., Schwey, D., and Amazouz, S. (2005). Sur tomates, la cartographie des indices de galles. Phytoma 584: 40-43.

Mathur, V.K., and Prasad, S.K. (1971). Occurrence and distribution of *Hirschmanniella oryzae* in the Indian Union with description of *H. mangaloriensis* n.sp. Indian Journal of Nematology 1:220–226.

Mathur, V.K., and Prasad, S.K. (1973). Control of *Hirschmanniella oryzae* associated with paddy. Indian Journal of Nematology 3:54–60.

Mathur, B.N., Handa, D.K., Swarup, G., Sethi, C.L., Sharma, G.L., Yadav, B.D. (1980). On the loss estimation and chemical control of molya disease of wheat caused by *Heterodera avenae* in India. Indian Journal Nematology 16: 152-159.

Mayr, E. and Ashlock, P. (1991). Principles of systematic zoology. 2nd ed.McGraw-Hill, New York, USA, p. 428.

McBeth, C.W. (1954). Some practical aspects of soil fumigation. Plant Disease Reporter Supplement 227: 95-97.

McBeth, C.W., Taylor, A.L. and A.L. Smith (1941). Note on staining nematodes in root tissue. Proc. Helminthol. Soc. Wash. 8: 26.

McGeachie, I., and Rahman, L. (1983). Ufra disease: a review and a new approach to control. Tropical Pest Management 29: 325–332.

McLean, K.S. and Lawrence, G.W. (1993). Interrelationship of *Heterodera glycines* and *Fusarium solani* in sudden-death syndrome of soybean. Journal of Nematology **25**: 434-9.

McSorley, R. (1987). Extraction of nematodes and sampling methods. In: Principles and Practice of Nematode Control in Crops (R.H. Brown, B.R. Kerry, eds.), pp. 13-41. Academic Press, Australia.

McSorley, R. and Gallaher, R.N. (1993). Effect of crop-rotation and tillage on nematode densities in tropical corn. J. Nematol. 25: 814-819.

Meadows, R. (2011). Nematodes Go Viral. PLoS Biol 9(1): e1001011. doi:10.1371/journal.pbio.1001011.

Meagher, J.W. (1972). Cereal cyst nematode (Heterodera avenae Woll). Studies on ecology and content in Victoria. Technical Bulletin 24. Victoria, Australia, Department of Agriculture. 50 pp.

Meher, H. C., K. K. Kaushal, E. Khan, and S. H. Naved. (1998). Use of esterase phenotypes of females for precise diagnosis of four *Heterodera* species. Indian Journal of Nematology 28:81–84.

Mehta, S.K., and Bajaj, H.K. (2005). Response of Resistant Germplasm to Different Races/Populations of Pigeonpea Cyst Nematode, Heterodera cajani. International Chickpea and Pigeonpea News Letter 12: 47–49

Mekete, Tesfamariam (2010). First parasitic nematodes reported in biofuel crops. University of Illinois at Urbana-Champaign. Science Daily. Retrieved from http://www.sciencedaily.com- /releases/2010/03/100317161956.htm

Mekete, T., Grey, M. and Niblack, T. (2009). Distribution, morphological description, and molecular characterization of *Xiphinema and Longidorous* spp. associated with plants (*Miscanthus* spp. and *Panicum virgatum*) used for biofuels. GCB Bioenergy 1: 257-266.

Meldal, B.H.M., Debenham, N.J., De Ley, P., De Ley, I.T., Vanfleteren, J.R., Vierstraete, A.R., Bert, W., Borgonie, G, Moens, T., Tyler P.A., Austen, M.A., Blaxter, M.L., Rogers, A.D., and Lambshead P.J.D. (2007). An improved molecular phylogeny of the Nematoda with special emphasis on marine taxa. Mol. Phylogenet. Evol. 42: 622-636.

Melakeberhan, H. and J. Dey (2003).Competition between *Heterodera glycines* and *Meloidogyne incognita* or *Pratylenchus penetrans*: Independent Infection Rate Measurements. J Nematol. 35(1): 1-6

Meldal, B. H. M., N. J. Debenham, P. De Ley, I. Tandingan De Ley, J. Vanfleteren, A. Vierstraete, W. Bert, G. Borgonie, T. Moens, P. A. Tyler, M. C. Austen, M. Blaxter, A. D. Rogers, and P. J. D. Lambshead (2007). An improved molecular phylogeny of the Nematoda with special emphasis on marine taxa. Molecular Phylogenetics and Evolution **42**:622636.

Meléndez, P.L. and Powell, N.T. (1970). The Pythium-root knot nematode complex in flue-cured tobacco. Phytopathology, 60: 1303

Melton, T.A. (1995). Disease management. In: 1995 Flue-cured Information. NC Agric. Ext. Serv. AG-187, Raleigh, NC (USA), pp. 85-111.

Merny, G. (1966). Biologie *d'Heterodera oryzae* Luc and Berdon, 1961. II. Rôle des masses d'oeufs dans la dynamique des populations et la conservation de l'espèce. Annls Epiphyt., 17: 445-449.

Merny, G. and Netscher, C. (1976). *Heterodera gambiensis* n.sp. (Nematoda: Tylenchida) parasite du mil et du sorgho en Gambie, Cah. ORSTOM Sér. Biol., 11: 209.

Mesel, I.D., Lee, H.J., Vanhove, S., Vincx, M. and Vanreusel. A. (2006). Species diversity and distribution within deep sea nematode genus, *Acantholaimus* on the continental shelf and slope in Antartica. Polar Biol. 29: 860-871.

Mateille, T., Duponnois, R. and Diop, M.T. (1995). Influence of abiotic soil factors and the host plant on the infection of photoparasitic nematodes of the genus *Meloidogyne* by the actinomycete parasitoid *Pasteuria penetrans*. Agronomie 15: 581-591.

Meyer, S. L. F., R. N. Huettel, and R. M. Sayre (1990). Isolation of fungi from *Heterodera glycines* and in vitro bioassays for their antagonism to eggs. Journal of Nematology 22: 532-537.

Meyer, S. L. F., and R. N. Huettel (1996). Application of a sex pheromone, pheromone analogs, and *Verticillium lecanii* for management of *Heterodera glycines*. Journal of Nematology 28: 36-42.

Meyer, S. L. F., and R.J. Meyer (1995). Effects of a mutant strain and a wild type strain of *Verticillium lecanii* on *Heterodera glycines* populations in the greenhouse. Journal of Nematology 27: 409-417.

Meyer, S. L. F., and R.J. Meyer (1996). Greenhouse studies comparing strains of the fungus *Verticillium lecanii* for activity against the nematode *Heterodera glycines*. Fundamental and Applied Nematology 19:305-308.

Mian, I.H.and Rodríguez-Kábana, R.(1982). Organic amendments with high tannin and phenolic contents for control of *Meloidogyne arenaria* in infested soil. Nematropica 12: 221-234.

Miller, P.M. (1975). Effect of the tobacco cyst nematode, *Heterodera tabacum*, on the severity of *Verticillium* and *Fusarium* wilts of tomato. Phytopathology 65: 81-2.

Miller, D.N., Bryant, J.E., Madsen, E.L, and Ghiorse, W.C. (1999). Evaluation and optimization of DNA extraction and purification procedures for soil and sediment samples. *Applied and Environmental Microbiology* 65: 4715–4724.

Milligan, S.B., Bodeau, J., Yaghoobi, J., Kaloshian, I., Zabel, P. and Williamson, V.M. (1998). The root knot nematode resistance gene Mi from tomato is a member of the leucine zipper, nucleotide binding, leucine-rich repeat family of plant genes. Plant Cell 10:1307-19.

Mishra, S.D., Dhawan, S.C., Ali, S.S., Haseeb, A. and Lingaraju, S. (2005). "Integrated management of plant nematodes/.soil pathogens in pulses based cropping systems." Final Report NATP (RPPS-2) pp.-61.

Mohandas, C., Rao, Y.S., and Sahu, S.C. (1981). Cultural control of rice root nematodes with *Sphenoclea zeylanica*. Proceedings of Indian Academy of Sciences 90:373–376.

Mokabli, A., Valette, S. and Rivoal, R. (2001). Différenciationb de quelques espèces de nématodes à kystes des cereals et des graminées par électrophorèse sur gel d'acétate de cellulose. *Nematologia Mediterranea* 29, 103-108.

Monfort, W.S., T.L. Kirkpatrick, D.L. Long, and S. Rideout (2006). Efficacy of a novel nematicidal seed treatment against *Meloidogyne incognita* on cotton. Journal of Nematology 38:245-249.

Mor, M. and Cohn. E. (1989). New nematode pathogens in Israel-Meloidogyne on wheat and Hoplolaimus on cotton. Phytoparasitica 17:221.

Morand, S., Legendre, P., Gardner, S. L., and Hugot, J.P. (1996). Body size evolution of Oxyurid (Nematoda) parasites: the role of hosts. *Oecologia*. 107: 274-282.

Mulk, M.M. (1976). *Meloidogyne graminicola.* In: *CIH Descriptions of Plant-parasitic Nematodes,* Set 6, No. 87. Commonwealth Agricultural Bureaux, Farnham Royal, UK.

Muller, C. (1884). Mittheilungen uber die unseren Kulturpflanzen schadlichen, das Geschlecht *Heterodera* bildenden Wurmer. *Landwirtschaftliche Jarhbucher, Berlin* 13: 1-42.

Mullin, P. G., T. S. Harris, and T. O. Powers. 2005. Phylogenetic relationships of Nygolaimina and Dorylaimina (Nematoda : Dorylaimida) inferred from small subunit ribosomal DNA sequences. Nematology 7:5979.

Mulvey, R.H. and Anderson, R.V. (1980). Description and relationships of a new root-knot nematode, *Meloidogyne sewelli* n. sp. (Nematoda: Meloidogynidae) from Canada and a new host record for the genus. Canada. Can. J. Zool. 58: 1551-1556.

Mulvey, R.H., Townshend, J.L. and Potter, J.W. (1975). *Meloidogyne microtyla* sp. novo from southwestern Ontario, Canada. Can. J. Zool. 53: 1528-1536.

Musser, G. G., and Carleton, M. D. (1993). Family Muridae. In: Mammal Species of the World: A Taxonomic and Geographic *Reference*, (D. E. Wilson and D. M. Reeder, eds.), pp. 501-755. Smithsonian Institution Press, Washington, DC.

Myint, Y.Y. (1981). Country report on root-knot nematode in Burma. Pages 163–170 *in* Proceedings of III International *Meloidogyne* Project Research and Planning Conference on Root-knot Nematodes, *Meloidogyne* spp., Region VI, Jakarta, Indonesia. USA: North Carolina State University and the United States Agency for International Development.

Nadler, S. (2002). Species delimitation and nematode biodiversity: phylogenies rule. Nematology, 4: 615-625.

Nadler, S., De Ley, P., Mundo-Ocampo, M., Smythe, A., Stock, S., Bumbarger, D., Adams, B., De Ley, I., Holovachov, O. and Baldwin, J. (2006). Phylogeny of Cephalobina (Nematoda): Molecular evidence for recurrent evolution of problem and incongruence with traditional classifications. Mol. Phylogenet. Evol. 40: 696-711

Nagesh, M., Parvatha Reddy, P. and Ramchander, N. (1998). Integrated management of *Meloidogyne incognita* and *Fusarium oxysporum* f. sp. *gladioli* in gladiolus using antagonistic fungi and neem cake. Proceedings of 3rd International Symposium of Afro-Asian Nematologists, SBI, Coimbatore, India, pp. 263-266.

Natarajan, N., Alan Cork, N. Boomathi, R. Pandi, S. Velavan and G. Dhakshnamoorthy (2006). Cold aqueous extracts of African marigold, *Tagetes erecta* for control tomato root knot nematode, *Meloidogyne incognita.* Crop Protection 25: 1210-1213

Narayanaswamy, B.C., Gowda, D.N. and Setty, K.G.H. (1982). *Heterodera gambiensis* (Nematoda: Heteroderidae), a new cyst forming nematode on fingermillet from Karnataka, India, J. Soil Boil. Ecol., 2:90.

Navas, A., Castagnone-Sereno, P., Blazquez, J. and Esperrago, G. (2001). Genetic structure and diversity within local populations of *Meloidogyne* (Nematoda: *Meloidogynidae*). Nematology, 3: 243-253.

Neal, J. C. (1888). Discussion of the affects of root-knot nematodes. Fla. Agr. Exp. Sta. Bull. 2 pp. 19-21.

Neal, J. C. (1889). The root knot disease of the peach, orange and other plants in Florida, due to the work of *Anguillula.* U.S. Dep., Agric. Div. Entomol. Bull. No.20, 31pp.

Needham, T. (1743). A letter concerning certain chalky tubulous concretions called malm; with some microscopical observations on the farina of the red lily, and of worms discovered in smutty corn. Philos. Trans. Roy. Soc. 42: 634-641.

Negron, J.A., and Acosta, N. (1989). The *Fusarium oxysporum* f.sp. *coffeae-Meloidogyne incognita* complex in bourbon coffee. Nematropica 19: 161–8.

Neher, D.A. and Campbell, C.L. (1994). Nematode communities and microbial biomass in soils with annual and perennial crops. Appl. Soil Ecol., 1:17-28.

Neilson, R. and Boag, B. (1996). The predicted impact of possible climatic change on virus-vector nematodes in Great Britain. Eur. J. Plant Pathol. 102:193–99

Nelson, F.K., Albert, P.S., and Riddle, D.L. (1983). Fine structure of the *Caenorhabditis elegans* secretory–excretory system. J. Ultra. Struct. Res. *82*, 156–171.

Netscher, C. (1969). L'ovogenèse e t la reproduction chez *Heterodera oryzae* et *H. sacchari* (Nematoda: Heteroderidae), Nematologica, 15:l0.

Netscher, C. and Pernes, J. (1971). Etude concernant l'influence de la constitution génétique sur la longueur des larves d' *Heterodera oryzae.* Nematologica, 17:336.

Netscher, C., Luc, M. and Merny, G. (1964). Description du mâle *d' Heterodera sacchari* Luc and Merny, 1963) Nematologica, 15 : 156.

Niblack, T. L. (1992). The race concept. Pp. 73–86 *in* R. D. Riggs and J. A. Wrather, eds. Biology and management of the soybean cyst nematode. St. Paul, MN: APS Press.

Nickle, W. R (1991). Manual of Agricultural Nematology. Marcel Dekker, New York.

Nielsen, C., Scharff, N., and Eibye, J. D. (1996). Cladistic analyses of the animal kingdom. *Biol. J. Linnean Soc.* 57, 385-410.

Niu, J.H., Heng Jian, Jian-Mei Xu, Yang-Dong Guo and Qian Liu (2010). RNAi technology extends its reach: Engineering plant resistance against harmful eukaryotes. African J. Biotechnology 9: 7573-7582

Noguera, R. (1982). Atteraciones en la produccion de rishitia en racces y tilosas en tallos en la interaccion *Meloidogyne-Fusarium* en plantas de tomate. Agronomia Tropical, 32, 303-8.

Noguera, R. and Smits, B.G. (1982). Variaciones en la microflora de la rizosfera del tomate infectado con Meloidogyne incognita. Agronomica Tropical, 32:147-54.

Noling, J.W. and Gilreath, J.P. (2002). Weed and nematode management: Simultaneous considerations. In: Gobenauf, G.L. (Ed.), Annu. Int. Res. Conf. on Methyl Bromide Alternatives and Emissions Reductions. Orlando, FL (USA).

Nordbring-Hertz, B., H.B. Jansson and A. Tunlid, (2002). Nematophagous Fungi: Encyclopedia of Life Sciences. Macmillan Publishers Ltd., London.

Nordbring-Hertz, B., H.B. Jansson and A. Tunlid, (2006). *Encyclopedia of Life Sciences* (Wiley, Chichester, UK).

Norton, D. C. (1978). Ecology of Plant-Parasitic Nematodes. John Wiley and Sons, New York. 268 pp.

Norton, D.C. (1987). Plant nematode communities. Int. J. nematol. 17: 215-222.

Nussbaumer, A.D., Bright, M., Baranyi, C., Beisser, C.J., and Ott, J.A. (2004). Attachment mechanism in a highly specific association between ectosymbiotic bacteria and marine nematodes. Aqua. Microb. Ecol. *34*, 239–246.

O'Brien, P.C. (1983). A further study on the host range of *Pratylenchus thornei.* Australian Plant Pathology 12:1–3.

O'Brien, D. G. and Prentice, E. G. (1930). *The West of Scotland Agricultural College Plant Husbandry* 2 : 63.

Odihirin,R.A. (1975). Occurrence of *Heterodera* cyst nematodes (Nematoda: Heteroderidae) on wild grasses in Southern Nigeria, OCC. Public. Nigerian Soc. P1. Protect., 1:24.

Odihirin, B.A. (1977). Irrigation water as a means for the dissemination of the sugar cane cyst nematode, *Heterodera sacchari* at Bacita Sugar Estate, Oec. Publ. Nigerian Soc. P1. Protect., 2:58.

Oka, Y. (2010). Mechanisms of nematode suppression by organic soil amendments—A review. Appl. Soil Ecol. 44: 101-115.

Oka, Y., S. Nacar, E. Putieusky, U. Ravid, Y. Zohara, and Y. Spiegal (2000). Nematicidal activity of essential oils and their components against the root knot nematode. Phytopathology 90 (7): 710—715.

Oka, Y., Shapira, N. and Fine, P. (2007). Control of root-knot nematodes in organic farming systems by organic amendments and soil solarization. Crop Prot. 26: 1556-1565

Okeniyi, M. O.1. Fademi O.A.l., Orisajo, S. B1, Adio, S.O.1, Otunoye, H. A.1. and Adekunle, O.V. (2010). Effect of botanical extracts on root-knot nematode (*Meloidogyne incognita*) infection and growth of cacao seedlings. J. Appl. Biosci. 36: 2346 - 2352

Okorie, C.C., C.C. Ononuju and I.A. Okwujiako (2011). Management of *Meloidegyne incognita* with *Pleurotus ostreatus* and *P. tuberregium* in soybean. *Int. J. Agric. Biol.*, 00: 401–405

Olatinwo, R., Borneman, J., and Becker, J. O. (2006). Induction of beet cyst nematode suppressiveness by the fungi *Dactylella oviparasitica* and *Fusarium oxysporum* in field microplots. Phytopathology, 96:855-859.

Ornat, C. and Sorribas, F.J. (2008). Integrated management of root-knot nematodes in Mediterranean horticultural crops. In: Cianco, A., Mukerji, K.G. (Eds.), Integrated Management and Biocontrol of Vegetable and Grain Crops Nematodes. Springer, Dordrecht (NLD), (Integrated management of plant pests and diseases, vol. 2), pp. 295-319.

Orton W.A. and Gilbert. W.W. (1912). Control of cotton wilt and root knot nematode. Circ. U.S. Bur. Pl. Ind., 92.

Osei K., Moss R., Nafeo A., Addico R., Agyemang A., Danso Y. and Asante J.S (2011). Management of plant parasitic nematodes with antagonistic plants in the forest-savanna transitional zone of Ghana. J.Appl. Biosci. 37: 2491 - 2495

Ou, S.H. (1972). *Rice diseases.* Kew, Common. Mycol. Inst., 368 p.

Padovan, A.C.B., Sanson, G.F.O., Brunstein, A. and Briones, M.R.S. (2005). Fungi evolution revisited: application of the penalized likelihood method to a Bayesian fungal phylogeny provides a new perspective on phylogenetic relationships and divergence dates of ascomycota groups. J.Mol Evol. 60:726–735.

Page, S.L.J., Bridge, J., Cox, P. and Rahman, L. (1979). Root and soil parasitic nematodes of deep water rice areas in Bangladesh. Intnl. Rice Res. Newsl., 4 : 10-11.

Panwar, M.S. and Rao, Y.S. 1998. Status of phytonematodes as pests of rice. Pages 49–81 *in* Nematode diseases in plants (Trivedi, P.C., ed.). New Delhi, India: CBS Publishers and Distributors.

Pak, J., and Fire, A. (2007). Distinct populations of primary and secondary effectors during RNAi in *C. elegans*. Science 315: 241–244.

Palanichamy, L (1973). Nematode problems of coffee in India. Indian Coffee: 37: 99-100.

Pàn, C.S. (1985). Studies on plant-parasitic nematodes on economically important crops in Fujian. III. Description of *Meloidogyne fujianensis* n. sp. (Nematoda: Meloidogynidae) infesting *Citrus* in Nanjing County. Acta Zoologica Sinica 31: 263-268.

Paolomaes Rius, J.E., Vovlas, N., Troccoli, A, Liebanas, G., Landa, B.B. and Castillo, P. (2007). A new rootknot nematode parasitizing sea rocket from Spanish Mediterranean coastal dunes: *Meloidogyne dunensis* n. sp. (Nematoda: Meloidogynidae). J. Nematol. 39: 190-202.

Patnaik, N.C. (1969). Pathogenicity of *Meloidogyne graminicola* (Golden and Birchfield, 1965) in rice. (Abstr.). A1l India Nenzatology Symposium, New Delhi, August 21-22: 12.

Pedigo, L. P. (1996). Entomology and Pest Management. Second Edition. 1996. Prentice-Hall Pub., Englewood Cliffs, NJ. 679 pp.

Pedigo, L. P., S. H. Hutchins and L. G. Higley. (1986). Economic injury levels in theory and practice. Annu. Rev. Entomol. 31:341—368.

Perry, R. N. (1987). Host induced hatching of phytoparasitic nematode eggs. *In:* Veech,]. and Dickson, D. (Eds). *Vistas on nemalology*. Hyatsville, MD, USA, Soc. Nematologists Inc. p. 159-164.

Perry, R. N. (1989). Dormancy and hatching of nematode eggs. *Parasif. Today*, 5: 377-383.

Perry, R. N., Clarke, A. and Beane. (1980). Hatching of *Heterodera goettingiana in vitro*. Nematologica 26: 493-495.

Perry, R. N., and Moens, M. (2011). Survival of parasitic nematodes outside the host. Pp. 1–27 in R.N. Perry, and D.A. Wharton, eds. Molecular and physiological basis of nematode survival. Wallingford, UK: CAB International.

Perry, R.N., Subbotin, S.A. and Moens, M. (2007). Molecular diagnostics of plant-parasitic nematodes. In 'Biotechnology and plant disease management.' (Eds ZK Punja, S De Boer, HI Sanfacon) pp. 195-226. (CAB International: Wallingford, UK)

Perry, R. N., Zunke, U. and Wyss, U. (1989). Observations on the response of the dorsal and subventral oesophageal glands of *Globodera rostochiensis* ro hatching stimulation. Revue Nématol., 12: 91-96.

Petersen, J.J. (1984). Nematode parasites of mosquitoes. Pages 797–820 *in* Plant and insect nematology (Nickle, W. R., ed.). New York, USA: Marcel Decker.

Petersen, D., Zijlstra, C., Wishart, J., Blok, V. and Vrain, T. (1997). Specific probes efficiently distinguish root-knot nematode species using signature sequence in the ribosomal intergenetic spacer. Fundam. Appl. Nematol. 20: 619-626.

Pfister, D.H. (1997). Castor, Pollux and life histories of fungi. Mycologia 89:1–23

Pham, T.B.7 (1990). Gall nematodes of vegetables and potatoes in Da Lat (Tei Nguen Plateau, Vietnam) and description of *Meloidogyne cynariensis*, a parasite of artichokes. Zoologichesky Zhurna 69 (4): 128-131.

Philis, I. (1988). Occurrence of *Heterodera latipons* on barley in Cyprus. Nemat. Med., 16: 223.

Pitcher, R.S. (1965). Interrelationships of nematodes and other pathogens of plants. Helminthological Abstracts, 34, 1-17.

Poage, Michael A., John E. Barrett, Ross A. Virginia and Diana H. Wall (2008). The Influence of Soil Geochemistry on Nematode Distribution, McMurdo Dry Valleys, Antarctica. Arctic, Antarctic, and Alpine Research, 40 (1): 119–128

Poinar, G.O., Jr. (1979). Nematodes for biological control of insects. Boca Raton, Florida, USA: CRC Press.

Poghossian, E.E. (1971). *Hypsoperine megriensis* n. sp. (Nematoda: Heteroderidae) in the Armenian SSR. *Doklady Akademii Nauk ArmyanskoT SSR* 53, 306-312.

Porter, D.M. and Powell, N.T. (1967). Influence of certain *Meloidogyne* species on Fusarium wilt development in flue-cured tobacco. Phytopathology, 57: 282-5.

Potts, F. A. (1932). The phylum Nematoda. Chap. VII in Borradaile, L.A., and Potts, F.A., *et al.,* The invertebrata. A manual for the use of students, pp. 214227, figs. 166 171. Cambridge and New York.

Powell, N.T. (1971). Interactions between nematodes and fungi in disease complexes. Annual Review of Phytopathology 9: 253–74.

Powell, N.T. (1971). Interaction of plant parasitic nematodes with other disease causing agents, in Plant Parasitic Nematodes, Vol. II (eds B.M. Zuckerman, W.F. Mai and R.A. Rohde), Academic Press, New York, pp. 119-36.

Powell N.T. (1979). Internal synergisms among organisms inducing disease. Pp. 113-133. In: Plant Disease, Vol. IV (Horsfall J.G. and Cowling E.B., eds.). Academic Press, New York, USA.

Powers, L.E., Menghi Ho, D.W. Freckman, and R.A. Virginia (1998). Distribution, Community Structure, and Microhabitats of Soil Invertebrates along an Elevational Gradient in Taylor Valley. Arctic, Antarctic, and Alpine Research 30 :133-141.

Powers, T.O. and Harris, T.S. (1993). A polymerase chain reaction method for identification of five major *Meloidogyne* species. *J. Nematol.* 25: 1–6.

Powers T., Todd T., Burnell A., Murray P., Fleming C., Szalanski A., Adams B., and Harris T. (I997). The rDNA internal transcribed spacer region as a taxonomic marker for nematodes. J. Nematol. 29: 441-450.

Powers, T.O., Mullin, P.G., Harris, T.S., Sutton, L.A., and Higgins, R.S. (2005). Incorporating molecular identification of *Meloidogyne* spp. into a large-scale regional nematode survey. *Journal of Nematology* 37: 226–235.

Pramer, D. (1964). Nematode trapping fungi. Science 144: 382–388.

Prasad D. (1997). Assessment of groundnut yield loss due to plant parasitic nematodes. Ann. Plant Prot. Sci. 4: 25–28.

Prasad, S.K. and Chawla, M.L. (1965). The potato root eelworm *Heterodera rostochiensis,* on the Indian Agricultural Research Institute, Indian J. Entomol. 23: 127.

Prasad, J.S., Panwar, M.S. and Y.S Rao. (1986). Screening of some rice cultivars against the root knot nematode, *Meloidogyne graminicola*. Ind. J. Nematol., 16:112-113.

Prot, J.C. and Matias, D.M. (1995). Effects of water regime on the distribution of *Meloidogyne graminicola* and other root-parasitic nematodes in a rice field toposequence and pathogenicity of *M. graminicola* on rice cultivar UPL Ri5. Nematologica 41,: 219-228.

Qiao, K., Shi, X., Wang, H., Ji, X. and Wang, K. (2011). Managing root-knot nematodes and weeds with 1,3-dichloropropene as an alternative to methyl bromide in cucumber crops in China. J Agric Food Chem. 59 (6): 2362-7.

Quarles, William. (1993). Rapeseed green manure controls nematodes. The IPM Practitioner. April. p. 15

Quarles, William. (2005). 2005 directory of least toxic pest control products. The IPM Practitioner, Vol. 26, No. 11/12. p.17.

Quénéhervé, P. (1990). Spatial arrangement of nematodes around the banana plant in the Ivory Coast: related comments on the interaction among concomitant phytophagous nematodes. Acta Ecologica, 11 (6): 875-886.

Radwan, M. A., M. M. Abu-Elamayem, S. M. Kassem, and E. K. El-Maadawy (2006). Soil amendment with dried weed leaves as non-chemical approach for the management of *Meloidogyne incognita* infecting tomato. Agricultural and applied biological sciences. 71(4):25-32.

Raffaelli, D.G., and Mason, D.F. (1981). Pollution monitoring with meiofauna using the ratio of nematodes to copepods. Marine Pollution Bulletin 12:158–163.

Rammah, A. and Hirschmann, H. (1988). *Meloidogyne mayaguensis* n. sp. (Meloidogynidae), a root-knot nematode from Puerto Rico. Journal of Nematology 20: 58-69.

Rammah, A. and Hirschmann, H.(1990). *Meloidogyne morocciensis* n. sp. (Meloidogyninae), a root-knot nematode from Morocco. Journal of Nematology 22: 279-291.

Rau, G.J. and Fassuliotis, G. (1965). *Hypsoperine spartinae* sp. n., a gall forming nematode on the roots of smooth cordgrass. Proceedings of the Helminthological Society of Washington 32: 159-162.

Rao,Y.S. and Israel, P. (1971). Studies on nematodes of rice and rice soils. Influence of soil chemical properties on the activity of *Meloidogyne graminicola*, the rice root-knot nematode. *Oryza*, 8 : 33-38.

Rao, Y.S. and Israel, P. (1972). Effect of temperature on hatching **of** eggs of the rice root-hot nematode *Meloidogyne graminicola. Oryza*, 9 : 73-75.

Rao, Y.S. and Jayaprakash, A. (1977). Leaf chlorosis due to infestation by a new cyst nematode. Int. Rice Res. Newsl., 2: 5.

Rao,Y.S. and Jayaprakash, A. (1978). *Heterodera oryzicola* n.sp. (Nematoda: Heteroderidae) a cyst nematode on rice (*Oryza sativa* L.) from Kerala State, India. Nematologica, 23:341.

Rao, Y.S., Prasad, J.S., Yadav, C.P. and Padalia, C.R. (1984). The influence of rotation crops in rice soils on the dynamics of parasitic nematodes. Biological Agriculture and Horticulture 2:69–78.

Rao, Y.S., Prasad, J.S., and Panwar, M.S. (1986). Nematode problems in rice: crop losses, symptomatology and management. Pages 279–299 *in* Plant parasitic nematodes of India (Swarup, G., and Dasgupta, D.R., eds.). New Delhi, India: Indian Agricultural Research Institute

Raski, D.J. (1954). Soil fumigation for the control of nematodes on grape replants. Plant Disease Reporter 38:811-817.

Rausch, R. L. (1983). The biology of avian parasites: helminths. *In: Avian Biology*, vol. VII (D. S. Farner, J. R. King, and K. C. Parkes, eds), pp. 367-442. Academic Press, NY.

Reaumur, Rene Antoine Ferchault de (1742). Memoires pour servir a l'historie des insectes. Vol. 6. Imprimerie Royale, Paris.

Redi, F. (1684). Page 253 in: Osservazion intorno agli animali viventi che si trovano negli animali viventi. 26 pls. Firenze.

Regnault-Roger, C., Philogène, B.J.R. and Vincent, C. (2002). Biopesticides d'origine végétale. Tec et Doc, Paris (FRA).

Reinhard, Karl J., Ulisses E. Confalonieri, Bernd Herrman, Luiz F. Ferreira and Adauto, J. G. De Araujo (1986). Recovery of parasite remains from coprolites and latrines: Aspects of Paleoparasitological Technique. *Homo* 37: 217-239.

Remillet, M., and Laumond, C. (1991). Spaerularioid nematodes of importance in agriculture. In: Nickle. W.R.(ed) Mannual of agricutural Nematology. Marcel Dekker, New York, pp 967-1024.

Reuven, M., Szmulewich, Y., Kolesnik, I., Gamliel, A., Zilberg, V., Mor, M., Cahlon, Y. and Ben-Yephet, Y. (2005). Methyl bromide alternatives for controlling fusarium wilt and root knot nematodes in carnations. Acta Hortic. 698: 99-104.

Rich, J.R. and Barker, K.R. (1984). Flowering Delay in Flue-cured Tobacco Infected with *Meloidogyne* Species. J. Nematol. 16:402-404.

Rich, J.R., Brito, J.A., Kaur, R.and Ferrell, J.A (2009). Weed species as hosts of *Meloidogyne*: a review. Nematropica 39: 157-185.

Rich, J.R.and Rahi, G.S. (1995). Suppression of *Meloidogyne javanica* and *M. incognita* on tomato with ground seed of castor, crotalaria, hairy indigo, and wheat. Nematropica 25: 159-164.

Riffle, J.W. (1963). *Meloidogyne ovalis* (Nematoda: Heteroderidae), a new species of root-knot nematode. Proceedings of the Helminthological Society of Washington 30: 287-292.

Riga, E., and G. Lazarovits. (2001). Development of an organic pesticide based on neem tree products. American Phytopathological Society / Mycological Society of America / Society of Nematology Joint Meeting. Abstracts of Presentations. Salt Lake City, Utah. Phytopatology 91: S141. Publication no.P2001-0096-SON.

Riggs, R.D. (1977). Worldwide distribution of soybean cyst nematode and its economic importance. Journal of Nematology 9: 34-39.

Riggs, R.D. and Schmitt, D.P. (1988). Complete characterization of the race scheme of *Heterodera glycines*. Journal of Nematology 20: 392-395.

Ringer, C.E., S. Sardanelli and L.R. Krusberg (1987). Investigations of the host range of the corn cyst nematode, *Heterodera zeae* from Maryland. Annl. Appl. Nematol., 1: 97–106

Ripoll, C., Favery, B., Lecomte, P., Van Damme, E., Peumans, W., Abad, P. and Jouanin, L. (2003). Evaluation of the ability of lectin from snowdrop (*Galanthus nivalis*) to protect plants against root-knot nematodes. Plant Science 164: 517–523.

Rivoal R., Valette S., Bekal S., Gauthier J.P., and Yahyaoui A. (2003). Genetic and phenotypic diversity in the graminaceous cyst nematode complex, inferred from PCR-RFLP of ribosomal DNA and morphometric analysis. European Journal of Plant Pathology 109: 227-241.

Rizvi, A.N. (2010). First record of three species of soil nematodes of the suborder Cephalobina from Ladakh region, Jammu and Kashmir, India. Journal of Threatened Taxa 2(11): 1286-1290.

Roberts, P.A. (1993). The future of nematology – Integration of new and improved management strategies. J. Nematol. 25: 383-394.

Rodríguez-Kábana, R., Morgan-Jones, G., and Chet, I. (1987). Biological-control of nematodes – soil amendments and microbial antagonists. Plant Soil 100: 237-247.

Romero, M.D., Andres, M.F., Lopez-brana, I. and Delibes, A. (1996). A pathogenic and biochemical comparison of two Spanish populations of the cereal cyst nematode. Nematologia Mediterranea 24, 235-244.

Roncadori, R. W. and R. S. Hussey. (1977). Interaction of the endomycorrhizal fungus Gigaspora margarita and root-knot nematode on cotton. Phytopathology 67:1507-1511.

Rosenbluth, J. (1968). Obliquely striated muscle. IV. Sarcoplasmic reticulum, contractile apparatus, and endomysium of the body muscle of a polychaete, Glycera, in relation to its speed. J. Cell Biol. 36: 245-259

Rossi, M., Goggin, F.., Milligan, S.B., Kaloshian, I., Ullman, D.E.and Williamson VM (1998). The nematode resistance gene Mi of tomato confers resistance against the potato aphid Proceedings of the National Academy of Sciences, USA 95:9750-54

Rosso, M.N., Vieira, P., de Almeida-Engler, J. and Castagnone-Sereno, P. (2011). Proteins secreted by root-knot nematodes accumulate in the extracellular compartment during root infection. Plant Signal Behav. 6: 8.

Rotimi M.O. and Moen M. (2002). The use of leaf extracts of some herbs in the control of *Meloidogyne incognita*. In: 30th Annual Conference Nigerian Society for Plant Protection, University of Agriculture, Abeokuta. September, 1st–4th 2002.

Roy, A.K. (1973). Reaction of some rice cultivars to the attack of *Meloidogyne graminicola. Indian J. Nematol.*, 3 : 72-73.

Ruanpanun, P., Laatsch, H., Tangchitsomkid, N. and Lumyong, S. (2011). Nematicidal activity of fervenulin isolated from a nematicidal actinomycete, Streptomyces sp. CMU-MH021, on *Meloidogyne incognita*. World J Microbiol Biotechnol. ; 27(6):1373-1380.

Rubner, A. (1996). Stud. Mycol. 39:1–134http://www.pnas.org/external-ref?access_num=10.2307/3761168&link_type=DOI

Rubtsova, T., Subbotin, S., Brown, D., and Moens, M. (2001). Description of *Longidorus sturhani* sp. n. (Nematoda: Longidoridae) and molecular characterisation of several longidorid species from Western Europe. Russ. J. Nematol. 9: 127-136.

Rudolphi, C. A. (1809). Entozoorum sive vermium intestinalium histoiria naturalis, Amsterdam.

Rudolphi, C. A. (1819). Entozoorum synopsis cut accedunt mantissa duplex et indices locu pletissümi. Berlin.

Rumpenhorst, H.J. (1985).Vergleichendeelektrophoretische Untersuchungen von Proteinen einiger Zystennematoden von Getreide und Gräsern. *Mitteilungen aus der Biologischen Bundesanstalt für Land- und Forstwirtschaft Berlin-Dahlem* 226, 64-74.

Runia, W.and Greenberger, A. (2005). Preliminary results of physical soil disinfestation by hot air. Acta Hortic. 698: 251-256.

Ruxandra I. Molnar, Gabi Bartelmes, Iris Dinkelacker, Hanh Wittel and Ralf J. Sommer (2011). Mutation Rates and Intraspecific Divergence of the Mitochondrial Genome of *Pristionchus pacificus*. Molecular Biology and Evolution 28:2317-2326.

Ryan, N. A., E. M. Duffy, A. C. Cassells and P. W. Jones (2000). The effect of mycorrhizal fungi on the hatch of potato cyst nematodes. Applied Soil Ecology 15 (2): 233-240

Sahoo, N.K., S. Ganguly and S. J. Eapen. (2000). Description of *Meloidogyne piperi* sp.n. (Nematoda: Meloidogynidae) isolated from the roots of *Piper nigrum* in South India. Indian J. Nematol. 30 (2) : 203-209.

Salen, H. and Sikora, R.A. (1984). Relationship between *Glomus fasciculatum* roots colonization of cotton and its effect on *Meloidogyne incognita*. Nematologica, 39:230-237

Samoiloff, M.R., and Bogaert, T. (1984). The use of nematodes in marine ecotoxicology. Pages 407–426 *in* Ecotoxicological testing for the marine environment: proceedings of an International Symposium, 12–14 Sep-1983, Ghent Belgium (Persoone, I.G., Jaspers, E., and Claus, C., eds.). Ghent, Belgium: State University of Ghent and Institute for Marine Scientific Research.

Sandground, J. (1922). A study of the life history and methods of control of the root knot nematode, *Heterodera radicicola*.South African J. Sci., 11: 399-418.

Sandground, J. (1923). *Oxyuris incognita* or *Heterodera radicicola*. J. Parasitol. 10:92-94.

Santos, M.S.N. de A (1968). *Meloidogyne ardenensis* n. sp. (Nematoda: Heteroderidae). a new British species of root-knot nematode. Nematologica 13 (1967): 593-598.

Sanyal, P.K. (2000). Screening for Indian isolates of predacious fungi for use in biological control against nematode parasites of ruminants. Vet. Res. Commun., 24: 55-62.

Sarah M. Nour, John R. Lawrence, Hong Zhu, George D. W. Swerhone, Martha Welsh, Tom W. Welacky, and Edward Topp (2003). Bacteria associated with cysts of the soybean cyst nematode (*Heterodera glycines*). Applied and Environmental Microbiology, 69: 607-615.

Sardanelli, S., L.R. Krusberg and A.M. Golden (1981). Corn cyst nematode, *H. zeae* in the United States. *Plant Dis.*, 65: 622

Sasser, J. N. (1990). Plant-parasitic Nematodes: The Farmer's Hidden Enemy. North Carolina State University Press, Raleigh, NC. p. 47.48.

Sarvanapriya, S. and M. Sivakumar (2005). Management of root knot nematode, *Meloidogyne incognita* on tomato using botanicals. Indian Journal of Natural Products and Resources 4:158-161.

Saxena, R. and Reddy, D.D.R. (1987). Crop losses in pigeonpea and mungbean by pigeonpea cyst nematode, *Heterodera cajani*. Indian J. Nematol. 17: 91-94

Sayre, R.M. and Starr, M.P. (1988). Bacterial diseases and antagonists of nematodes. In: Poinar, G.O., Jansson, H.B. (Eds.), Diseases of Nematodes, Vol. 1. CRC Press, Boca Raton, FL, pp. 69–101.

Schacht, H. (1859). Über einige Feinde der Rübenfelder. Zeit. Ver. Rubenzuckerindustrie Zolluer. 9:175-179.

Schenck, N.C., R. A. Kinloch, and D.W. Dickson. (1975). Interaction of endomycorrhizal fungi and root-knot nematode on soybean. Pp. 605-616 in F. E. Sanders, B. Mosse, and P. B. Tinker, eds. Endomycorrhizas. New York: Academic Press.

Schierenberg, E. (2005). Unusual cleavage and gastrulation in a freshwater nematode: developmental and phylogenetic implications. Dev. Genes Evol. *215*, 103–108.

Schmidt, A. (1871). Uber den Ruben-Nematoden (Heterodera Schachtii A. S.). Z. Zucklnd. Zoolversein 21:1-19.

Schneider, A. (1866). Monographie der Nematoden. Reimer, Berlin.

Schroeder, J., Thomas, S.H. and Murray, L. (1993). Yellow and purple nutsedge and chile peppers host southern root-knot nematode. Weed Sci. 41:150-156.

Schulte, F. (1989). Life history of *Rhabditis (Pelodera) orbitalis*- a larval parasite in the eye orbit of arvicolid and murid rodents. Proc. Helminthol. Soc. Wash. 56: 1-7.

Scopoli, G.A. (1777). Introductio ad historiam naturalum sistens genera lapidum plantarum etanimalium hactenus detecta caracteribus essentialibus donata, in tribus divisa, subinde ad leges naturae, Prague. 506 pp.

Seastedt, T.R., Jams, S.W., and Todd, T.C. (1988).Interactions among soil invertebrates, microbes and plant growth in the tall grass prairie. Agric. Ecosystem Environ,. 24:219-228.

Sethi, C.L. and Gaur, H.S. (1986). Nematode management : An overview. 425-445. **In:** Gopal Swarup and D.R. Dasgupta (Eds.), Plant Parasitic Nematode India, IARI, New Delhi, 497 p

Serfoji, P., Rajeshkumar, S. and Selvaraj, T. (2010). Management of root-knot nematode, *Meloidogyne incognita* on tomato CV Pusa Ruby. by using vermicompost, AM fungus, *Glomus aggregatum* and mycorrhiza helper bacterium, *Bacillus* coagulans. Journal of Agricultural Technology, 6(1): 37-45

Setterquist, R., Smith, G., Jones, R. and Fox, G. (1996). Diagnostic probes targeting the major sperm protein gene that may be useful in the molecular identification of nematodes. J. Nematol. 28: 414-421.

Shahina, F. and Mohammad A. Maqbool (1989). *Heterodera cynodontis n.* **sp.** (*Nematoda : Heteroderidae*) from *Cynodon dactylon* (*L.*) in Pakistan. *Revue Nérnatol.* 12: 395-400

Shagalina, L., Ivanova, T and Krall, E. (1985).Two new gall nematodes from the genus *Meloidogyne* (Nematoda: Meloidogynidae), parasites of trees and shrubs. Eesti NSV Teaduste Akadeemia Toimetised, Bioloogialne 34: 279-287.

Shakil, N. A.; D. Prasad; D. B. Saxena and A. K Gupta (2004). Nematicidal activity of essential oils of Artemisia annua against root-knot and reniform nematodes. Annals of Plant Protection Sciences, 12: 403-408.

Shapiro, D.I., Nyczepir, A.P., Lewis and E.E. (2006). Entomopathogenic nematodes and bacteria applications for control of the pecan root-knot nematode, *Meloidogyne partityla* in the greenhouse. Journal of Nematology 38: 449-454.

Sharma, N.K., Thapa, C.D., and Nath, A. (1981). Pathogenicity and identity of myceliophagous nematodes infesting *Agaricus bisporus* (Lang) Sing. in Himachal Pradesh (India). Indian J. Nematol.11:230-231.

Sharma, S.B., and Nene, Y.L. (1990). Effects of soil solarization on nematodes parasitic to chickpea and pigeonpea. Journal of Nematology 22: 658–664.

Sharma, S.B., Nene Y.L., Reddy, M.V. and McDonald, D. (1993). Effect of *Heterodera cajani* on biomass and grain yield of pigeon pea on vertisol in pot and field experiments. Plant Pathology 42: 163-167.

Sharma, S.B., Rego, T.G., Mohiuddin, M., and Nageswara Rao, V. (1996). Regulation of densities of *Heterodera cajani* and other plant parasitic nematodes in semi-arid tropical production systems. Journal of Nematology 28: 244–251.

Sharma S.B., Siddiqui M.R., Fazul Rahman P. Ali S.S. and Ansari M.A. (1998). Description of *Heterodera swarupi* sp. n. (Nematoda: Heteroderidae), a parasite of chickpea in India. International Journal of Nematology, 8 : 111-116.

Shi-Bin W., Liang L., Ju W., Yong J., and i-Wei J. (2008). PCR-DGGE analysis of nematode diversity in Cu-contaminated soil. Pedosphere 18: 621- 627.

Shreenivasa, K. R., K. Krishnappa and N.G. Ravichandra (2007). Interaction Effects of Arbuscular Mycorrhizal Fungus *Glomus fasciculatum* and Root-knot Nematode, *Meloidogyne incognita* on Growth and Phosphorous Uptake of Tomato. Karnataka J. Agric. Sci., 20(1): 57 - 61

Siddiqi, M.R. (1980). The origin and phylogeny of the nematode orders Tylenchida Thorne, 1949 and Aphelenchida, n. ord. Helminthological Abstracts - Series B *49,* 143–170.

Siddiqi, M. R. (1983). Phylogenetic relationships of the soil nematode orders Dorylaimida, Mononchida, Triplonchida and Alaimida, with a revised classification of the subclass Enoplia. Pakistan Journal of Nematology **1**:79110.

Siddiqi, M.R. and Booth, W. (1991) *Meloidogyne* (*Hypsoperine*) *mersa* sp. n. (Nematoda: Tylenchina) attacking *Sonneratia alba* trees in mangrove forest in Brunei Darussalam. Afro-Asian Journal of Nematology 1: 212-220.

Singh, S.P. (1969). A new plant parasitic nematode *Meloidogyne lucknowica* n. sp. from the root galls of *Lufta cylindrica* (sponge gourd) in India. *Zoologischer Anzeiger* 182: 259-270.

Sidhu, G. and Webster, J.W. (1977) Predisposition of tomato to the wilt fungus (*Fusarium oxysporum lycopersici*) by the root-knot nematode (*Meloidogyne incognita*). Nematologica, 23: 436-42.

Siebold von, C. T. E. (1842; 1843; 1848; 1850). Ueber die Fadenwurmer der Insekten. Entomol. Z. 3: 146-161; 4: 78-84; 9: 290-300; 11: 329-336

Sijmons, P. C., Atkinson, H. J. and Wyss, U. (1994). Parasitic strategies of root nematodes and associated host cell responses. Annual Review of Phytopathology, 32: 235 - 259.

Sijmons, P. C., Grundler, F. M. W., vonMende, N., Burrows, P. R. and Wyss, U. (1991). *Arabidopsis thalliana* as a new model host for plant-parasitic nematodes. The Plant Journal 1:245 - 254.

Sijen, T., Steiner, F.A., Thijssen, K.L.and Plasterk, R.H.A. (2007). Secondary siRNAs result from unprimed RNA synthesis and form a distinct class. Science 315: 244–247.

Sikora, R. A. (1979). Predisposition to *Meloidogyne* infection by the endotrophic mycorrhizal fungus *Glomus mosseae*. Pp. 399-404 in F. Lamberti and C. E. Taylor, eds. Root knot nematode (Meloidogyne species) systematics, biology and control. New York: Academic Press.

Sikora, R.A. (1988). Plant parasitic nematodes of wheat and barley in temperature and temperate semi-arid regions - a comparative analysis. In M.C. Saxena, R.A. Sikora and J.P.

Srivastava, eds. Nematodes parasitic to cereals and legumes in temperate semi-arid regions, p. 46-48. Aleppo, Syria, ICARDA.

Sikora, R.A. and Fernandez, E. (2005). Nematode parasites of vegetables. In: Luc, M., Sikora, R.A., and Bridge, J. (eds.) *Plant parasitic nematodes in Subtropical and Tropical Agriculture.* 2nd edn. CAB International, Wallingford, UK, pp 319-394.

Singh,S., Kumar,S., Bajaj, H.K., and Ganguly, A.K.(1998). Esterase and malate dehydrogenase patterns of races of *Heterodera cajani* and *H. zeae*. Indian J.Nematol. 28(2): 105-109.

Sivaprasad, P., Jacob, A. and George, B. (1990). Root knot nematode infestation and nodulation influenced by V A mycorrhizal association in cowpea. Indian Journal of Nematology, **20**:49-52

Skarbilovich, T.S. (1959) On the structure of systematics of nematodes order Tylenchida Thorne, 1949. *Acta Parasitologica Polonica* 7:117-132.

Sledge, E.B. and Golden, A.M. (1964). *Hypsoperine graminis* (Nematoda: Heteroderidae), a new genus and species of plant-parasitic nematode. Proceedings of the Helminthological Society of Washington 31:83-88.

Smitley, D.R., Warner, W.R., Bird, G.W. (1992). Inuence of irrigation and *Heterorhabditis bacteriophora* on plant-parasitic nematodes in turf. Journal of Nematology 24 (Suppl.), 637–641.

Sobczak and Wladyslaw Golinowski (2009). Structure of Cyst Nematode, Feeding Sites. Plant Cell Monographs, 15:153-187.

Sohlenius, B., Bostrom, S., and Sandor, A. (1988). Carbon and nitrogen budgets of nematodes in arable soil. Biol. Fertil. Soils 6:1-8.

Somasekhar, N., Denardo, E.A. B., and Grewal. P.S. (2002). Impact of inundative application of entomopathogenic nematodes on non-target nematode communities in turf grass ecosystem. J. Nematol. 32:461.

Sommer, R.J., Carta, L.K., Kim, S.Y. and Sternberg, P.W. (1996). Morphological, genetic and molecular description of *Pristionchus pacificus* sp.n. (Nematoda: Neodiplogastridae). Fundam. Appl. Nematol. **19**: 511–521

Soriano, I.R.S., Prot, J.C. and Matias, D.M. (2000). Expression of tolerance for *Meloidogyne graminicola* in rice cultivars as affected by soil type and flooding. Journal of Nematology 32, 309-317.

Spallanzani, L. (1769). Nouvelles recherches sur les decouvertes microscopiques, etc. Londres and Paris.

Spaull, V.W. (1977). *Meloidogyne propora* n. sp. (Nematoda: Meloidogynidae) from Aldabra atoll, western Indian ocean, with a note on M. *javanica* (Treub). Nematologica 23: 177-186.

Spurr, H.W., Jr. (1982). Mode of action of nematicides. Pp. 269-776 *In*: An Advanced Treatise on *Meloidogyne*: Vol. 1 Biology and Control. North Carolina State University Graphics. Raleigh, NC.

Srivastava, A.N. and C.L. Sethi, (1984). Relationship of initial populations of *H. zeae* with plant growth of maize and nematode reproduction. Indian J. Nematol., 14: 110–114

Srivastava. A.N. and Kaushal, K.K. (1986). Occurrence and distribution of *Heterodera zeae* and *H sorghi* in India." Nat. Conf.. Plant Parasitic Nematodes, New Delhi, p.34

Srivastava. A.N. and Kaushal, K.K. (1991). *Heterodera zeae* at high altitudes in Himachal Pradesh. Ind. J. Nematol. 21:163.

Stalin, C., Ramakrishnan, S.and Jonathan, E.I. (2007). Management of root knot nematode *Meloidogyne incognita* in bhumyamalaki (*Phyllanthus amarus*) and makoy (*Solanum nigrum*). Biomed 2: 119-122.

Starr J. L. and M. C. Black.(1995). Reproduction of *Meloidogyne arenaria, M. incognita,* and *M. javanica* on Sesame. Supplement to the Journal of Nematology 27:624-627.

Starr, J.L., Jeger, M.J., Martyn, R.D. and Schilling, K. (1989) Effects of *Meloidogyne incognita* and *Fusarium oxysporum f. sp. vasinfectum* on plant mortality and yield of cotton. Phytopathology, 79: 640-6.

Steeves, R.M., Todd, T.C., Essig, J.S. and Trick, H.N. (2006). Transgenic soybeans expressing siRNAs specific to a major sperm protein gene suppress *Heterodera glycines* reproduction. Functional Plant Biology **33**: 991–999.

Steinbuch, J.G. (1799). Das Grasalchen, *Vibrio agrostis*.Naturforscher 28:233-259.

Steiner, G. (1923). *Aplectana krassi* n. sp., eine in der Blattwespe Lyda sp. parasitierende Nematodenform, nebst Bomerkungen uber das Seitenorgan der parasitischen Nematoden. Zentralblatt fuer Bakteriologie Parasitenkunde Infektionskrankheiten und Hygiene Aeilung I Originale 59:14-18.

Steiner, G. (1929). *Neoaplectana glaseri* n. g., n. sp. (Oxyuridae) a new nemic parasite of the Japanese beetle (Popillia japonica Newm.). Journal of the Washington Academy of Science 19:436-440.

Stern, V. M., R. F. Smith, R. van den Bosch, and K. S. Hagen. (1959). The integrated control concept. Hilgardia 29:81-101.

Stevens, C., V.A. Khan, and A.Y. Tang (1990). Solar heating of soil with double plastic layers: a potential method of pest control. p. 163-68. In: Proceedings of the 22nd National Agricultural Plastics Congress. Nat. Ag. Plastics Assoc., Peoria, IL.

Stirling, G. R. (1991). Biological Control of Plant Parasitic Nematodes: Progress, Problems and Prospects (CAB International, Wallingford, UK).

Stirling, G.R. and Mankau, R. (1979). Mode of parasitism of *Meloidogyne* and other nematode eggs by *Dactylella oviparasitica*. J. Nematol. 11: 282-288.

Stirling, G., Griffin, D., Ophel-Keller, K., McKay, A., Hartley, D., Curran, J., Stirling, A., Monsour, C., Winch, J. and Hardie, B. (2004). Combining an initial risk assessment process with DNA assays to improve prediction of soil borne diseases caused by root-knot nematode (*Meloidogyne* spp.) and *Fusarium oxysporum* f. sp. *lycopersici* in the Queensland tomato industry. Australas. Plant Pathol. 33: 285-293.

Stock, S., Campbell, J., and Nadler, S. (2001). Phylogeny of *Steinernema* Travassos, 1927 (Cephalobina: Steinernematidae) inferred from ribosomal DNA sequences and morphological characters. J. Parasitol. 87: 877-889.

Stone, A.R. (1973a) *Heterodera pallida* n. sp. (Nematoda: Heteroderidae), a second species of potato cyst nematode. Nematologica 18, 591-606.

Stone, A.R. (1973b) *Heterodera pallida* and *Heterodera rostochiensis*. CIH Descriptions of Plantparasitic Nematodes No. 16 and 17. CAB International, Wallingford, UK.

Storey, G.W. and Evans, K, (1987). Interactions between *Globodera pallida* juveniles, *Verticillium dahliae*, three potato cultivars, with descriptions of associated histopathologies. *Plant Pathology* **36**, 192–200.

Strubell, A. (1888). Untersuchungen iiber den Bau und die Entwicklung des Riibennematoden *Heterodera schachtii* Schmidt. Bibliotheca Zool. Orig. Abh. Gesammt. Zool.. Heft. 2. 50 pp.

Sturhan, D. (1988). New host and geographical records of nematode parasitic bacteria of the *Pasteuria penetrans* group. Nematologica 34:350– 356

Sturhan, Dieter (2002). Notes on the genus *Cactodera* Krall and Krall, 1978 and proposal of *Betulodera betulae* gen. nov., comb. nov. (Nematoda: Heteroderidae). Nematology 4: 875-882

Sturhan, D.; Wouts, W. M.; Subbotin, S. A. (2007). An unusual cyst nematode from New Zealand, *Paradolichodera tenuissima* gen.n.,sp. n. (Tylenchida: Heteroderidae). Nematology 9 (4): 561-571

Sturhan, D. and Rumpenhorst, H.J. (1996). Untersuchungen über den *Heterodera avenae*-Komplex. Mitteilungen aus der BiologischenBundesanstaltfür Land- und Forstwirtschaft Berlin-Dahlem 317: 75-91.

Subbotin, S. A., D. Sturhan, V. N. Chizhov, N. Vovlas, and J. G. Baldwin. (2006). Phylogenetic analysis of Tylenchida Thorne, 1949 as inferred from D2 and D3 expansion fragments of the 28S rRNA gene sequences. Nematology 8: 455474.

Subbotin, S.A., Rumpenhorst, H.J. and Sturhan, D. (1996). Morphological and electrophoretic studies on populations of the *Heterodera avenae* complex from the former USSR. Russian Journal of Nematology 4, 29-38.

Subbotin, S., Halford P. and Perry R. (1999). Identification of populations of potato cyst nematodes from Russia using protein electrophoresis, rDNA-RFLP and RAPDS. Russ. J. Nematol. 7: 57-63.

Subbotin, S.A., Waeyenberge, L., Molokanova, I.A. and Moens, M. (1999). Identification of *Heterodera avenae* group species by morphometrics and rDNA-RFLPs. Nematology 1: 195-207.

Subbotin, S.A., Waeyenberge, L. and Moens, M. (2000). Identification of cyst forming nematodes of the genus *Heterodera* (Nematoda: Heteroderidae) based on the ribosomal DNA-RFLP. Nematology 2:153-164.

Subbotin, S.A., Vierstraete, A., De Ley, P., Rowe, J., Waeyenberge, L., Moens, M. and Vanfleteren, J.R. (2001). Phylogenetic relationships within the cyst-forming nematodes (Nematoda, Heteroderidae) based on analysis of sequences from the ITS regions of ribosomal DNA. Molecular Phylogenetics and Evolution 21: 1-16.

Subbotin S.A., Sturhan D., Chizhov V., Vovlas N. and Baldwin J. (2006). Phylogenetic analysis of Tylenchida Thorne, 1949 as inferred from D2 and D3 expansion fragments of the 28S rRNA gene sequences. Nematology, 8: 455-474.

Subbotin, S.A., Sturhan, D., Rumpenhorst, H.J. and Moens, M. (2002). Description of Australian cereal cyst nematode *Heterodera australis* sp. n. (Tylenchida: Heteroderidae). Russian Journal of Nematology 10, 139-148.

Subbotin, S.A., Sturhan, D., Rumpenhorst, H.J., Moens, M. (2003). Molecular and morphological characterisation of the Heterodera avenae species complex (Tylenchida: Heteroderidae). Nematology 5: 515-538.

Subbotin, S., Sturhan D., Vovlas N., Castillo P., Tanyi T., Moens, M. and Baldwin, J. (2007). Application of the secondary structure model of rRNA for phylogeny:

D2-D3 expansion segments of the LSU gene of plantparasitic nematodes from the family *Hoplolaimidae* Filipjev, 1934. Mol. Phylogenet. Evol. 43: 881-890.

Subbotin S., Vovlas N., Crozzoli R., Sturhan D., Lamberti F., Moens M. and Baldwin J. (2005). Phylogeny of Criconematina Siddiqi, 1980 (Nematoda: Tylenchida) based on morphology and D2-D3 expansion segments of the 28S-rRNA gene sequences with application of a secondary structure model. Nematology, 7: 927-944.

Sudhaus, W. and Fitch, D.H.A. (2001). Comparative studies on the phylogeny and systematics of the Rhabditidae (Nematoda). J. Nemat. 33: 1–72

Suhail, A. (2003). Effect of different organic amendments with *Paecilomyces lilacinus* for the management of soil nematodes. Archives of Phytopathology and Plant Protection 36 (2): 103 - 109

Sukul, N.C. (1992): Plant antagonistic to plant-parasitic nematode. Indian Review of Life Sciences, **12**: 23–52.

Sumner, D.R. and Minton, N.A. (1987). Interaction of Fusarium wilt and nematodes in soybean. Plant Disease, 71:20-23.

Suresh, C.K. and Bagyaraj, D. J. (1984). Interaction between vesicular-arbuscular mycorrhizae and a root-knot nematode and its effect on growth and chemical composition on tomato. Nematologia Mediterranean, **12**:31- 39

Suresh, C.K., Bagyaraj, D.J. and Reddy, D.D.R.(1985). Effect of vesicular-arbuscular mycorrhiza on survival, penetration and development of root-knot nematode in tomato. Plant Soil. **87**:305-308.

Swarup, G., R. L. Mathur, A. R. Seshadri, D. J. Raski and B. N. Mathur (1976). Response of wheat and barley to soil fumigation by D-D and DBCP against 'molya' disease caused by *Heterodera avenae*. Indian J. Nematol. 6:150-155.

Swarup, G., Sethi, C.L. Seshadri, A.R. and Kaushal, K.K. (1979). On the biotypes of *H. avenae,* the causal organism of 'molya' disease of wheat and barley in India. Indian J. Nematology 9:164-168.

Szczech, M., Rondomanski, W., Brzeski, M.W., Smolinska, U., and Kotowski, J.F. (1993). Suppressive effect of a commercial earthworm compost on some root infecting pathogens of cabbage and tomato. Biol. Agric. Hortic. 10: 47-52.

Tanino, K., Motomasa Takahashi, Yoshihide Tomata, Hiroshi Tokura, Taketo Uehara, Takashi Narabu and Masaaki Miyashita (2011). Total synthesis of solanoeclepin A. Nature Chemistry **3**:484–488.

Taylor C.E. (1990). Nematode interactions with other pathogens. *Annals of Applied Biology* 116: 405–16.

Taylor, A.L. and C.W. McBeth (1941). A practical method of using methyl bromide as a nematicide in the field. Proceedings of the Helminthological Society of Washington 8: 26-28.

Thacker J.R.M. (2002). An Introduction to Arthropod Pest Control. Cambridge University Press, Cambridge: 343.

Tchesunov, A.V., and Riemann, F. (1995). Arctic sea ice nematodes (Monhysteroidea), with descriptions of *Cryonema crassum* gen. n., sp. n. and *C. tenue* sp. n. Nematologica 41: 35–50.

Terenteva, TG. (1965). *Meloidogyne kirjanovae* n. sp. (Nematoda: Heteroderidae). Materialy Nauchnoi Konferentsii Vesouyuznoi ObscheJ Gel'mint 4: 277-281.

Teuchert, G. (1977). The ultrastructure of the marine gastrotrich, *Turbanella cornuta* Remane (Macrodasyoidea) and its functional and phylogenetical importance. Zoomorphologie, 88:189-246.

Thoden, T. C., Korthals, G. W. and Termorshuizen, A. J. (2011). Organic amendments and their influences on plant-parasitic and free-living nematodes: a promising method for nematode management? Nematology 13:133-153.

Thomason, I.J. (1987). Challenges facing nematology: environmental risks with nematicides and the need for new approaches. *In* J.A. Veech and D.W. Dickson, eds. *Vistas on nematology,* p. 469-476. Hyattsville, USA, Society of Nematologists.

Thorn, R.G. and A. Tsueneda (1993). Interactions between *Pleurotus* species, nematodes and bacteria on agar and in wood. *Trans. Mycol. Soc. Japan*, 34: 449–464

Thorne, G. (1961). Principles of Nematology. McGraw-Hill, New York.

Thorne, G. (1969). *Hypsoperine ottersoni* sp. n. (Nematoda: Heteroderidae) infesting Canary Grass, *Phalaris arundinaceae* (L.) Reed in Wisconsin. Proceedings of the Helminthological Society of Washington *36*, 98-102.

Thuy, T.T.T. (2010). Incidence and effect of *Meloidogyne incognta* (Nematoda-Meloidogyninae) on black pepper plants in Vietnam. Ph.D thesis submitted to Hanoi University.

Tietjen, J.H., and Lee, J.H. 1977. Feeding behaviour of marine nematodes. Pages 21–35 *in* Ecology of marine benthos (Coull, B.C., ed.). Columbia, South Carolina, USA: University of South Carolina Press.

Tigano, M.ST, Carneiro, R.M.D.G., Jeyaprakash, A., Dickson, OW. and Adams, B.J. (2005). Phylogeny of *Meloidogyne* spp. based on 185 rNA and the intergenic region of mitochondrial DNA sequences. *Nematology* 7: 851-862.

Timothy, C. P., and Robert, G. L.(1992). Mycorrhizal interactions with soil organisms, In: Handbook of Applied Mycology Soil and Plants. Vol. I (Eds. Dilip, K. Arora, Bharat Rai., K. G. Mukerji and Guy R. Knudsen), pp.77-130, Marcel Dekker Inc. USA, New York, 720 pp.

Timm, L., D. Pearson and B. Jaffee (2001). Nematode trapping fungi in conventionally and organically managed corn-tomato rotations. Mycologia, 93: 25-29.

Toida, Y. and Yaegashi, T. (1984). Description of *Meloidogyne suginamiensis* n. sp. (Nematoda: Meloidogynidae) from mulberry in Japan. Japanese Journal of Nematology 14:49-57.

Tom, T., Yamashita, David, R., Viglierchio and Richard V. Schmitt (1986). Responses of nematodes to nematicidal applications following extended exposures to subnematicidal stress. Revue Nématol., 9 (1) : 49-60

Toruan-Mathius, N., Pancoro, A. and Sudarmadji, D. (1995). Root characteristics and molecular polymorphism associated with resistance to Pratylenchus coffeae in Robusta coffee. Menara Perkebunan 63:43-51.

Treub, M. (1885). Onderzoekingen over Sereh-Ziek Suikkeriet gedaan in s'Lands Plantentium te Buitenzorg. Mededeelingen uit's Lands Plantentium, Batavia, 2:1-39.

Triantaphyllou A. C. (1985). Cytogenetics, cytotaxonomy and phylogeny of root-knot nematodes. In: An Advanced Treatise on *Meloidogyne*, Volume1 ed.J.N.Sasser and C.C.Carter,pp.113–26.Raleigh:NorthCarolinaState Univ. Graphics.

Triantaphyllou, A.C. (1993). Hermaphroditism in *Meloidogyne hapla*. J. Nematol. 25:15–26

Triffith, M. J. (1930). *J. of Helminthology* **7**: 19–48.

Trudgill, D.L. (1991). Resistance to and tolerance of plant-parasitic nematodes in plants. *Annu. Rev. Phytopathol.* 29:167–92.

Trudgill, D. L. (1995). An assessment of the relevance of thermal time relationships to nematology. Fundamental and Applied Nematology, 18: 407 - 417.

Tsay,T.T.; S.T. Wu and Y. Y. Lin. (2004). Evaluation of Asteraceae plants for control of *Meloidogyne incognita*. Journal of Nematology 36: 36-41.

Tyler. J. (1938). Egg out put of the root knot nematode. Proc. Helminthol. Soc. 5: 49-54.

Tyson, E. (1683). *Lumbricus teres*, or some anatomical observations on the round worm bred in human bodies. Phil. Tr., Lond. (146), v. 13:154-161

Uehara, T., Mizukubo, T., Kushida, A. and Momota, Y. (1998). Identification of *Pratylenchus coffeae* and *P. loosi* using specific primers for PCR amplification of ribosomal DNA. Nematologica, 44: 357-368.

Upavanavinoda (1935). Edited by G.P.Mojumdar. The Indian Research Institute, Calcutta,

Umarao, Rao, S.B., Kaushal, K.K. and Ganguly, A.K. (2004). Molecular genetic variation in Indian populations of cereal cyst nematode, *Heterodera avenae*: implications on virulence and breeding for disease resistance. In 'Proceedings of the second international group meeting, 23-26 September 2002, Agharkar Research Institute, Pune, India.' pp. 234-253.

Umarao, Rao, S.B., Kaushal, K.K., Ganguly, A.K., and Gaur, H.S. (2007). RAPD analysis of intra-specific variation in *Heterodera avenae* populations of India. International Journal of Nematology 17:29-34.

Umarao, and Sashi, V. (2008). Molecular characterization of Indian populations of *Heterodera filipjevi* using PCR-RFLP of rDNA. International Journal of Nematology 18:118-122

Urwin, P. E, Atkinson H. J., Waller, D.A. *and* McPherson M. J. (1995). Engineered oryzacystatin-I expressed in transgenic hairy roots confers resistance to *Globodera pallida*. *Plant Journal* **8**: 121–131.

Urwin, P.E., Green J.*and* Atkinson H.J.(2003). Expression of a plant cystatin confers partial resistance to *Globodera*, full resistance is achieved by pyramiding a cystatin with natural resistance. Molecular Breeding **12**: 263–269.

Urwin, P.E., Levesley, A., McPherson, M.J.*and* Atkinson, H.J. (2000). Transgenic resistance to the nematode *Rotylenchulus reniformis* conferred by *Arabidopsis thaliana* plants expressing proteinase inhibitors. Molecular Breeding 6: 257–264.

Urwin, P.E., Lilley, C.J. *and* Atkinson, H.J. (2002). Ingestion of double-stranded RNA by pre parasitic juvenile cyst nematodes leads to RNA interference. Molecular Plant–Microbe Interactions **15**: 747–752.

Urwin, P.E., Lilley, C.J., McPherson, M..J. *and* Atkinson, H.J. (1997a). Resistance to both cyst- and root-knot nematodes conferred by transgenic *Arabidopsis* expressing a modified plant cystatin. Plant Journal **12**: 455–461.

Urwin, P.E., Lilley, C.J., McPherson, M.J. *and* Atkinson, H.J. (1997b). Characterisation of two cDNAs encoding cysteine proteases from the soybean cyst nematode *Heterodera glycines.* Parasitology **114**: 605–613.

Urwin, P.E., McPherson, M.J. *and* Atkinson, H.J. (1998). Enhanced transgenic plant resistance to nematodes by dual proteinase inhibitor constructs. Planta **204**: 472–479.

Usha, K. (1980). *Studies on the cyst nematodes of rice in Kerala.* M Sc Thesis, Kerala Agricultural University, 80 p.

Uziel, A.,and R. Sikora. (1992). Use of non-target isolates of the entomopathogen *Verticillium lecanii* (Zimm.) Viegas to control the potato cyst nematode, *Globodera pallida* (Stone). Nematologica 38:123-130.

Van Berkum, J.A. and Seshadri, A.R. (1970). Some important nematode problems in India. In 10th Int. Nematology Symp., Pescara, Italy, p. 136-137.

Van Bezooijen, J. (2006). Methods and Techniques for Nematology, Wageningen,112pp.

Van Damme, V. Hoedekie, A. and Viaene, N. (2005). Long-term efficacy of *Pochonia chlamydosporia* for management of *Meloidogyne javanica* in glasshouse crops. Nematology 7: 727-736.

Van der Vossen, E.A.G., van der Voort, J., Kanyuka, K., Bendahmane, A. and Sandbrink, H. (2000). Homologues of a single resistance-gene cluster in potato confer resistance to distinct pathogens: a virus and a nematode. Plant J. 23:567-76.

Van Der Beek, J., Los, J., and Pijanacker, L. (1998). Cytology of parthogenesis of ve *Meloidogyne* species. Fundam. Appl. Nematol. 21:393–99

Vanfleteren J.R. and Vierstraete A.R. (1999). Insertional RNA editing in metazoan mitochondria: The cytochrome b gene in the nematode *Teratocephalus lirellus.* RNA 5: 622–624.

Van Gundy, S.D., Gustavo Perez, Jose, B., Stelzy, L.H., and Thomson, I.J. (1974). A pest management approach to the control of *Pratylenchus thornei* on wheat in Mexico. Journal of Nematology 6:107–116.

Vanholme B, De Meutter J, Tytgat T, Van Montagu M, Coomans A, and Gheysen G. (2004). Secretions of plant-parasitic nematodes: a molecular update. Gene 332:13–27

Vanholme, B., Mitreva, M., Van Criekinge W, Logghe M, Bird D, McCarter JP, and Gheysen G. (2006). Detection of putative secreted proteins in the plant-parasitic nematode, *Heterodera schachtii*. Parasitol. Res. 98:414–424.

Varaprasad, K.S., Sharma, S.B., Loknathan, T.R. (1997). Nematode constraints to pigeonpea and chickpea in Vidarbha region of Maharashtra in India. International Journal of Nematology 7: 152-157.

Vejdovsky, F. (1886). Zur Morphologic der Gordüden. Z. Wissensch. Zoo1. 43: 369-433.

Verdejo-Lucas, S., Sorribas, F.J., Ornat, C., and Galeano, M.(2003). Evaluating *Pochonia chlamydosporia* in a double-cropping system of lettuce and tomato in plastic houses infested with *Meloidogyne javanica*. Plant Pathol. 52: 521-528.

Verhoef, H.A. and Brussard, L. (1990). Decomposition and nitrogen mineralization in natural and agro-ecosystems.The contribution of soil animals. Biogeochemistry, 11: 175-211.

Viaene, N.M.and Abawi, G.S. (1998). Management of *Meloidogyne hapla* on lettuce in organic soil with sudangrass as a cover crop. Plant Dis. 82: 945-952.

Vincenzo Candido, Trifone D'Addabbo, Martino Basile, Donato Castronuovo and Vito Miccolis (2008). Greenhouse soil solarization: effect on weeds, nematodes and yield of tomato and melon Agron. Sustain. Dev. 28 : 221-230

Vishnudasan, D., Tripathi, M.N., Rao, U., Khurana, P. (2005). Assessment of nematode resistance in wheat transgenic plants expressing potato proteinase inhibitor (*PIN2*) gene. Transgenic Research **14**: 665–675.

Vovlas, N. and R.N. Inserra (1996). Distribution and parasitism of root knot nematodes in Citrus. Nematol. Circ., 217. Fla. Dept. Agric. and Consumer Services.

Wachira, P.M., Kimenju. J.W., Okoth, S.A, and Mibey, R.K. (2009). Stimulation of Nematode-Destroying Fungi by Organic Amendments Applied in Management of Plant Parasitic Nematode. Asian Journal of Plant Sciences.

Waeyenberge, L., Ryss, A., Moens, M., Pinochet, J., and Vrain, T. (2000). Molecular characterization of 18 *Pratylenchus* species using rDNA Restriction Fragment Length Polymorphism. Nematology, 2: 135-142.

Waite, I., O'Donnell, A., Harrison, A., Davies, J., Colvan, S., Ekschmitt, K., Dogan, H., Wolters, V., Bongers, T., Bongers, M., Bakonyi, G., Nagy, P., Papatheodorou, E., Stamou, G., and Bostrom, S. (2003). Design and evaluation of nematode 18S rDNA primers for PCR and denaturing gradient gel electrophoresis (DGGE) of soil community DNA. Soil Biol. Biochem. 35: 165-173.

Walia, R.K. (2004). Biopesticides for the management of nematodes - Is it a practical reality? pp 53-54. **In:** National Symposium on Paradigms in Nematological Research for Biodynamic Farming, Bangalore, November, 17-19. pp: 53-54

Walia, R.K. and Bajaj, H.K. (1988). Further studies on existence of races in Pigeon pea cyst nematode, *Heterodera cajani*. International Journal of Nematology 18: 269–272.

Walia, K. and H.K. Bajaj, (2000). Morphological and morphometrical variations in two races of *H. cajani* Koshy. *Indian J. Nematol.*, 30: 124–128

Wallace, H.R. (1965). The ecology and control of the cereal root nematode. J. Austr. Inst. Agric. Sci., 31: 178-186

Wallace, H.R. (1978). The diagnosis of plant diseases of complex etiology. Annual Review of Phytopathology, 16:379-402.

Wang, D., Kumar, S., and Hedges, B. (1999). Divergence time estimates for the early history of animal phyla and the origin of plants, animals and fungi. Proc. Royal Soc. London Ser. B 266, 163-171.

Wang, K.H., R. McSorley and R.N. Gallaher (2004). Effect of *Crotalaria juncea* amendment on squash infected with *Meloidogyne incognita*. J. Nematol., 36: 290-296.

Wasilewska, L. (1971). Nematodes of the dunes in the Kampinos Forest: II. Community structure based on number of individuals, state of biomass and respiratory metabolism. Ekol Pol. 19: 651-688.

Wasilewska, L. (1989). Impact of human activities on nematodes, in Charholm, C. and Bergstrom, L. Eds. Ecology of Arable Land, Kluwer Academic, Dordrecht, The Netherlands, 123-132.

Wasilewska, L., Oloffs, P.C., and Webster, J.M. (1975). Effects of carbofuran and PCB on development of bacteriophagous nematode *Acrobeloides nanus*. Canadian Journal of Zoology 53:1709–1715.

Waterhouse, P.M., Graham, M.W., Wang, M.B.(1998). Virus resistance and gene silencing in plants can be induced by simultaneous expression of sense and antisense RNA. Proceedings of the National Academy of Sciences, USA 95: 13959–13964.

Watson, J. R. (1915). Root-knot on Tomato and Its Control. Fla. Agr. Exp. Sta. Bull. 125, pp. 62-64

Wei, Z.M., and S.V. Beer (1996). Harpin from *Erwinia amylovora* induces plant resistance. Acta Horticulturae 411:223-225.

Weir, G. H. and Bonavia, D. (1985). Coprolitos y dieta del preceramico tardio de la costa Peruana. Bull. French Inst. Andean Stud. 14, 85-140.

Wescott, S.W.III and Barker,K.R. (1976). Interaction of *Acrobeloides buetschlii* and *Rhizobium legminosarum* on Wando pea. Phytopathology ^6: 468-472.

Wharton, D.A. (1986). A functional biology of nematodes. Baltimore, Maryland, USA: The Johns Hopkins University Press. 192 pp.

Wharton, D.A. (2002). Life at the Limits, Organisms in Extreme Environments. Cambridge University Press, Cambridge. 307 pp.

Whitehead, AG. (1960). The root-knot nematodes of East Africa. I. *Meloidogyne africana* sp. n., a parasite of arabica coffee (*Coffeae arabica* L.). Nematologica 4:272-278.

Whitehead, A. G. (1968). Taxonomy of *Meloidogyne* (Nematoda: Heteroderidae) with descriptions of four new species. Transaction of the Zoological Society of London. 31:263-401.

Whitehead, A.G. (1968). Nematodea. In: Le Pelley R.H. (ed.) Pests of Coffee. Longmans, Green and Co. Ltd., London and Harlow.

Whittaker, R.H. (1969). New concepts of kingdoms of organisms. Science 163:150–160

Wider, T.L., and G.S. Abawi. (2000). Mechanism of suppression of *Meloidogyne hapla* and its damage by a green manure of Sudan grass. Plant Disease. 84: 562-568.

Wieczorek, K., Golecki, B. and Gerdes, L. (2006). Expansins are involved in the formation of nematode-induced syncytia in roots of *Arabidopsis thaliana*. Plant J. 48: 98–112.

Wieczorek, K., Hofmann, J., Blöchl, A., Szakasits, D., Bohlmann, H. and Grundler, F.M.W. (2008). Arabidopsis endo-1,4-beta-glucanases are involved in the formation of root syncytia induced by *Heterodera schachtii*. Plant J. 53: 336–351

Will. K. and Rubinoff, D. (2004). Myth of the molecule: DNA barcodes for species cannot replace morphology for identification and classification. Cladistics 20: 47-55.

Williams, Greg, and Pat Williams (1990a). Sesame residues vs. harmful nematodes. Hort. Ideas. March. p. 35.

Williams, Greg and Pat Williams (1990b). Some plant nutrients repel harmful nematodes. Hort. Ideas. June. p. 63.

Williams, Greg, and Pat Williams (1993). Wheat vs. nematodes causing peach tree short life. Hort. Ideas. July. p. 76.

Williams, Mary S.R. and Supapan Seraphin (1998). Heavy metal biomineralization in free-living nematodes, *Panagrolaimus* spp. Materials Science and Engineering C, 6:47-51.

Winston, W.M., Sutherlin, M., Wright, A.J., Feinberg, H., and Hunter, C.P. (2007). *Caenorhabditis elegans* SID-2 is required for environmental RNA interference. Proceedings of the National Academy of Sciences, USA 104: 10565–10570.

Winter, M.D., McPherson, M.J, and Atkinson H.J. (2002). Neuronal uptake of pesticides disrupts chemosensory cells of nematodes. Parasitology 125: 561–565.

Woese, C.R., Kandler, O. and Wheelis, M.L. (1990). Towards a natural system of organisms: proposal for the domains Archaea, Bacteria, and Eucarya. Proc. Natl. Acad. Sci. USA. 87: 4576–4579.

Womersley, C. (1980). The effect of different periods of dehydration/rehydration periods upon the ability of second stage larvae of Anguina tritici to survive desiccation at 0 per cent relative humidity. Ann. Appl. Biol. 95: 221–224.

Wyss, U., Grundler, F.M.W. and Munch, A. (1992). The parasitic behaviour of second stage juveniles of*Meloidogyne incognita* in roots of *Arabidopsis thaliana*. Nematologica, 38, 98 - 111.

Yadav, B.C., Veluthambi, K. and Subramaniam, K. (2006). Host-generated double stranded RNA induces RNAi in plant-parasitic nematodes and protects the host from infection. Molecular and Biochemical Parasitology 148: 219–222.

Yang, H., Powell, N.T. and Barker, K.R. (1976). The Influence of *Trichoderma harzianum* on the root-knot Fusarium wilt complex in cotton. J.Nematol. 8(1):81-6.

Yang, Y., E. Yang, A. Zhiqiang and X. Liu, (2007). Evolution of nematode-trapping cells of predatory fungi of the Orbiliaceae based on evidence from rRNA-encoding DNA and multiprotein sequences. Proc. Natl. Acad. Sci., 104: 8379-8384

Yang, B.J. and Eisenback, J.D. (1983). *Meloidogyne enterolobii* n. sp. (Meloidogynidae), a root-knot nematode parasitizing pacara earpod tree in China. Journal of Nematology 15: 381-391.

Yang, B.J., Wang, O.L. and Feng, R.Z. (1988). *Meloidogyne kongi* n. sp. (Nematoda: Meloidogynidae) a root-knot nematode parasitizing *Citrus* sp. in Guangxi, China. Journal of Jiangsu Agricultural College 7:1-9.

Yang, B.J., Hu, K.J. and Xu, B.W (1988). A new species of root-knot nematode *Meloidogyne lini* n. sp. parasitizing rice. Journal of Yunnan Agricultural University 3: 11-17.

Yang, B.J., Hu, K.J., Chen, H. and Zhu, W. (1990). A new species of root-knot nematode *Meloidogyne jianyangensis* n. sp. parasitizing mandarin orange. *Acta Phytopathologica Sinica* 20,259-264.

Yanoviak, S.P., M. Kaspari, R. Dudley and G. Poinar Jr. (2008). Parasite-induced fruit mimicry in a tropical canopy ant. The *American Naturalist* 171: 536–544.

Yeates, G.W. (1987). Nematode feeding and activity: The importance of development stages. Biol. Fertil. Soils. 3; 143-146.

Yeates, G.W., and Wardle, D.A.(1996). Nematodes as predator and prey: relationship to biological control and soil processes. Pedobiologia.40: 43-50.

Yeates, G.W. (1979). Soil nematodes in terrestrial ecosystems. Journal of Nematology 11:213–229.

Yeates, G.W., Bongers, T., De Geode, R.G.M., Freckman, D.W., and Georgieva, S.S. (1993). Feeding habits in soil nematode families and genera - An outline for social ecologists. Journal of Nematology 25:315–331.

Yepsen, Roger B. Jr. (ed.) (1984). The Encyclopedia of Natural Insect and Disease Control. Rev. ed. Rodale Press, Emmaus, PA. p. 267—271.

Y. Seshagiri Rao,J. Satyanarayana Prasad and A.V. Surya Rao (1984). Interaction of the cyst and root-knot nematodes in rsots of rice. Revue Nématol. **7** (2): 117-120.

Yucel, S., I.H. Elekcioglu, A. Uludag, C. Can, M.A., Sogut, A. Ozarslandan and E. Aksoy, (2002). The second year results of Methyl Bromide alternatives in the Eastern Mediterranean. Proceedings of 2002 Annual International Research

Conference on Methyl Bromide Alternatives and Emissions Reductions, Nov. 5-8, Orlando, Florida, USA., pp: 1-4.

Yushin, V.V., Yoshida, M., and Spiridonov, S.E.(2003). Self-moving spermatophores: spermatozoan dimorphism in *Steinernema* (Steinernematidae, Rhabditida). Russ. J. Nematol. 11, 151–152.

Zhang, S.S. (1993). *Meloidogyne mingnanica* n. sp. (Meloidogynidae) parasitizing citrus in China. Journal of Fujian Agricultural University 22 (Suppl.): 69-76.

Zhang, S.S. and Weng, Z.M. (1991). Identification of root-knot nematode species in Fujian. Journal of Fujian Agricultural College 20: 158-160.

Zhang, S.S., Gao, R.X. and Weng, Z.M. (1990). *Meloidogyne citri* n. sp. (Meloidogynidae), a new root-knot nematode parasitizing citrus in China. Journal of Fujian Agricultural College 19: 305-311.

Zhang, Y.M. (1983). A root-knot nematode, *Meloidogyne sinensis* n. sp., from potatoes in China. Acta Scientiarum Naturaum Universitatis Shandong 2: 88-95.

Zhang, Y.M. and Su, C.D. (1986). A new species of the genus *Meloidogyne* from Shandong Province, China (Tylenchida: Meloidogynidae). Journal of Shandong University 21: 95-103.

Zheng, L., Lin, M. and Zheng, M. (1990). Occurrence and identification of a new disease of the citrus, Donghai root-knot nematode - *Meloidogyne donghaiensis* sp. novo in coast soil of Fujian, China. *Journal* of *Fujian Academy* of *Agricultural Sciences* 5: 56-63.

Zijlstra, C., Donkers-Venne, D., and Fargette, M. (2000). Identification of *Meloidogyne incognita, M, javanica* and *M. arenaria* using sequence characterized amplified region (SCAR) based PCR assays. Nematology 2: 847-853.

Zolda, P. (2006). Nematode communities of grazed and ungrazed semi-natural steppe grasslands in Eastern Austria. Pedobiol., 50: 11-22

Zrzavy, J., Mihulka, S., Kepka, P., Bezdek, A., and Tietz, D.(1998). Phylogeny of the Metazoa based on morphological and 18S ribosomal DNA evidence. Cladistics 14, 249-40

Zuckerman, B. (1987). Nematodes as models to study biological aging. Pages 414–423 *in* Vistas on nematology: A commemoration of the Twentyfifth Anniversary of the Society of Nematologists (Veech, J.A., and. Dickson, D.W., eds.). Hyattsville, Maryland, USA: Society of Nematologists, Inc.

Xue, B., Baillie, D., Beckenbach, K., and Webster, J. (1992). DNA hybridization probes for studying the affinities of three *Meloidogyne* populations. Fundam. Appl. Nematol. 15: 35-41.

Zunke, U., Rossner, J., and Wyss, U. (1986). Parasitierungsverhalten von *Aphelenchoides hamatus* an funf verschiedenen phytopathogenen pilzen und an *Agaricus campestris*. Nematologica 32:194–201.

Zunke, U., and J. D. Eisenback. (1998). Morphology and ultrastructure, p. 31-56. *In* S. B. Sharma (ed.), The cyst nematodes. Chapman and Hall, London, United Kingdom.

Annexure

Books on Nematodes 2002 Onwards

2002

1. Plant Resistance to Parasitic Nematodes J.L. Star, R. Cook and J. Bridge (Eds). ISBN: 0851994660/2002/CABI Publishing/ 258pp/ £65
2. The Biology of Nematodes D.L. Lee (Ed). ISBN: 0415272114/ Taylor and Francis/ 2002/ 365pp/ £125
3. Nematode Control in Crop M. Mashkoor Alam and Neeta Sharma. ISBN: 8185860203/ International Book Distributors, Booksellers and Publishers/ 2002/ 432pp/£35.50
4. The Behavioural Ecology of Parasites E.E. Lewis, J.F. Campbell and M.V.K. Sukhdeo (Eds)ISBN: 0851996159/CABI Publishing/2002/358pp/£65.00
5. Tylenchida: Parasites of Plants and Insects, 2nd Edition M R Siddiqi ISBN: 0851992021/ CABI Publishing/2002/ 833pp/£145
6. The plant-parasitic Nematode Genus *Meloidogyne* Göldi, 1892 (Tylenchida) in Europe.G. Karssen. ISBN: 9004127909/ E.J. Brill/2002/ 160pp/ £45
7. Plant Resistance to Parasitic Nematodes/J L Starr, R Cook and J Bridge. ISBN: 0851994660/Oxford Press University/2002/256pp/£49.95
8. Identification Guides for the Most Common Genera of Plant-Parasitic Nematodes J. D. Eisenback. Mactode Publications, 3510 Indian Meadow Drive, Blacksburg, VA 24060, USA
9. Nematology Laboratory Investigations, Vol. 1 - Morphology and Taxonomy J. D. Eisenback. Mactode Publications, 3510 Indian Meadow Drive, Blacksburg, VA 24060, USA

2003

10. An Introduction to Nematodes: Entomophilic Nematology/R. Perry and D. Wright (Eds), Pensoft Publishers, distributed by NHBS/ 2003/ £33.50
11. Monograph of Soil Nematodes from Coastal Douglas-fir Forests in British Columbia Panesar and Marshall

2004

12. Nematode Behaviour/R. Gaugler and A.L. Bilgrami (Eds), ISBN: 0861998189/ CABI Publishing/2004/400pp/£ 75
13. Nematology: Advances and Perspectives. Vol I. Nematode Morphology, Physiology and Ecology, Z.X. Chen, S.Y. Chen and D.W. Dickson (Eds). ISBN: 0851996450/CABI Publishing/2004/ hardcover/636pp/£95
14. Nematology: Advances and Perspectives Vol II. Nematode Management and Utilization/Z.X. Chen, S.Y. Chen and D.W. Dickson (Eds). ISBN: 0851996469/ CABI Publishing/2004/597pp//£75
15. The Pinewood nematode, *Bursaphelenchus xylophilus*/Series: NEMATOLOGY MONOGRAPHS AND PERSPECTIVES 1; Proceedings of an International Workshop, University of Evora Portugal, August 20-22, 2001/M. Mota and P. Vieira (Eds). ISBN: 9004132678/ E.J. Brill/ 2004/292 pp/£86

2005

16. Plant Parasitic Nematodes in Subtropical and Tropical Agriculture R. A. Sikora and J. Bridge (Eds). ISBN: 0851997279/ CABI Publishing/ hardcover 2nd edition/ 896pp/£100
17. Free-living Nematodes of Hungary (Nematoda Errantia). Vol 1. Pedozoologica Hungarica. Taxonomic, zoogeopgraphic and faunistic studies on the soil animals, No 3,Istvan Andrássy. ISBN: 9637093907/ Hungarian Natural History Museum and Systematic Zoology Research Group of Hungarian Academy of Sciences./2005/518pp

2006

18. Functional and Detailed Morphology of the Tylenchida (Nematoda) E. Geraert. ISBN: 9004148957/E.J. Brill/2006/200pp/£113
19. Life and Work of Dr Johannes Govertus de Man (1850-1930) A Crustacea and Nematoda specialist. G. Karssen. ISNB: 9004149694/E.J. Brill/2006/ 118pp/£83
20. Criconematina (Nematoda: Tylenchida)/Series: FAUNA OF NEW ZEALAND 55/W.M. Wouts. ISBN: 0478093810:Manaaki Whenua/2006/ 232pp/£82
21. Nematode Parasites of Birds (including Poultry) from South Asia M.L. Sood. IDBN: 8181890159/ 2006/ International Book Distributors, Booksellers and Publishers/ 824pp/ £100

22. Freshwater Nematodes: Ecology and Taxonomy/Eyualem-Abebe, W. Traunspurger and I. Andrassy (Eds). ISBN: 0851990096/ CABI Publishing/ 2006/752pp/£125
23. Plant Nematology/R.N. Perry and M. Moens (Eds). ISBN: 1845930568/ CABI Publishing/2006/ 447pp/ £55

2007

24. Plant Nematodes of Agricultural Importance/J. Bridge and J. Starr. ISBN: 9781840760637/ Manson Publishing/ 2007/ 128pp, illustrations in colour/ £42.50
25. Integrated Management and Biocontrol of Vegetable and Grain Crops Nematodes/Series: INTERGRATED MANAGEMENT OF PLANT PESTS AND DISEASES 2/A; Ciancio and K.G. Mukerji (Eds). ISBN: 9781402060625/ Spinger-Verlag/ 2007/ 360pp/ £130.50
26. Management of Nematode and Insect-borne Plant Diseases G. Saxena and K.G. Mukerji. ISBN: 9781560221357/ Haworth Press/2007/ 29pp/ £46.99
27. Free-living Nematodes of Hungary Nematoda Errantia). Vol 2.. Pedozoologica Hungarica. Taxonomic, zoogeopgraphic and faunistic studies on the soil animals, No 4. Istvan Andrássy. ISBN: 9637093982/ Hungarian Natural History Museum and Systematic Zoology Research Group of Hungarian Academy of Sciences./2007/496pp

2008

28. Tylenchidae of the World/E. Geraert. ISBN: 9789038213552/ Academia Press/2008/ 540pp/£45.50
29. Nematodes as Biological Control Agents/P. S. Grewal; R. Ehlers and D.I. Shapiro-Ilan (Eds). ISBN: 9781845934545/softcover 2008/ CABI Publishing/505pp/£39.95 (Original publication 2005)
30. Plant Nematodes. Methodology, Morphology, Systematics, Biology and Ecology/M. R. Khan. ISBN: 978-1-57808-533-0/ August 2008/360 pp/ £ 49/Science Publishers
31. Diseases of Horticultural Crops: Nematode Problems and their Management P. Parvatha Reddy. ISBN: 9788172335434/ Scientific Publishers/2008/ 380pp/ £86
32. Cell Biology of Plant Nematode Parasitism/Series: PLANT CELL MONOGRAPHS 15/R. H. Berg and C.G. Taylor (eds). ISBN: 9783540852131/ Springer-Verlag/2008/ 270pp/ £99
33. Plant-Parasitic Nematodes of Coffee/R. M. Souza (Ed). ISBN: 978-1-4020-8719-6/Springer/355 pp/approx. $239.00

2009

34. Root-knot Nematodes/Edited by Roland N. Perry, Rothamsted Research, UK, Maurice Moens, Institute for Agricultural and Fisheries Research,

Belgium, and James L. Starr, Texas A&M University, USA. ISBN: 978 1 84593 492 7/CABI Publishing/496pp//£99.50 / $189.05 / •139.3

35. Life and Work of PROF. DR LUCIEN DE CONINCK. Biologist, humanist and freemason./W. Decraemer, A. Coomans and E. Geraert. ISBN: 978 90 382 1499 3/ September 2009/ 169pp/• 15/Available at J. STORY-SCIENTIA Scientific Booksellers, Sint Kwintensberg 87, B 9000 Gent, Belgium.
36. Nematodes as Environmental Indicators/M. Wilson and T. Khakouli-Duarte (Eds). ISBN: 9781845933852, CABI-publishing/2009/352pp/£85
37. Keys to the Nematode Parasites of Vertebrates: Archival Volume R.C. Anderson, A.G. Chabaud and S. Willmott (Eds). ISBN: 9781845935726, CABI publishing/2009/463pp/£95
38. Free-living nematodes of Hungary (Nematoda errantia), III István Andrássy. Published in the series Pedozoologica Hungarica/ Hungarian National History Museum and Systematic Zoology Research Group of the Hungarian Academy of Sciences, Budapest

2010

39. Systematics of Cyst Nematodes (Nematoda: Heteroderinae), Part A SA Subbotin, M Mundo-Ocampo and JG Baldwin. In series: NEMATOLOGY MONOGRAPHS AND PERSPECTIVES 8A, 351 pages, figs, tabs.ISBN-13: 9789004162259/ E.J. Brill/ Hardcover/£140.00 / $216 / •163 approx.
40. Systematics of Cyst Nematodes (Nematoda: Heteroderinae), Part B SA Subbotin, M Mundo-Ocampo and JG Baldwin. /In series: NEMATOLOGY MONOGRAPHS AND PERSPECTIVES 8B, 512 pages, figs, tabs./ISBN-13: 9789004164345/ E J Brill / l Hardcover /£150.00 / $232 / •175 approx.
41. Mononchida: The Predatory Soil Nematodes Wasim Ahmad and M Shamim Jairajpuri/In series: NEMATOLOGY MONOGRAPHS AND PERSPECTIVES 7, 298 pages, E J Brill/ISBN-13: 9789004174641/ E.J. Brill / Hardcover l 2010 / £136.00 / $210 / •159 approx.

2011

42. Crustacea, Platyhelminthes, Nematoda, Annelida, Rotifera and Tardigrada of the Seychelles islands.Gerlach, J. (Editor) Soft cover 160 printed pages 24 x 16 cm, 98 photos, 175 line illustrations/ISBN 978-0-9558636-9-1. Price £40.00
43. The Evolutionary History of Nematodes: As Revealed in Stone, Amber and Mummies/George Poinar. In series: NEMATOLOGY MONOGRAPHS AND PERSPECTIVES 9, 332 pages/ISBN-13: 9789004175211 / Hardcover / E J Brill / £136.00 / $210 / •159 approx.
44. Molecular and Physiological basis of Nematode Survival Edited by R.N. Perry and D.A. Wharton. ISBN 978 1 84593 687 7/CABI Publishing/ 352pp/Paperback/£95.00/$ 180.00/•135

Index

K

L

M

N

O

P

T

U

V

W

X

Y

Z